Fundamental Optical Design

Fundamental Optical Design

Michael J. Kidger

SPIE PRESS

Bellingham, Washington USA

Library of Congress Cataloging-in-Publication Data

Kidger, Michael J.
Fundamental Optical Design / Michael J. Kidger.
p. cm. – (SPIE monograph ; PM92)
Includes bibliographical references and index.
ISBN 0-8194-3915-0
1. Geometrical optics. I. Title. II. Series

QC381 .K53 2001
535'.32—dc21

2001042915
CIP

Published by

SPIE
P.O. Box 10
Bellingham, Washington 98227-0010 USA
Phone: +1 360.676.3290
Fax: +1 360.647.1445
Email: Books@spie.org
Web: http://spie.org

Second soft cover printing 2019
Soft cover ISBN: 978-0-8194-9599-0

Printed in the United States of America

CONTENTS

Chapter 7 Principles of Lens Design / 139

Chapter 8 Achromatic Doublet Objectives / 167

Chapter 9 Petzval Lenses and Telephoto Objectives / 183

Chapter 10 Triplets / 199

Chapter 11 Eyepieces and Afocal Systems / 209

FOREWORD

In preparing this Foreword to *Fundamental Optical Design*, the first of two volumes of work by my late husband, Dr. Michael Kidger, I thought it fitting that the reader should know more about the "man." You may ask, "What was he like?" What was it about him that made so many of his friends and colleagues send in letters of tribute with donations, so that we were able to establish the Michael Kidger Memorial Scholarship annual award?

Michael and I met in 1955, when he was still a student. Such was his academic ability that in his final school year he was awarded a Royal Scholarship to Imperial College, London, and also a County Major Scholarship from the County of Worcestershire, UK, where he lived.

Michael's interest and enthusiasm for lens design was stimulated by Professor Walter Welford, who lectured in lens design at Imperial College as part of the undergraduate course. After graduating, Michael continued his research at Imperial College and was awarded an MSc. He left Imperial College in 1960 to work with Rank Taylor Hobson in Leicester. At Rank he worked on a range of optical products, but particularly on zoom and fixed focus camera lenses. Later in his life he developed an interest in cameras, building up a comprehensive collection, as well as a selection of screw lenses.

Michael continued his interest in flying gliders, which he had started at the Lasham Gliding Club when he was an undergraduate. While on vacation at the Cambridge Gliding Club in 1963, during what seemed to be a week of long hot summer days, it was his delight to fly and to take advantage of the weather and resulting thermals.

In these early years for Michael, the Applied Optics Section of the Physics Department at Imperial College included a group who individually and collectively were clever and inspiring. They had considerable insight into the future developments in computer-aided optical design and in the development and applications of lens design. This prestigious and dedicated group of academic professionals included Professor Charles Wynne, Professor Harold H. Hopkins, Professor Walter Welford, and Professor W. D. Wright.

In 1963 Michael was invited to join this group as a research assistant to develop raytracing programs. Using an IBM 7090 computer he rewrote the group's damped least squares software in FORTRAN, adapting programs written by his colleagues to investigate double Gauss designs. He was awarded his Ph.D. for this work in 1971. Subsequently he was appointed a lecturer in the Physics Department of Imperial College. He held this appointment for over 20 years.

Michael's interests went well beyond his professional expertise. During weekends he took time to make and fly model gliders. He built a model engine, thus following in the footsteps of his father and grandfather. As a young boy he would ride on his bicycle around the countryside and visit various train stations to take photographs of the trains. An album of early steam engines exists as a permanent record of these trips.

As the power and speed of microcomputers increased, software developed to take advantage of the new technology. Consequently the FORTRAN programs were rewritten into HPBasic under Michael's direction. In 1978, after pressure from Michael, the optics group at Imperial College obtained a Hewlett Packard 9845. Then came the HP85 portable computer, upon which Michael rewrote many of the large programs. He demonstrated his programs at the 1980 International Lens Design Conference in Oakland, California, where his work excited many of those attending the meeting.

While all these developments were taking place, Michael continued with his lectures at Imperial College. In addition to a considerable amount of undergraduate teaching, he also taught graduate optics for the MSc program. Then, by persuading the college to equip the first-year laboratories with HP85s, he completely changed the teaching of computing to first-year undergraduates. As a result of this innovation students were freed from the burden of unnecessary handwritten notes. They took their first step forward in the use of computers as a fundamental tool in the pursuit of their future profession. For this generation of students and those to follow, Michael had championed a major step forward in their competitive and professional development.

Michael found that there was considerable interest in his lens design program. To his surprise, companies and individuals wanted to buy it, but as he said, "I wrote it to use for myself, not to sell." However, in 1982 after careful discussion we decided to establish Kidger Optics Ltd. As part of our work with the company we started to exhibit widely with SPIE—The International Society for Optical Engineering, both in the United States and internationally. As we traveled worldwide we often met Michael's former students, who enjoyed exchanging news with him.

Although we had formed a business, it was in teaching that Michael really came alive. Not only did he possess an unending wealth of information about lens design—there never was a question he could not answer—he was a natural teacher. He brought to his teaching kindness and humor, with a considerable intuitive understanding of the problems of the student, of whatever age.

Michael began the lens design courses whilst still at Imperial College early in the 1980s, supported by Dr. Walter Welford. He continued the courses after the formation of Kidger Optics, subsequently giving them all over the world. The lectures he gave covered a full spectrum of knowledge in all aspects of lens design. Some were at a "basic" level, some at a "higher" level. He believed profoundly that the purpose of the courses was to teach the student about lens design and not "to sell SIGMA," his computer-aided optical design program. These were five-day courses incorporating a mixture of teaching followed by instruction, with practical use of examples.

Michael was the kind of person who "got on with the job." If his work was innovative or remarkable to others, for him it was no more than he should have done.

Throughout his life Michael had a great love of cricket. It was his obsession once the cricket season was under way, and particularly the cricket Test Series, when it seemed that each room in our home had a radio or TV switched on, so that not a piece of the game should be missed if he moved from one room to the next. When he was traveling it was my job to fax copies of the various newspaper reports to him, wherever he was. His love of cricket led to his love of Australia and the Australian people. Somehow, he managed to fit in a lens design course in Australia whenever there was a Test Match. In Sydney he worked with his friend and colleague Dr W. H. Steel, giving lens design courses. As a result of one of the courses he met a fellow cricket enthusiast, Richard Walmsley. They watched cricket at the Adelaide cricket ground, developing and enjoying a continuing friendship over the years.

Michael enjoyed music—all things musical had been central to the Kidger lives, and he shared his love of opera with his son David. With his daughter Julia, her husband Ian, and our family dogs, he enjoyed visits to the country fairs. These were special times for him.

He had a remarkable wealth of knowledge. There never seemed to be any question or issue that we could not discuss, or on which he was not well informed. He was a patient, extremely kind and clever man, with a quiet sense of humor, who though reluctant to speak of his own abilities was quick to recognize the abilities of others.

I hope that this Foreword has brought you closer to the "man." To give you another "picture" and further understanding, I am pleased to include David Williamson's letter that he sent to me shortly after Michael died. This tribute brings together so much that so many of his colleagues and students wrote to me in their tributes to Michael. David writes:

"As you may know, Michael was a great influence in my choice of lens design as a career, as well as my teacher in the mysteries of geometrical optics and aberration theory. Looking back, I can see how lucky I was to have studied at Imperial College then—it has certainly served me well over the years. Even now, when my bosses and customers in the U.S. are telling me I am using too many lenses or mirrors with impossible tolerances, I think back to how Michael may have dealt with the situation—a wry smile and a few well-chosen words. Rarely am I as calm, clear or succinct as he, but the guidance is still there! Of course the simplicity and clarity of his teachings have helped countless people over the years."

We live in an exciting age of tremendous technological developments. Those men and women who have chosen a career in lens design have a special gift with which to work, enabling them to be strong contributors to the great advances that are being made in science and technology. In working with exciting and innovative new equipment, we should remember that, as in the past, it is the man or woman and their capabilities that interpret and provide the required results; the equipment, the computer, the software, are the tools for this creation and innovation. This book is about learning the many parts of lens design that you need to know to enable you to achieve what you want in your research and your workplace. Michael was always available to answer any questions; his book will continue to answer your questions. He would want you to enjoy everything about lens design, to use your knowledge together with the latest technology to enable you to find the very best results, to go forward as he did.

You are now his students—I know he would wish you every success.

In conclusion, I want to recognize the long and dedicated efforts of Don O'Shea—President of SPIE, 2000—for his work on this book. I also want to recognize David Williamson, a former student of Michael's, without whom this book would not have been possible. David worked tirelessly over very many months reviewing, editing, and incorporating the heart of Michael's work into the manuscript. I would also like to acknowledge the help of the SPIE staff, in particular Rick Hermann. I want to acknowledge the work of Brian Blandford who has carefully reviewed the text, correcting and adding material. I would also like to thank colleagues and friends for their assistance.

.

Tina E. Kidger
September 2001

PREFACE

This volume is based on Michael Kidger's short course for SPIE entitled "Fundamental Optical Design." It reviews basic geometrical optics and third-order aberration theory, using the nomenclature and sign conventions of the Optical Design Group at Imperial College in the 1960s given in W.T. Welford's book, *Aberrations of Optical Systems*.[1]

Michael's courses for SPIE were abbreviated forms of workshops that he taught for Kidger Optics. In these short courses, Michael concentrated on the application of this theory to the design of a variety of simple optical systems, with students spending about half of the course time working on these lenses with Michael's optical design code, SIGMA, under his supervision. This book attempts to re-create, for the reader, the teaching style and practical work that made these courses so popular with students all around the world. The design examples include prescriptions and aberration data generated by SIGMA, although the interested reader will find sufficient data to further explore the designs with any available optical design software.

With the advent of the PC in the 1980s, such software has become much more accessible to engineers and scientists without formal training in optics. Michael's short courses were aimed at such newcomers to what can seem, at first, a very daunting field. He always emphasized the need to understand why a particular lens works (or, more commonly, does not work!), rather than to blindly hope that the optimization code will find a miraculous, practical solution. In fact, the earlier Imperial College courses on optical design did not encourage access to an optimization program (in the 1960s and 1970s, residing on mainframe computers) until well after the student had a thorough understanding of geometrical optics and third-order aberration theory. This more academic approach, as part of a master's course in applied optics, encouraged a certain intangible "feel" for the subject that had the potential to develop into a more intuitive approach as experience was gained with a wider variety of optical systems, some of which might be novel or innovative.

In spite of the enormous improvements in computing power, optical design remains a discipline that is best developed in apprenticeship to a master practitioner. Traditionally, this takes years, but Michael gave many students a brief taste of this mysterious process of osmosis for a few days or hours. In the fast-paced modern world, this is becoming increasingly rare and precious, and it is hoped that this book at least provides a glimpse of it for posterity.

One of the reasons optical design is not easily learned from a textbook is that, in many cases, it is a necessary but not sufficient condition that the third-order aberrations are correctable. When fifth- and higher-order aberrations dominate, as they do at the larger apertures and field sizes often required, analytical dissection of the problem starts to fail, and optimization codes, experience and intuition become the designer's principal "tools of the trade." In the commercial world, there is also the need to find a design that is manufacturable, often with conflicting requirements of low cost and high performance. A second volume, *Intermediate Optical Design*, will explore some of these issues—applied to more complex designs—but the intermediate material will remain firmly grounded in the foundations of this first volume.

Most of the material in these volumes originates either directly from Michael's course notes, or from the unfinished book that he was working on. The editing process has included, for completeness, the addition of some material, while at the same time trying to retain Michael's original intent and style. In the first volume, this includes a brief discussion of pupil aberrations, some additional visual optical designs, as well as some catadioptric astronomical telescopes. Our hope is that Michael would be pleased with the result!

David M. Williamson
September 2001

[1]W. T. Welford, *Aberrations of Optical Systems*, Adam Hilger (1986).

List of Symbols

A	ni (refraction invariant)
B	thin-lens conjugate variable (or magnification variable)
c	curvature of a surface (= $1/R$)
C	thin-lens shape variable (or bending variable)
C_1	Seidel coefficient of longitudinal chromatic aberration or "axial color"
C_2	Seidel coefficient of transverse chromatic aberration or "lateral color"
d	axial distance between two surfaces
D	distance between two surfaces, measured along an exact ray
e	eccentricity of a conicoid
ε	$(1-e^2)$ for a conicoid
E	defined by $E \cdot H = \bar{h}/h$ ($\bar{h}$ is the paraxial chief ray height)
f, f'	focal length
h	paraxial ray height
H	Lagrange invariant (= $n\,u\,\eta$)
i	paraxial angle of incidence
I	exact angle of incidence
k	conic constant of a conicoid = $-e^2 = \varepsilon - 1$
K	power of a surface or system
l	Object distance, measured from surface (or lens) to the object
L, M, N	direction cosines of exact ray
n	refractive index
q	ratio of n / n'
R	radius of a surface (= $1 / c$)
S_1	Seidel spherical aberration coefficient
S_2	Seidel coma coefficient
S_3	Seidel astigmatism coefficient
S_4	Seidel field curvature coefficient (Petzval sum = S_4/H^2)
S_5	Seidel distortion coefficient
u	paraxial ray angle
V	Abbe V-value = $(n_d - 1) / (n_F - n_C)$
W	wavefront aberration
x, y, z	ray coordinates at a surface
x, y, z	ray coordinates at the vertex plane of a surface
η, ξ	coordinates in the object space

Note that a primed quantity refers to the image space.

Any quantity associated with a chief ray is denoted by a bar, e.g., $\bar{h}, \bar{u}$

CHAPTER 1
GEOMETRICAL OPTICS

In this chapter we will introduce most of the basic concepts of geometrical optics, although it is likely that most readers will be familiar with these concepts from a study of more elementary texts.

Although all of the basic principles of geometrical optics can be derived from a knowledge of the wave nature of light, we will not follow this approach here. Except in special cases, an understanding of the rigorous derivations of these principles is not helpful in lens design.

However, we will discuss the limitations of geometrical optics when they are relevant to lens design, because there are situations in which an understanding of physical optics is absolutely essential. In lens design, as in other branches of applied science, it is most helpful to use the simplest approximation that can be used for any given task.

Geometrical optics can be considered to describe, with a high degree of accuracy, the properties of lenses as the wavelength of the radiation, λ, approaches zero. In this situation, diffraction effects disappear. So geometrical optics will be quite accurate for the design of short-wavelength x-ray imaging systems (if the wavelength is short enough). On the other hand, geometrical optics is rarely completely adequate for the design of thermal imaging systems operating at wavelengths around 10 μm. In the visible waveband, some lenses can be designed and evaluated completely by using geometrical optics, while the evaluation of other lenses must use physical optics. However, in almost all cases, lenses are actually designed using the results of geometrical optics.

1.1 Coordinate system and notation

In this book we discuss primarily the design of centered optical systems. We define the optical axis of a lens to be the z-axis, with the y-axis in the plane of the diagram, in Fig. 1.1. The x-axis is orthogonal to the y- and z-axes; in a right-handed coordinate system the x-axis is positive into the diagram. In the case of lenses that are centered, the z-axis represents the common optical axis of the refracting and reflecting surfaces.

In many equations in geometrical optics, we are concerned with quantities that are affected by refraction or reflection at a surface or at a lens. In these cases, we represent quantities after refraction or reflection as primed quantities; for example, we shall see that we write n' for the refractive index after a surface in Eq. (1.1) below.

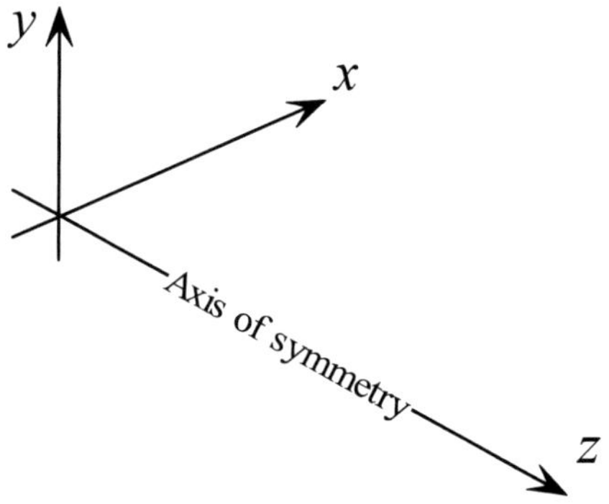

Figure 1.1. The coordinate system.

1.2 The rectilinear propagation of light

One of the most obvious properties of light, which is seen very clearly by observation of the path of a laser beam, is that it propagates in straight lines. In reality this is an approximation. As physical optics predicts and experiment confirms, any beam of light diverges to an extent determined by the beam width and the wavelength.

Furthermore, the rectilinear propagation of light is dependent on the uniformity of the medium through which it is passing. The classical example of a situation in which the medium is not uniform is the atmosphere; the existence of mirages and the compression of the image of the sun, when it is very near the horizon, are both due to nonuniformity of the atmosphere. More recent examples of nonuniform materials are gradient-index lenses.

Despite these reservations, the geometrical optics presented here will assume rectilinear propagation of light, embodied in the concept of the light ray. Many results that are discussed in Chapters 1 through 7 will demonstrate the usefulness of this approximation.

1.3 Snell's law

The inception of optical design, in my opinion, occurred in 1621. In that year, Snell formulated the law of refraction, which states that if the angle between an incident ray and the surface normal at the point of incidence, called the **angle of incidence**, is I; and if the **angle of refraction**, the angle between the refracted ray and the normal, is I'; then the angles are related by the equation

$$n \sin I = n' \sin I'. \tag{1.1}$$

In addition, Snell's law states that the incident ray, the refracted ray and the normal to the surface at the point of incidence are all in the same plane (Fig. 1.2). The quantities n and n' are the refractive indices of the two materials. While

Eq. (1.1) can be taken as the definition of refractive index, it should also be noted that the refractive index of a material is more fundamentally defined as

$$n = \frac{c}{v}, \tag{1.2}$$

where c is the velocity of light in a vacuum and v is the velocity of light in the material.

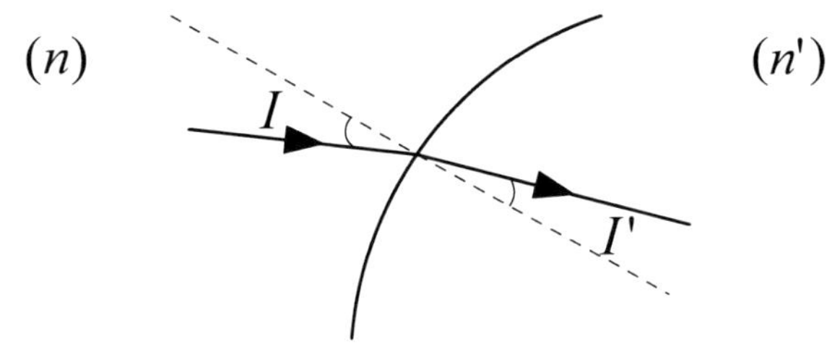

Figure 1.2. Snell's law (the law of refraction).

Since, at any surface, the ratio of the refractive indices determines the refracted ray angle, it is convenient to write

$$q = \frac{n}{n'}, \tag{1.3}$$

so that Snell's law is simplified to

$$\sin I' = q \sin I. \tag{1.4}$$

In the case of reflection (Fig. 1.3), the angle of the reflected ray is equal to the angle of incidence. Because of the sign convention for angles in ray tracing, the two angles have opposite sign. Therefore, the law of reflection is given as

$$I' = -I. \tag{1.5}$$

In lens design, it is quite usual to treat reflection as a special case of refraction, by assuming that

$$n' = -n \text{ or } q = -1. \tag{1.6}$$

This device is very useful in designing centered systems with reflecting surfaces, as the formulae for refraction can be applied unchanged to reflective surfaces, provided we adopt the convention that the refractive index changes sign after each reflection. After an even number of reflections, when the rays are traveling in the same sense as their initial direction, the refractive index will be positive; after an odd number of reflections, the refractive index will be assumed to be negative.

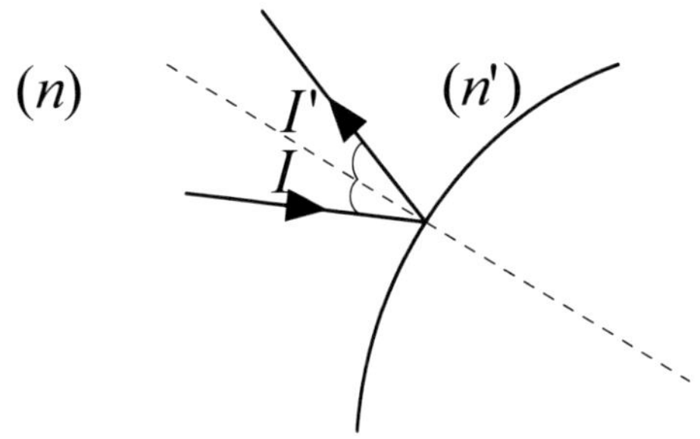

Figure 1.3. The law of reflection.

In the case of complex decentered systems, such as systems with several folding mirrors, this convention can become very confusing, and it is probably then more convenient to treat reflection as a separate case, keeping all refractive indices positive.

1.4 Fermat's principle

Fermat's principle is one of the most important theorems of geometrical optics. While it is not used directly in practical lens design (unlike Snell's law) it is used to derive results that would be impossible, or more difficult, to derive in other ways.

It may be stated as follows.

Figure 1.4 shows a physically possible path for a ray from A to D, and let the lengths of the segments along the ray be d_1, d_2, d_3.

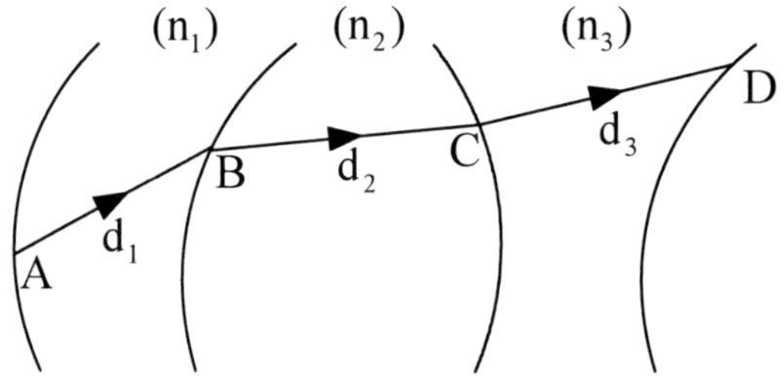

Figure 1.4. Optical path length.

We then define the optical path length in any medium to be the product of the distance traveled and the refractive index:

$$\text{Optical path length} = [\text{ABCD}] = \sum n_i d_i, \tag{1.7}$$

where the square brackets are used to distinguish the optical path length from a geometrical distance.

Fermat's principle states that the optical path length along a physically possible ray is stationary. For example, take the simple case of a plane refracting surface, as shown in Fig. 1.5.

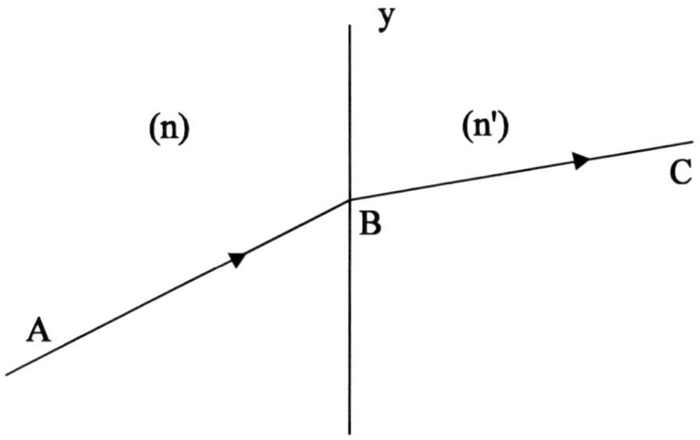

Figure 1.5. An example of Fermat's principle.

Here we have a ray passing through two points A and C; it is assumed to intersect the plane refracting surface at B. Fermat's principle states that if we write an expression for the optical path length as a function of y, and then differentiate with respect to y, the point where the differential is zero will represent the point B.

1.5 Rays and wavefronts—the theorem of Malus and Dupin

We have encountered the concepts of rays and optical path length, and we must now define the wavefront. In geometrical optics we define a wavefront as follows:

A wavefront is a surface of constant optical path length, from a point in the object.

In other words, if we trace several rays from a source at a point A, as shown in Fig. 1.6, then the points B_1, B_2, B_3, etc., all represent points that have the same optical path length from A. In the case of these points it is clear that since they are in the same medium as A, the locus of the points B_1, B_2, B_3, etc., is a surface of constant optical path length and therefore a **wavefront**. It is a sphere centered on A. If, however, we take the points C_1, C_2, C_3, etc., which are not in the same medium as A, the wavefront that is described by these points is not, in general, a sphere.

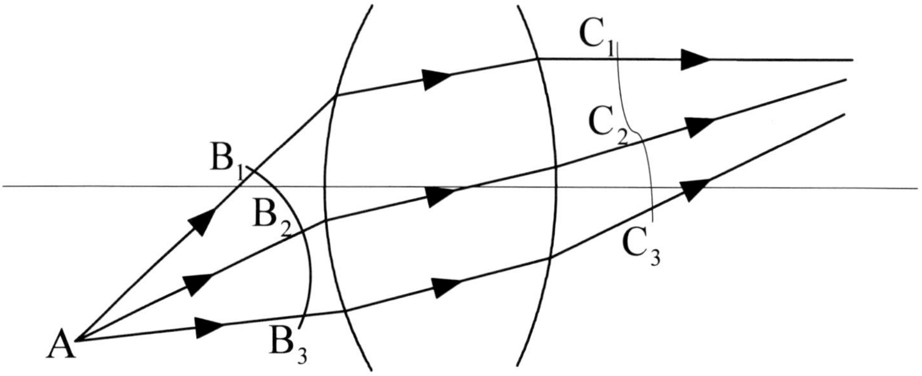

Figure 1.6. Wavefronts.

Note that this definition of a wavefront depends purely on geometrical optics, and we have not considered physical optics at all. In fact, in most cases, this **geometrical** wavefront does correspond to a surface of constant phase, as determined by **physical optics**; the major exception to this correspondence is the case of a Gaussian laser beam, with a very low convergence angle.

The theorem of Malus and Dupin states that these geometrical wavefronts are orthogonal to the rays from the point A; in other words, the rays are always perpendicular to the geometrical wavefront.[1]

There is an exception to this. In the case of a nonisotropic material, such as a birefringent crystal, the refractive index depends on the direction of propagation of the ray, and in this case the rays are not, in general, normals to the wavefront.

The major use of this theorem in practical lens design is to assist in understanding the relationship between transverse ray aberrations and wavefront aberrations, which we will discuss later.

1.6 Stops and pupils

The diagram below (Fig. 1.7) shows a simple lens system, with a stop between two lenses. If this stop limits the size of the beam from an axial point, it is known as the aperture stop.

Since the diameter of the beam passing through an optical system is always limited by something, every optical system has an aperture stop of some sort. In some cases, the aperture stop is a separate entity, as in the diagram. In other cases the aperture stop may simply be part of the lens mount.

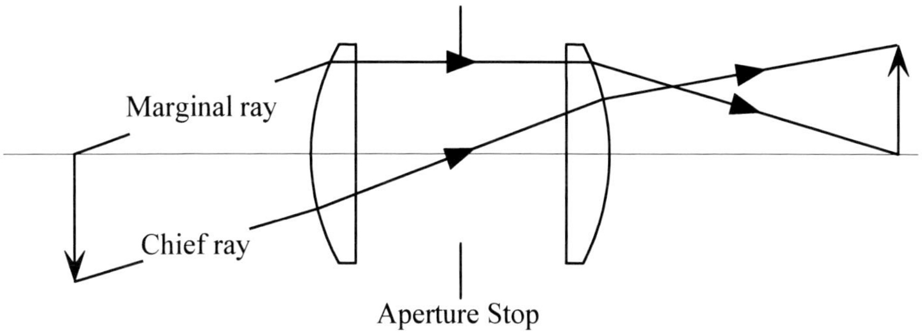

Figure 1.7. Simple lens system and aperture stop.

1.6.1 Marginal and chief rays

In Fig. 1.7, the ray that passes from the center of the object, at the maximum aperture of the lens, is normally known as the **marginal ray**. It therefore passes through the edge of the aperture stop. Conventionally, this ray is in the *y-z* plane, usually called the **meridian plane**.

The **chief ray** is defined to be the ray from an off-axis point in the object passing through the center of the aperture stop; although there can be an infinite number of such rays, we can usually assume, at least for centered systems, that the chief ray is also restricted to the meridian plane.

1.6.2 Entrance and exit pupils

In the case of a lens with an internal aperture stop, as in Fig. 1.7, the concepts of entrance pupil and exit pupil are particularly important. If we view the lens in Fig. 1.7 from the object space, we see an image of the aperture stop, known as the **entrance pupil**. Its position is found by the construction shown in Fig. 1.8.

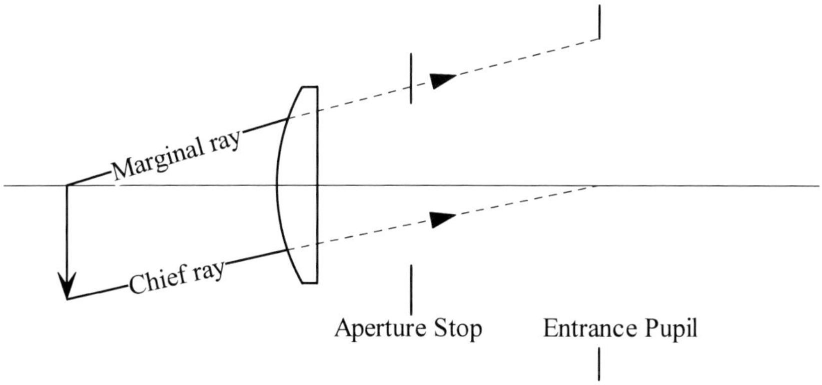

Figure 1.8. Construction to find the entrance pupil position.

In the image space, the image of the aperture stop is known as the **exit pupil**, and its position is found using a similar construction.

1.6.3 Field stops

In addition to the aperture stop, which limits the size of the image-forming beam, all optical systems also have a **field stop**, which limits the size of the image or the object. This field stop is normally at, or close to, the object or image surface.

In some cases, the function of the field stop is very clear. For example, in the case of eyepieces used in microscopes or binoculars, the size of the field of view is limited by a physical stop at the plane of the intermediate image. In the case of a camera, the field stop is effectively the "gate" that defines the area of the film that is exposed.

1.7 Surfaces

The interface between media of different refractive indices may be described by any number of geometrical shapes. Here we will consider only some of the more common forms.

1.7.1 Spheres

Since many lenses are constructed from surfaces that are nominally spherical in form, it is convenient at this point to discuss several forms of the equations that can be used to represent a spherical surface.

The equation of a sphere, of radius R, and passing through the origin is given by

$$R^2 = x^2 + y^2 + (z - R)^2 . \tag{1.8}$$

Writing $r^2 = x^2 + y^2$, for convenience

$$R^2 = r^2 + (z - R)^2 , \tag{1.9}$$

or

$$z^2 - 2zR + r^2 = 0 , \tag{1.10}$$

giving

$$z = R \pm \sqrt{R^2 - r^2} . \tag{1.11}$$

For the first intersection point (as shown in Fig. 1.9) we need the negative square root, giving

$$z = R - \sqrt{R^2 - r^2}\,. \tag{1.12}$$

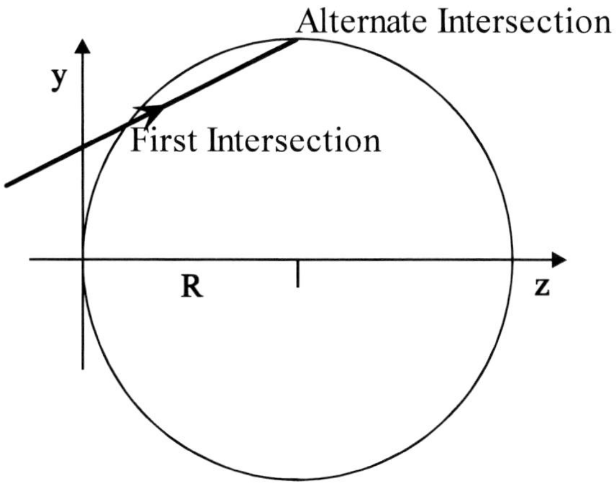

Figure 1.9. The sphere.

Since R is often large compared with r, we would be determining z by the difference between two large numbers.

To overcome this, we first replace Eq. (1.11) by

$$z = R\left[1 - \sqrt{1 - \left(\frac{r}{R}\right)^2}\right], \tag{1.13}$$

which always gives us the solution nearest to the vertex of the surface, and we also replace the radius of curvature of the surface, R, with $1/c$, where c is the curvature of the surface. Then we have

$$z = \frac{1 - \sqrt{1 - (cr)^2}}{c}\,. \tag{1.14}$$

By simple algebra we can also write this equation as

$$z = \frac{cr^2}{1 + \sqrt{1 - (cr)^2}}\,. \tag{1.15}$$

It is sometimes useful to have an expression for z in terms of a power series in r^2. Returning to Eq. (1.14),

$$\begin{aligned} zc &= 1-\sqrt{1-c^2r^2} \\ &= 1-\left(1-\frac{1}{2}c^2r^2-\frac{1}{8}c^4r^4-\frac{1}{16}c^6r^6-....\right) \end{aligned} \tag{1.16}$$

and therefore

$$z=\frac{1}{2}c\ r^2+\frac{1}{8}c^3r^4+\frac{1}{16}c^5r^6+.... \tag{1.17}$$

While we would never calculate the z-coordinate of a surface from Eq. (1.17), it will be useful later.

1.7.2 Quadrics of revolution (paraboloids, ellipsoids, hyperboloids)

If we take the case of an ellipsoid (Fig. 1.10) as a simple example, its equation can be written in the form

$$\frac{(z-a)^2}{a^2}+\frac{y^2}{b^2}=1, \tag{1.18}$$

where a and b are the semi-major axes of the ellipsoid, and

$$b^2=a^2(1-e^2)=a^2\varepsilon, \tag{1.19}$$

where e is the eccentricity of the ellipsoid, and ε is defined to be $(1-e^2)$. Clearly, $\varepsilon=b^2/a^2$. Equation (1.18) can then be transformed into

$$\varepsilon z^2-2a\varepsilon z+r^2=0 \tag{1.20}$$

in a similar manner to Eq. (1.10).

It may be shown that the vertex radius of the surface, R, is equal to $b^2/a=a\varepsilon$, so following a similar procedure to that described above, we get

$$z=\frac{cr^2}{1+\sqrt{1-\varepsilon(cr)^2}}, \tag{1.21}$$

where $c = \frac{1}{R} = \frac{1}{a\varepsilon}$.

In most lens design programs, this is rewritten as

$$z = \frac{cr^2}{1 + \sqrt{1 - (1+k)c^2r^2}}. \tag{1.22}$$

where k is referred to as the **conic constant,** and

$$k = \varepsilon - 1 = \frac{b^2}{a^2} - 1 = -e^2. \tag{1.23}$$

Note that, although we have started from the simple case of the ellipsoid, in order to relate to the simple High School equation for the ellipsoid, Eqs. (1.20) and (1.21) are general equations describing any conicoid. The significance of e, ε, and k are therefore as follows:

	$k > 0$	$\varepsilon > 1$	oblate ellipsoid
$e = 0$	$k = 0$	$\varepsilon = 1$	sphere
$0 < e < 1$	$-1 < k < 0$	$0 < \varepsilon < 1$	prolate ellipsoid
$e = 1$	$k = -1$	$\varepsilon = 0$	paraboloid
$e > 1$	$k < -1$	$\varepsilon < 0$	hyperboloid

If we expand Eq. (1.22) as a power series, we get a similar result to Eq. (1.17), but with slight differences due to the extra term, ε:

$$z = \frac{1}{2}c\,r^2 + \frac{1}{8}(1+k)c^3r^4 + \frac{1}{16}(1+k)^2c^5r^6 + \ldots. \tag{1.24}$$

We therefore see that the differences between a sphere and a conicoid appear as terms proportional to r^4, r^6, and r^8, but the term in r^2 is the same in both cases.

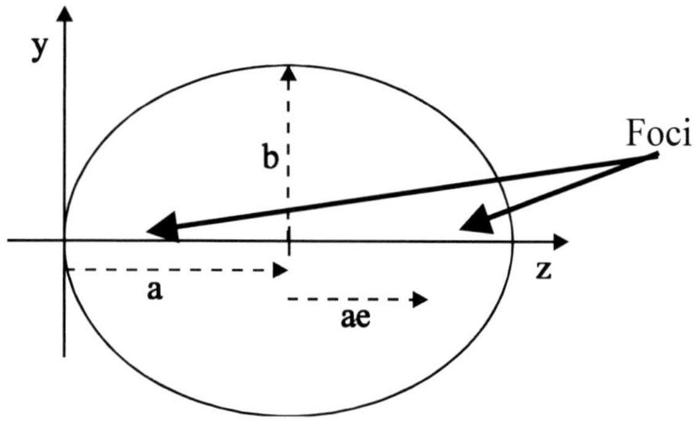

Figure 1.10. A quadric of revolution (in this case, a prolate ellipsoid).

1.7.3 Oblate ellipsoid

In this case, the surface is generated by rotation of an ellipse about its **minor** axis (Fig. 1.11). Since its foci are not on the optical axis, z, it is a not a conic section.

Following the same procedure as in Eqs. (1.18) to (1.23) (and keeping a as the major axis, and b as the minor axis), the equation of the ellipse is

$$\frac{(z-b)^2}{b^2}+\frac{y^2}{a^2}=1. \tag{1.25}$$

Let

$$b^2=a^2\left(1-e^2\right) \text{ or } a^2=\phi b^2; \tag{1.26}$$

then

$$\phi z^2-2b\phi z+r^2=0, \tag{1.27}$$

and

$$z=\frac{cr^2}{1-\sqrt{1-\phi c^2 r^2}}, \tag{1.28}$$

where

$$\phi=1+k=\frac{1}{1-e^2}, \text{ so } k=\frac{1}{1-e^2}-1=\frac{e^2}{1-e^2}. \tag{1.29}$$

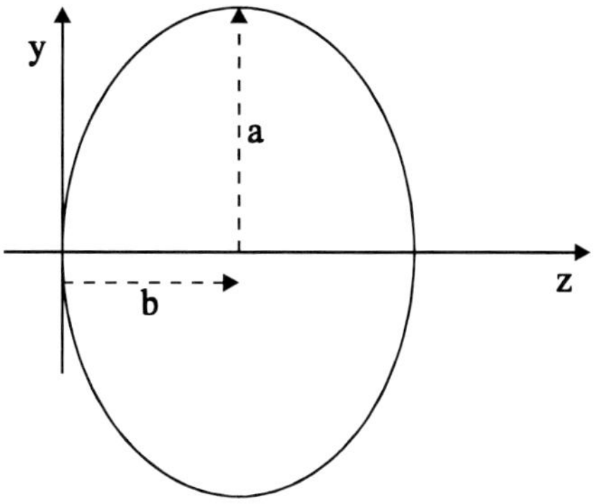

Figure 1.11. Oblate ellipsoid.

1.7.4 The hyperbola

The equation of a hyperbola can be written in the form

$$\frac{(z-a)^2}{a^2}-\frac{r^2}{b^2}=1, \tag{1.30}$$

where a and b are the semi-major axes of the ellipsoid;

$$b^2=a^2(e^2-1)=-a^2\varepsilon, \tag{1.31}$$

where ε is defined to be $(1-e^2)$; and e is the eccentricity of the hyperboloid. Clearly, $\varepsilon=-b^2/a^2$.

Equation (1.31) can then be transformed into

$$\varepsilon z^2-2a\varepsilon z+r^2=0 \tag{1.32}$$

in a similar manner to Eq. (1.10). This is the same equation as the equation for an ellipsoid [Eq. (1.20)], and therefore all of the results that apply to an ellipsoid also apply to a hyperboloid. Of course, for a (prolate) ellipsoid, $-1<k<0$ and for a hyperboloid, $k<-1$. Specifically, from Eq. (1.32),

$$\begin{aligned} z &= a\left(1-\sqrt{1-\frac{r^2}{\varepsilon a^2}}\right) \\ &= a\left[1-\left(1-\frac{r^2}{2\varepsilon a^2}-\frac{r^4}{8\varepsilon^2 a^4}-\frac{r^6}{16\varepsilon^3 a^6}---\right)\right] \\ &= \frac{r^2}{2\varepsilon a}+\frac{r^4}{8\varepsilon^2 a^3}+\frac{r^6}{16\varepsilon^3 a^5}. \end{aligned} \tag{1.33}$$

Comparing the first term with Eq. (1.17), it can be seen that the vertex curvature

$$c=\frac{1}{\varepsilon a} \text{ or } R=\varepsilon a. \tag{1.34}$$

Therefore, we get Eq. (1.24) again, but for the hyperboloid:

$$z=\frac{1}{2}c\,r^2+\frac{1}{8}(1+k)c^3r^4+\frac{1}{16}(1+k)^2c^5r^6+.... \tag{1.24}$$

Given a and b,

$$k=\varepsilon-1=-\left(1+\frac{b^2}{a^2}\right) \text{ and } c=\frac{1}{\varepsilon a}=-\frac{a}{b^2}. \tag{1.35}$$

1.7.5 Axicon

An axicon is a conical surface whose axis of symmetry is the optical axis (the z-axis). This can be represented very effectively as a hyperboloid, with a very large curvature at the vertex of the surface. If c becomes very large, Eq. (1.22) becomes

$$z=\frac{c\cdot r^2}{\sqrt{-(1+k)c^2r^2}}=\frac{r}{\sqrt{-k-1}}=\frac{r}{\sqrt{e^2-1}}. \tag{1.36}$$

Therefore, the cone half-angle, θ, shown in Fig. 1.12, is given by

$$\tan\theta=\frac{r}{z}=\sqrt{e^2-1}$$

and

$$\cos^2\theta=\frac{1}{e^2}=-\frac{1}{k} \text{ or } k=-\frac{1}{\cos^2\theta}. \tag{1.37}$$

Therefore, an axicon can be represented by a hyperboloid, with a conic constant given by Eq. (1.37) and a large vertex curvature. Figure 1.12 is a drawing of several axicons with different curvatures, but all with the same cone half-angle θ = 30 deg, such that

$$k=\frac{-1}{\cos^2(30)}=\frac{-1}{0.75}=-1.33333.$$

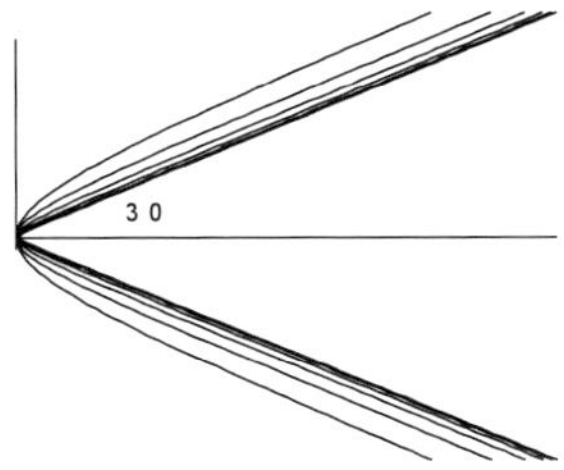

Figure 1.12. Several axicons, with $k = -1.3333$ ($\theta = 30$ deg) and different curvatures.

References

1. W. T. Welford, *Aberrations of Optical Systems*, Adam Hilger (1986).

CHAPTER 2
PARAXIAL OPTICS

In this chapter, we will consider the properties of optical systems in the region close to the optical axis, usually known as the "paraxial region." The treatment of light in this paraxial region is referred to as "paraxial optics."

In many texts this region is referred to as the "Gaussian region" and the properties of light in this region are known as "Gaussian optics." However, in modern optics many scientists and optical engineers working with laser beams with a Gaussian profile also make use of the term "Gaussian optics." Therefore, we will use the terms "paraxial optics" and "paraxial region" in order to eliminate any confusion.

Although we limit our discussion to a small area enclosing the optical axis, paraxial optics is very important. In any well-corrected lens, the rays from outside the paraxial region *must* intersect the image surface at points very close to the paraxial image points, and we therefore find that many concepts that are strictly defined for the paraxial region are applicable outside of the paraxial region as well.

2.1 Paraxial rays

The paraxial approximation enables us to compute some of the most basic properties of an optical system with a minimum of calculation. With the advent of electronic computers, it might be thought that such an approximate treatment would not be needed. After all, why calculate approximately what can be easily computed with great precision? It is because the paraxial treatment can provide the designer with insight into the performance of the system with a modest effort.

2.1.1 The sign convention

The progress of geometrical optics has not been helped by the existence of several different sign conventions. Before we begin our discussion of paraxial optics, it is necessary to define our conventions for ray positions and angles.

In Fig. 2.1 below we have a ray that intersects the optical axis on the *right-hand* side of the vertex of the surface. We measure ray intersection distances from the surface vertex, the point where the surface intersects the optical axis, to the ray intersection point, as shown. We use the normal Cartesian sign convention. Therefore, the z-coordinate of the ray intersection point will be positive as shown. In addition, the y-coordinate of the ray at the surface will be positive, as the ray intersects the surface *above* the optical axis.

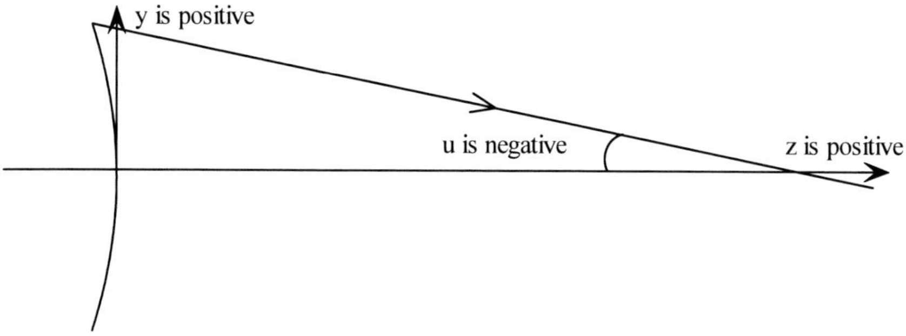

Figure 2.1. Sign conventions.

The ray angle in the figure is defined to be negative, as the slope of the ray is negative, using the normal conventions of coordinate geometry. Note that none of the signs depend on the direction of propagation of the ray; if the ray travels from right to left along the same path, the ray coordinates and angle are unaffected.

2.1.2 The paraxial region

As we have seen in Chapter 1, a spherical surface can be completely described by a power series of the form

$$z = \frac{1}{2}cr^2 + \frac{1}{8}c^3r^4 + \frac{1}{16}c^5r^6. \quad (1.17)$$

We define the paraxial region to be the region where any refracting or reflecting surface can be described by the first term in this power series:

$$z = \frac{1}{2}cr^2. \quad (2.1)$$

In addition, we limit the extent of the object and image surfaces to terms proportional to the square of the object and image sizes. Further, we assume that any angle that we may encounter is sufficiently small so that its sine is equal to the angle itself in radians, and its cosine is equal to unity. Under these conditions, it may be shown that the paraxial region is a region in which every plane in the object space is imaged perfectly on a plane in the image space.

2.2 The cardinal points

The cardinal points are the focal points, principal points and nodal points. There is a pair of each, defined by two paraxial rays traveling through the lens from the object and image directions. As we shall see, for optical systems whose objects and images are in air (the most common situation), the nodal points coincide with the principal points. These two points and the two foci are often called the four cardinal points of the lens.

2.2.1 Principal points

In Fig. 2.2 we show a thick, centered optical system, forming an image of an axial object point at infinity. Within the paraxial region this system will form a perfect image at a point F′. We extend the ray in the object space and the image space until they intersect at a point P_1'. The plane through P_1' that intersects normal to the optical axis at P′ is known as the **second principal plane**. P′ is the **second principal point,** and the axial image point F′ is the **second principal focus.**

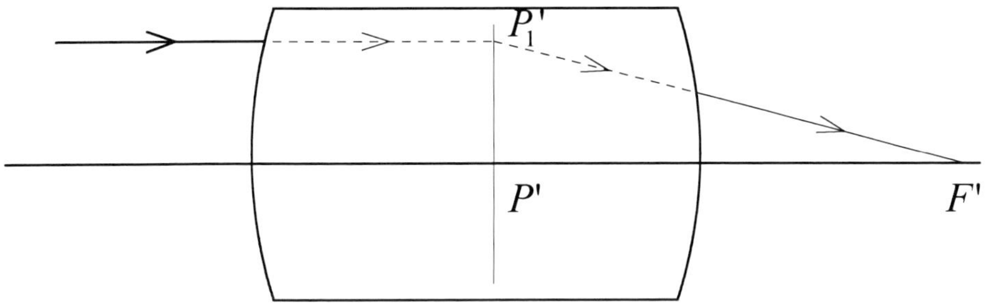

Figure 2.2. The second principal point (P'_1) and principal focus (F′).

We also define corresponding quantities for a second ray, shown in Fig. 2.3, parallel to the axis in the image space; and these quantities are naturally known as the **first principal point** (P), the **first principal plane** (P_1P), and the **first principal focus** (F).

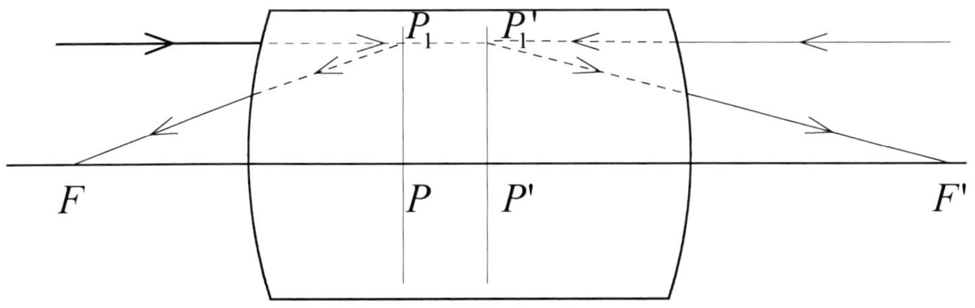

Figure 2.3. The two principal planes of unit magnification

If we consider the case where the two rays discussed above are at the same distance from the axis, as in Fig. 2.3, we can see that the rays both pass through

P_1 and P_1', showing that P_1' is the image of P_1. In other words, P_1 and P_1' are conjugate points, and they are clearly points of unit magnification; the planes containing P_1 and P_1' (the principal planes) are therefore planes of unit magnification.

2.2.2 Nodal points

If we consider a parallel fan of rays that enter a thick lens, as in Fig. 2.4, we can see that it is possible to find one ray that exits the lens at an angle parallel to its initial angle. This is ray *a* in Fig. 2.4. The points where ray *a* crosses the optical axis are the **nodal points** (N and N′) of the lens.

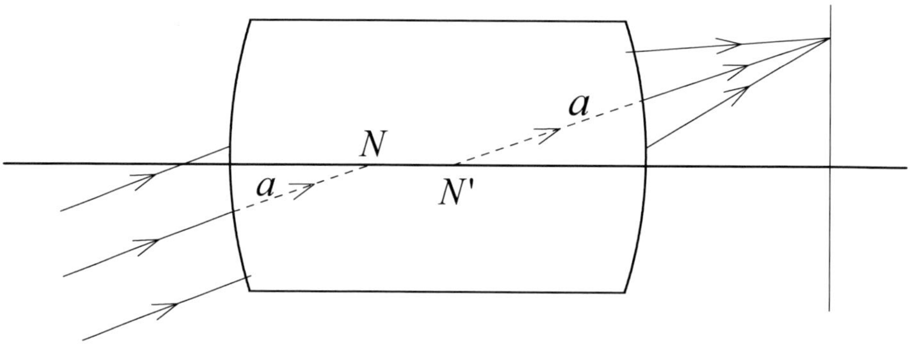

Figure 2.4. The two nodal points.

The position of the nodal points can be found by using the construction shown in Fig. 2.5. Consider ray *b*, which has been chosen to be parallel to the axis in the image space. This ray must pass through the first principal focus, F, and it also passes through the off-axis point, F_1', in the second focal plane. We also consider ray *c*, which has been drawn so that the segment of *c* in the image space is parallel to the segment of *b* in the object space. Since all rays that pass through F_1' must be parallel to each other in the object space, the object-space segment of ray *c* must be parallel to its image-space segment, as indicated in the diagram. Therefore, the points N and N′ are the nodal points, and it is clear that

$$PF = F'N' = f. \tag{2.2}$$

Similarly,

$$P'F' = FN = f'. \tag{2.3}$$

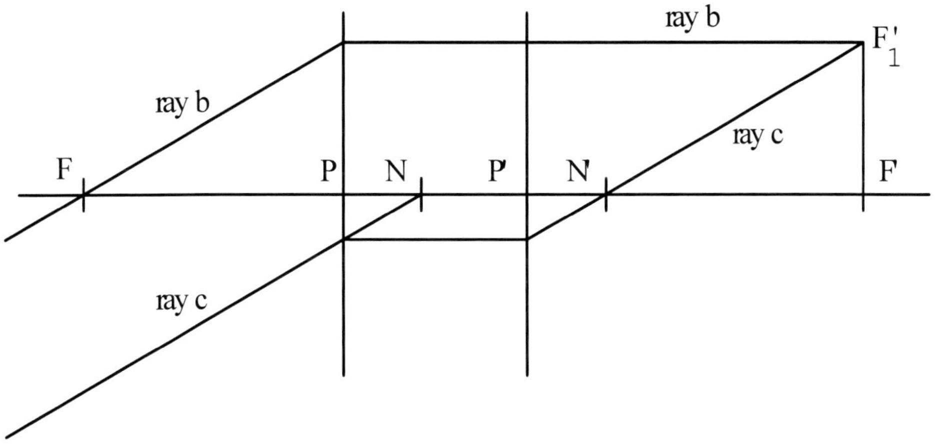

Figure 2.5. Determination of nodal point positions.

2.3 Paraxial properties of a single surface

We now discuss the properties of a single refracting surface. In Fig. 2.6, we consider a paraxial ray from an object point O, which forms an image at O′, having passed through the refracting surface at S. We have a wavefront PA in the object space and a wavefront SB in the image space. AQ is the plane through the vertex of the surface. The object distance is l and the image distance is l'; u is the paraxial ray angle AOP, and u′ is the paraxial ray angle BO′S. The height of the ray is h. We deliberately do not specify whether this is the ray height at P or Q or S, because in the paraxial approximation it is of no importance.

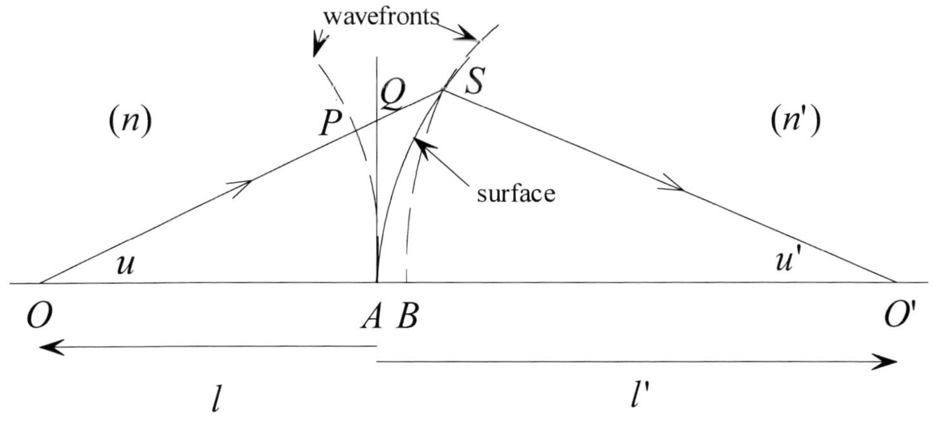

Figure 2.6. A paraxial ray.

Since O and O′ are conjugate points (object and image), it is clear that we have, in the object space, a spherical wavefront diverging from O and, in the image space, a spherical wavefront converging on O′. Therefore, the optical paths of a ray passing along the axis and a ray through S must be equal:

$$[\mathrm{OABO'}] = [\mathrm{OPQSO'}] \tag{2.4}$$

Since

$$[\mathrm{OA}] = [\mathrm{OP}] \text{ and } [\mathrm{BO'}] = [\mathrm{SO'}], \tag{2.5}$$

it follows that

$$[\mathrm{AB}] = [\mathrm{PQS}] = [\mathrm{PQ}] + [\mathrm{QS}]. \tag{2.6}$$

To the paraxial approximation,

$$\mathrm{QS} = \frac{1}{2}ch^2, \tag{2.7}$$

and

$$\mathrm{PQ} = -\frac{1}{2}\frac{h^2}{l}. \tag{2.8}$$

So, the optical path length from P to S is

$$[\mathrm{PQS}] = \frac{1}{2}nh^2\left(c - \frac{1}{l}\right). \tag{2.9}$$

Similarly,

$$[\mathrm{AB}] = \frac{1}{2}n'h^2\left(c - \frac{1}{l'}\right). \tag{2.10}$$

So it follows that, since wavefronts are surfaces of equal path length,

$$nh^2\left(c - \frac{1}{l}\right) = n'h^2\left(c - \frac{1}{l'}\right) \tag{2.11}$$

or

$$\frac{n'}{l'} = \frac{n}{l} + (n' - n)c. \tag{2.12}$$

Note the signs of these quantities. In Fig. 2.6, l' is positive, since the image is on the *right* of the surface. l is negative, since the object is on the *left* of the surface. Because the center of curvature of the surface is on the *right* of the surface, $R > 0$ and therefore $c > 0$.

Equation (2.12), which is analogous to the simple equation often given in high school textbooks on physics:

$$\frac{1}{f} = \frac{1}{u} + \frac{1}{v}, \tag{2.13}$$

gives us the relationship between the object distance and the paraxial image distance, measuring these distances from the vertex of the refracting surface.

2.4 Paraxial ray tracing

Returning to Eq. (2.12), if we multiply both sides of the equation by h, we have

$$\frac{n'h}{l'} = \frac{nh}{l} + (n' - n)hc; \tag{2.14}$$

and if we substitute

$$u' = -\frac{h}{l'} \text{ and } u = -\frac{h}{l}, \tag{2.15}$$

we get

$$n'u' = nu - hc(n' - n). \tag{2.16}$$

These quantities, u and u', represent the paraxial ray angles. Note that in Fig. 2.6, u is positive and u' is negative; we therefore need the minus signs in Eq. (2.15) in order to keep the sign convention consistent.

In order to calculate the complete path of a paraxial ray through a lens, we need one further relationship.

If we have a paraxial ray, with a height h_i at surface i, and an angle u_i' after refraction (or reflection) at surface i, the image distance of this ray, l_i', is given by

$$l_i' = -\frac{h_i}{u_i'}. \tag{2.17}$$

It should then be clear that the object distance for the next surface, surface $(i+1)$, will be given by

$$l_{i+1} = l_i' - d\,. \tag{2.18}$$

From Fig. 2.7 and using Eq. (2.17) again we have

$$h_{i+1} = h_i + u_i' d\,. \tag{2.19}$$

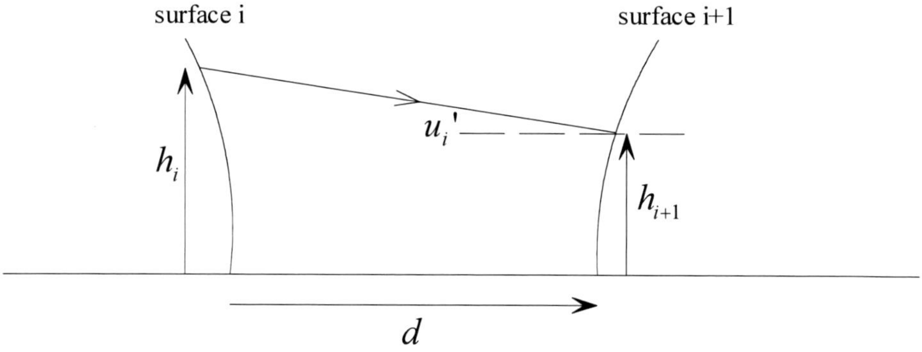

Figure 2.7. Paraxial ray transfer.

To trace a paraxial ray, therefore, we use Eq. (2.16),

$$n'u' = nu - hc(n' - n)\,, \tag{2.16}$$

and Eq. (2.19) alternately. Equation (2.16) is used to calculate the value of u' after refraction at the first surface; Eq. (2.19) is then used to calculate the value of h at the second surface. This process is repeated for subsequent surfaces until we have found the values of h and u' at the last (kth) surface of the lens. We then calculate the back focal length, *bfl*, from Eq. (2.19), by setting $h_{k+1} = 0$,

$$bfl = -\frac{h_k}{u_k'}\,. \tag{2.20}$$

In the case of an object at infinity we can also calculate the second focal length, f', from the construction shown in Fig. 2.8, by using the formula

$$f' = -\frac{h_1}{u_k'}\,. \tag{2.21}$$

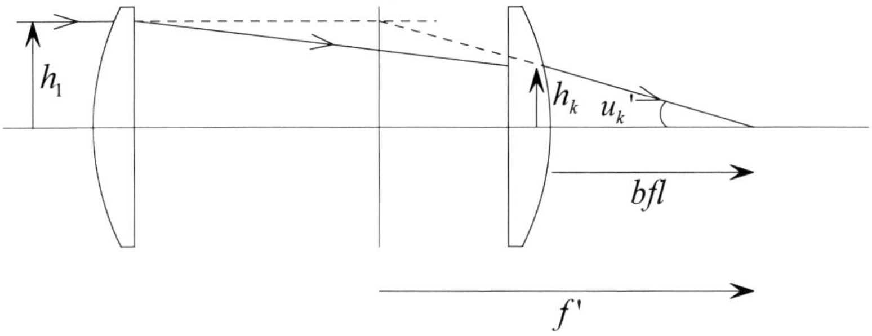

Figure 2.8. Calculation of the back focal length and second focal length.

2.4.1 Discussion of the use of paraxial ray trace equations

It is important to note that Eqs. (2.16) and (2.19) are derived from Eqs. (2.12) and (2.18):

$$\frac{n'}{l'} = \frac{n}{l} + (n' - n)c\,, \tag{2.12}$$

$$l_{i+1} = l_i' - d\,. \tag{2.18}$$

Since these were derived from the paraxial approximation, they are only valid for rays very close to the axis. We have derived the paraxial ray trace Eqs. (2.16) and (2.19) by multiplying Eqs. (2.12) and (2.18) by h and u, respectively, where h and u are the ray height and angle; in this context, Eqs. (2.16) and (2.19) only represent the path of a ray through a lens accurately if h and u are small.

If we use the paraxial ray trace equations to calculate the back focal length and second focal length, it should be clear that h can have any value, consistent with the object position being correct, since in Eq. (2.20), h_k and u_k' are both proportional to the initial value of h. The essential point here is that Eqs. (2.16) and (2.19) are linear equations, involving only the first powers of h and u.

It is therefore common practice to trace a paraxial marginal ray through a lens with a value of h corresponding to the full aperture of the lens, although this is almost always well outside the paraxial region. The terms "back focal length" and "focal length" in fact refer to paraxial quantities. Values derived by the method outlined here will be completely accurate, although the values of h and u from which they are calculated will only be an approximation to the real values of the ray height and angle.

2.5 The Lagrange invariant

Suppose we trace two paraxial rays through a lens, as shown in Fig. 2.9. For the purposes of illustration we will assume that one ray is a paraxial marginal ray and the other is a paraxial chief ray. Just in case these terms cause any confusion, we emphasize that a paraxial marginal ray is a ray passing through the edge of the aperture stop, traced using the paraxial ray trace equations. Similarly, a paraxial chief ray is a ray passing through the center of the aperture stop, usually at the full field of view of the lens, but again traced using the paraxial ray trace equations.

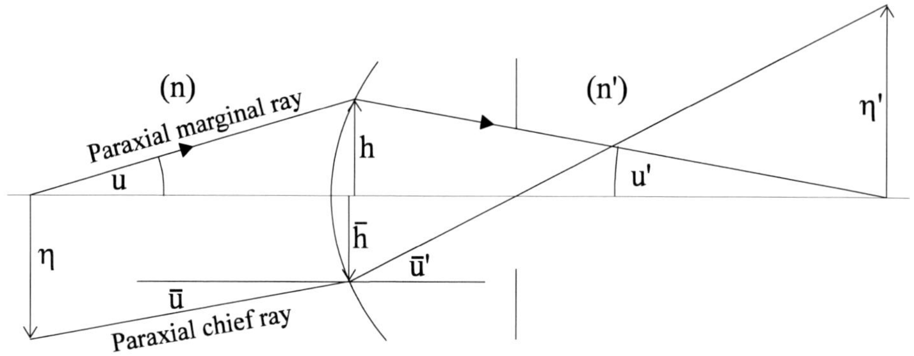

Figure 2.9. The Lagrange invariant.

At any given surface the paraxial marginal ray is traced using

$$n'u' = nu - hc(n' - n), \tag{2.16}$$

and the paraxial chief ray is traced using the same equation, but writing $\bar{u}$ for the chief ray angle and $\bar{h}$ for the chief ray height:

$$n'\bar{u}' = n\bar{u} - \bar{h}c(n' - n). \tag{2.22}$$

Eliminating $c(n'- n)$ we have

$$(nu - n'u')\bar{h} = (n\bar{u} - n'\bar{u}')h, \tag{2.23}$$

or

$$n(u\bar{h} - \bar{u}h) = n'(u'\bar{h} - \bar{u}'h) = H. \tag{2.24}$$

This quantity is known as the **Lagrange invariant**, and is denoted by *H*. So far, we have shown that *H* is invariant on refraction at a surface, but we will now show that it is also invariant on transfer between two surfaces.

For the paraxial marginal ray,

$$h_{i+1} = h_i + u_i' d_i , \tag{2.19}$$

and for the paraxial chief ray,

$$\bar{h}_{i+1} = \bar{h}_i + \bar{u}'_i d_i . \tag{2.25}$$

Eliminating *d*, we have

$$(h_i - h_{i+1})\bar{u}'_i = (\bar{h}_i - \bar{h}_{i+1})u_i' , \tag{2.26}$$

or

$$u_i' \bar{h}_i - \bar{u}_i' h_i = u_i' \bar{h}_{i+1} - \bar{u}'_i h_{i+1} , \tag{2.27}$$

and multiplying by the refractive index *n*, we have

$$n(u_i' \bar{h}_i - \bar{u}_i' h_i) = n(u_i' \bar{h}_{i+1} - \bar{u}_i' h_{i+1}) = H . \tag{2.28}$$

Equation (2.28) shows that *H* is also invariant after a transfer from one surface to another. Since it is invariant both during refraction and also during a transfer between surfaces, it is identical at all surfaces and spaces in a lens.

We do not have to use an axial ray and a chief ray to prove this result. In fact, any two different rays could have been used to show the invariance of *H*. In practice, however, *H* is usually computed from the paraxial marginal ray and the paraxial chief ray, so this convention has been followed here.

The Lagrange invariant is one of the most important concepts in paraxial optics. It can be shown that the square of the Lagrange invariant is proportional to the total energy transmitted by a lens, assuming that the object radiates uniformly. As we shall see later, it is also fundamental in the calculation of third-order aberrations, as well as being related to the number of linear pixels resolved by a perfect lens.

2.5.1 Transverse (lateral) magnification

At the object surface, the paraxial marginal ray height is, by definition, zero, and the chief ray height equals the object radius, η. Therefore, the Lagrange invariant simplifies to

$$H = n(u\eta - \bar{u}h) = n(u\eta - \bar{u}0) = nu\eta. \tag{2.29}$$

Similarly, at the image surface, *H* simplifies to

$$H = n'\left(u'\eta' - \bar{u}'0\right) = n'u'\eta' \tag{2.30}$$

and we have the simple result:

$$H = nu\eta = n'u'\eta' \; . \tag{2.31}$$

This result is extremely important because it relates the transverse magnification, m, which we define as

$$m = \frac{\eta'}{\eta}, \tag{2.32}$$

to the paraxial marginal ray angles, u and u', by the following relationship:

$$m = \frac{nu}{n'u'} \; . \tag{2.33}$$

In the common case of an object at infinity, u is zero and therefore the transverse magnification is also zero.

We also have to remember that in most finite-conjugate systems the image is inverted. This of course means that the magnification is negative, and this is the situation shown in Fig. 2.10.

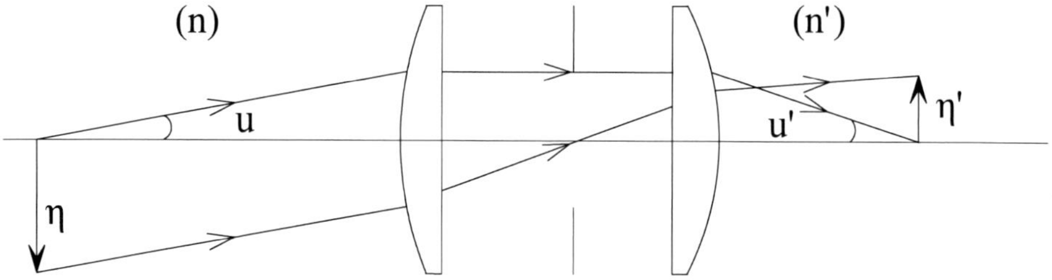

Figure 2.10. Transverse magnification.

2.5.2 Afocal systems and angular magnification

For the case when both the object and image are at infinity, the transverse magnification, as calculated by Eq. (2.32), becomes indeterminate. A system of this type is known as an **afocal** system; examples of common afocal systems are laser beam expanders and binoculars. (Actually, binoculars are often adjusted by the user for physiological reasons, to place the image at a finite distance in front of the eye, but for the purposes of this discussion we will assume that the image is at infinity.)

In the case of an afocal system we are not usually interested in the transverse magnification, but in the angular magnification, since it is this quantity that determines the angular subtense of the image. Since $u = 0$ in the object space, and $u' = 0$ in the image space, the Lagrange invariant simplifies to

$$H = -n\bar{u}h = -n'\bar{u}'h' \cdot \tag{2.34}$$

We then have

$$\text{Angular magnification} = \frac{\bar{u}'}{\bar{u}} = \frac{nh}{n'h'} . \tag{2.35}$$

In most afocal systems the object space and the image space are air, and this then further simplifies to

$$\text{Angular magnification} = \frac{\bar{u}'}{\bar{u}} = \frac{h}{h'} . \tag{2.36}$$

Since an afocal system produces an image at infinity, we can see from the definitions of principal points and the principal focus that the focal length will be infinite, and therefore the power is zero. However, the system does form a finite image of a finite object point, as we can see from Fig. 2.12, where A is an axial object point and A′ is the corresponding axial image point. Since the rays in Fig. 2.11 and 2.12 are identical, we can see that if the object height is h, the image height is h', where these quantities are the radii of the entrance pupil and the exit pupil, respectively. Therefore, the transverse magnification for a zero-power system, when used at finite conjugates, is given by

$$\text{Magnification} = \frac{1}{\text{Angular magnification}} . \tag{2.37}$$

Figure 2.11. Afocal system and angular magnification.

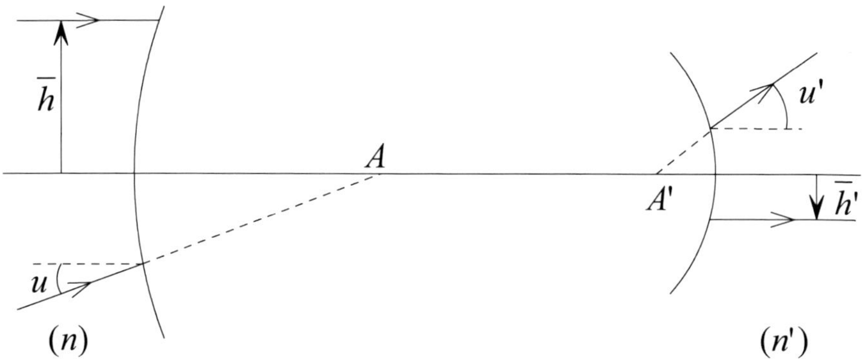

Figure 2.12. Zero-power (afocal) system at finite conjugates.

This is important in the design of binoculars. For example, if we have an instrument with an entrance pupil diameter of 30 mm, which is a common value, and if the angular magnification is 8, the diameter of the exit pupil must be 30/8 mm = 3.75 mm. Since the exit pupil of the instrument must be close to the pupil of the user's eye, we can see that the Lagrange invariant gives us some very important information.

2.6 Newton's conjugate distance equation

We now consider conjugate distance equations, which give us the position of the image in terms of the object position.

In Fig. 2.13, the ray $O_1FP_1O_1'$ is chosen to pass through the first principal focus, and it is therefore parallel to the optical axis in the image space. Similarly, the ray $O_1P_1'F'O_1'$ is parallel to the axis in the object space, and it must therefore pass through F′. Then, by similar triangles, we can see that

$$\frac{\eta}{z} = -\frac{\eta'}{f} \text{ and } \frac{\eta'}{z'} = -\frac{\eta}{f'}, \tag{2.38}$$

and therefore m, the magnification, is given by

$$m = \frac{\eta'}{\eta} = -\frac{f}{z} = -\frac{z'}{f'}, \tag{2.39}$$

so we have

$$zz' = ff'. \tag{2.40}$$

This is Newton's conjugate distance equation; it is rarely used in this form, but we can derive a more useful equation from it as follows.

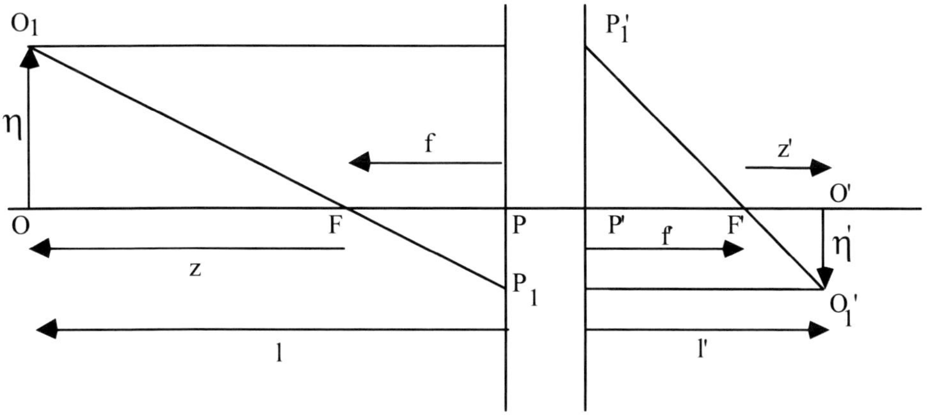

Figure 2.13. Newton's conjugate distance equation.

From Fig. 2.13 we can see that

$$z = l - f \quad \text{and} \quad z' = l' - f', \tag{2.41}$$

and therefore

$$\begin{aligned} m &= -\frac{f}{z} = -\frac{f}{l-f} \\ &= -\frac{z'}{f'} = -\frac{l'-f'}{f'}. \end{aligned} \tag{2.42}$$

From this,

$$ff' = (l-f)(l'-f'), \tag{2.43}$$

and therefore

$$1 = \frac{f'}{l'} + \frac{f}{l}, \tag{2.44}$$

or

$$l' - f' = \frac{l'f}{l}. \tag{2.45}$$

Then,

$$m = -\frac{l' - f'}{f'} = -\frac{l'f}{lf'} . \tag{2.46}$$

From Eqs.(2.33) and (2.15):

$$m = \frac{nu}{n'u'} = \frac{nl'}{n'l} , \tag{2.47}$$

and by comparing Eqs. (2.46) and (2.47) we see that

$$\frac{n'}{f'} = \frac{-n}{f} . \tag{2.48}$$

This quantity,

$$K = \frac{n'}{f'} = \frac{n}{-f} , \tag{2.49}$$

is known as the power of a system. The equivalent focal length (efl) is defined to be $1/K$, and is the (second) focal length of a system if the image space medium is air.

The conjugate distance Eq. (2.44) then becomes

$$\frac{n'}{l'} = \frac{n}{l} + K \tag{2.50}$$

or

$$n'u' = nu - hK . \tag{2.51}$$

2.7 Further discussion of the cardinal points

The relationships discussed earlier indicate that if we know F, F′, P, P′, then we can also find the positions of N and N′; these relationships were determined purely from geometry and the definitions of these quantities. But, we were not able to derive the relationship between f and f' purely by geometry. We need the Lagrange invariant, in the form of Eq. (2.47), in order to derive the important result

$$\frac{n'}{f'} = \frac{n}{f}. \tag{2.48}$$

With this result, we can continue the discussion of nodal points and principal points that was shown in Fig. 2.5. For lenses in air the refractive indices in the object space and the image space are both equal to unity, and, in this case,

$$-f = f'. \tag{2.52}$$

Therefore, the positions of P and N in Fig. 2.5 coincide, and P′ and N′ also coincide. We can now see, in general terms, where these nodal points and principal points will be, for some simple cases.

First, consider a thick equi-convex lens. If the lens is equi-convex, the ray that determines the nodal points will pass through the exact center of the lens. From Fig. 2.14 we can see that the ray angles in air are larger than the ray angles inside the lens, and it follows that N and N′ (and therefore P and P′) are as shown.

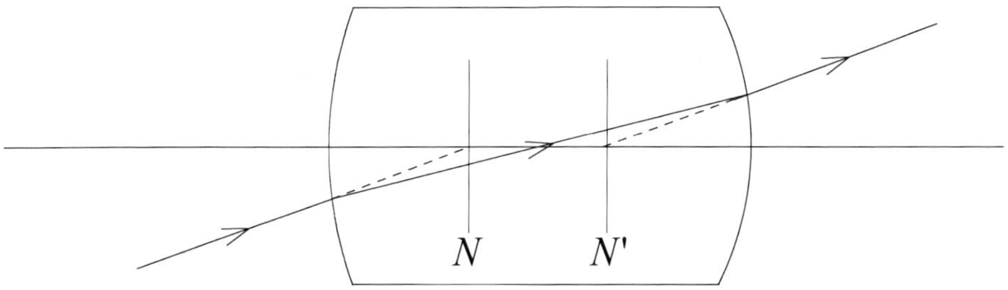

Figure 2.14. Nodal points for a single thick lens.

Now consider two separated thin lenses. For simplicity, consider the case when the two lenses have equal powers. Again, the nodal points will be determined by a ray that passes through the exact center of the system, and we can see from Fig. 2.15 that in this case N′ is to the *left* of N.

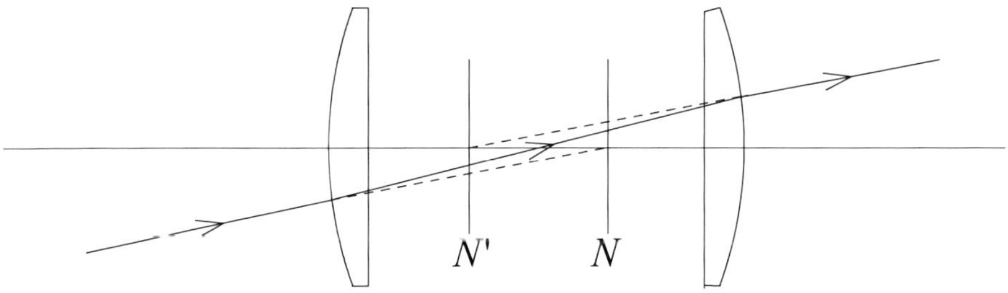

Figure 2.15. Nodal points for two separated lenses.

2.7.1 The combination of two lenses

When we wish to determine the focal length of a lens or the image position, we can do this by carrying out a paraxial ray trace using Eqs. (2.16) and (2.19). Even on a slow computer this is a very short calculation. We can also derive some useful results algebraically from the paraxial ray trace equations.

Suppose we have two thin lenses (they need not be lenses—they can also be surfaces, or optical systems as represented by their principal plane positions) such as shown in Fig. 2.16. Assuming that the powers of the lenses (or surfaces) are K_1 and K_2, and that the separation between them is d, we can trace, algebraically, a paraxial ray through the system as follows. If a parallel ray ($u_1 = 0$) is incident at a height h_1 at lens 1, its angle after refraction will be given by Eq. (2.16):

$$n_1'u_1' = n_1u_1 - h_1K_1 = -h_1K_1 . \tag{2.53}$$

At lens 2 we therefore have, using Eq. (2.19),

$$h_2 = h_1 + du_1' = h_1\left(1 - \frac{K_1}{n_1'}\right). \tag{2.54}$$

Then, after refraction at lens 2,

$$\begin{aligned} n_2'u_2' &= n_2u_2 - h_2K_2 \\ &= n_1'u_1' - h_2K_2 \\ &= -h_1K_1 - h_1\left(1 - \frac{K_1}{n_1'}\right)K_2 \\ &= -h_1\left(K_1 + K_2 - \frac{dK_1K_2}{n_1'}\right). \end{aligned} \tag{2.55}$$

However, the final ray angle, u_2', is given by

$$n_2'u_2' = -h_1K ,$$

where K is the power of the complete system, and it therefore follows that the power of the overall system is given by

$$K = K_1 + K_2 - \frac{dK_1K_2}{n_1'}. \tag{2.56}$$

Although it is of less interest, we can also, using the same method, show that the positions of the two principal planes are given by

$$\delta = \frac{dK_2n_1}{n_1'K} \tag{2.57}$$

and

$$\delta' = -\frac{dK_1n_2'}{n_1'K}, \tag{2.58}$$

where δ and δ' are as shown in Fig. 2.16.

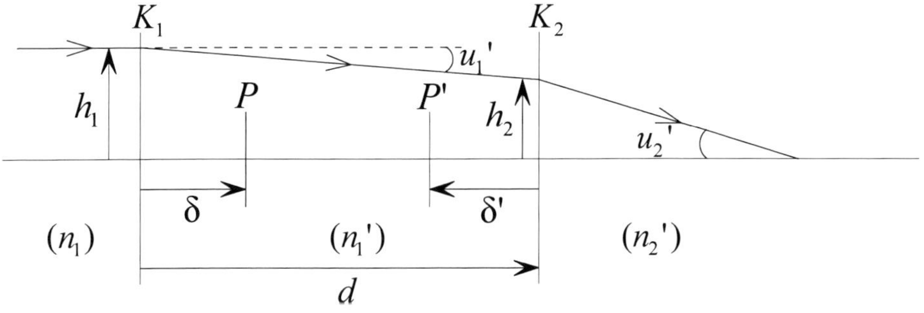

Figure 2.16. Two separated lenses.

2.7.2 The thick lens

The equations given in the previous section can be used to derive some properties of a thick lens (Fig. 2.17). We have already seen that the principal points of a single surface are at the vertex of the surface, and we can therefore use Eq. (2.5).

We write for surface 1,

$$K_1 = (n-1)c_1, \tag{2.59}$$

and for surface 2,

$$K_2 = -(n-1)c_2. \tag{2.60}$$

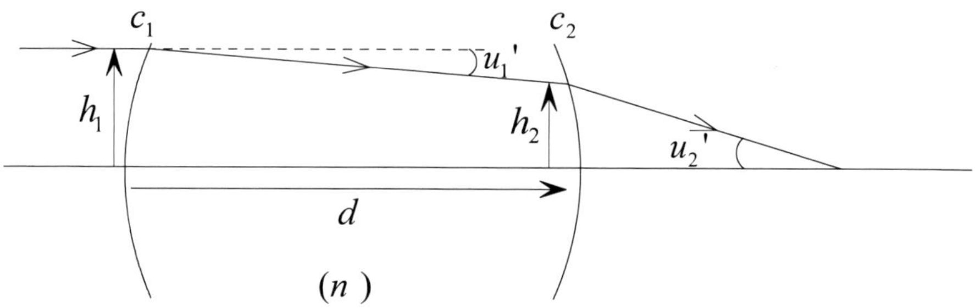

Figure 2.17. The thick lens.

d is simply the axial lens thickness, and therefore

$$\begin{aligned} K &= K_1 + K_2 - \frac{dK_1K_2}{n} \\ &= (n-1)\left[c_1 - c_2 + \frac{(n-1)dc_1c_2}{n}\right]. \end{aligned} \tag{2.61}$$

The principal plane positions are given by

$$\delta = -dc_2\left(\frac{n-1}{n}\right)f' \tag{2.62}$$

and

$$\delta' = -dc_1\left(\frac{n-1}{n}\right)f'. \tag{2.63}$$

As an example, let us take the simple case of a thick unit-power plano-convex lens with a refractive index of 1.5.

Let $K = 1$, then $c_1 = 2.0$

and

$$\delta' = -d2\frac{0.5}{1.5}1 = -\frac{2}{3}d$$

and

$$\delta = 0.$$

Therefore, the principal planes are positioned as shown in Fig. 2.18, and PP′, the distance between them, is 1/3 *d*.

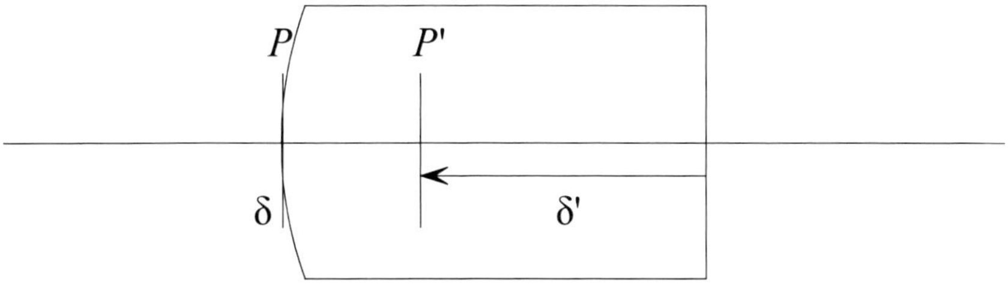

Figure 2.18. Principal planes of a thick plano-convex lens.

Taking the case of an equi-convex lens in which $c_1 = -c_2$, we assume arbitrarily that $d = 0.25$. Then, from Eq. (2.61), $c_1 = -c_2 = 1.04555$, to give a focal length of 1.

Then,

$$\delta = -\delta' = 0.3485\, d\,,$$

and it follows that

$$PP' = d\,[1-2\,(0.3485)] = 0.303\, d.$$

This case is shown in Fig. 2.19. The important point to notice is that the distance PP′, sometimes known as the **interstitium**,[2] or **hiatus**[3] is almost the same in both cases.

Remembering that for a lens in air the nodal points and the principal planes coincide, it can be seen that for the plano-convex case, P is at the vertex of surface 1, since this is the only position that permits a ray to pass through the lens with no change in its angle (although there is a lateral deviation). Similarly, in the case of the equi-convex lens, the nodal points must be symmetrically arranged.

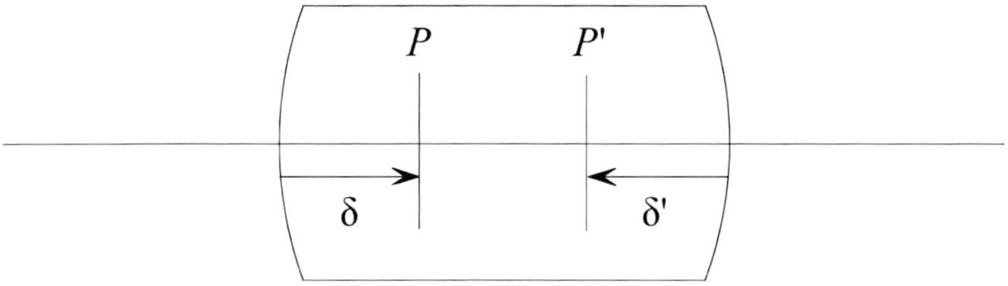

Figure 2.19. Principal planes of a thick biconvex lens.

2.7.3 System of several elements

As we have already mentioned, the paraxial properties of lenses are almost always determined by paraxial ray tracing, in practice. However, we can derive one more result that is sometimes useful analytically.

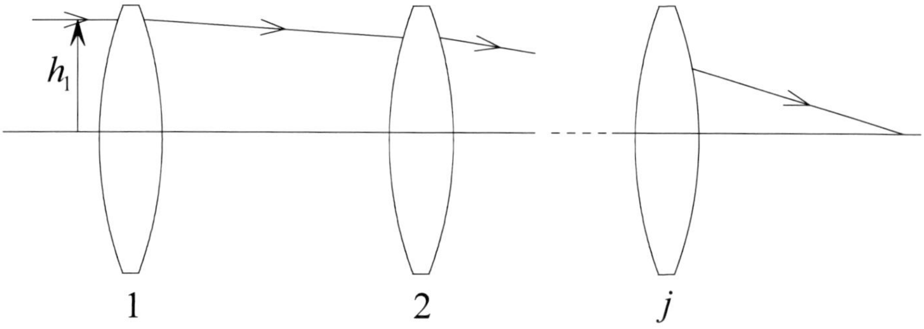

Figure 2.20. System of several elements.

Suppose we have several elements, as shown in Fig. 2.20. These elements do not have to be lenses—they can be surfaces, but the result that we shall derive is most useful for the case when the elements are, in fact, thin lenses. If we trace a paraxial ray through the system, initially parallel to the optical axis, and with an initial ray height of h_1, we can find expressions for the angles after refraction at each element as follows:

$$\begin{aligned} n_1'u_1' &= -h_1K_1, \\ n_2'u_2' &= n_2u_2 - h_2K_2, \end{aligned} \tag{2.64}$$

until the last element is reached:

$$n_j'u_j' = n_ju_j - h_jK_j .$$

Adding these together, and noting that $n_1'u_1' = n_2u_2$, we obtain

$$\begin{aligned} n_j'u_j' &= -(h_1K_1 + h_2K_2 + \ldots + h_jK_j) \\ &= -h_1K, \end{aligned} \tag{2.65}$$

where K is the power of the complete system. Dividing by h_1, it therefore follows that

$$K = K_1 + \frac{h_2 K_2}{h_1} + \frac{h_3 K_3}{h_1} + ... + \frac{h_i K_i}{h_1}. \tag{2.66}$$

Thus, the contribution of any element to the overall system is dependent on both the power of component and the height of the incident parallel ray at the component.

2.8 The refraction invariant, *A*

If we consider a spherical surface of curvature c, with a paraxial marginal ray having an angle u incident at a height h, it can be seen from Fig. 2.21 that the paraxial angle of incidence is given by

$$i = hc + u. \tag{2.67}$$

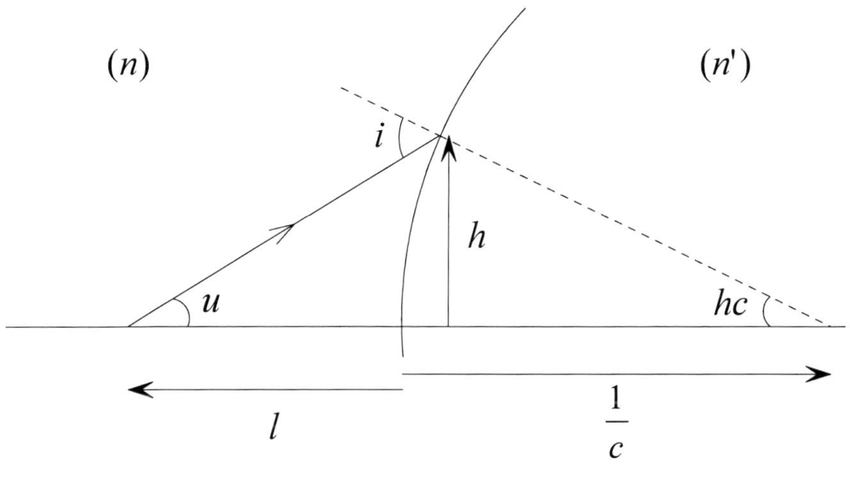

Figure 2.21. The refraction invariant, *A*.

In the paraxial approximation we replace the sines of angles by the angles in radians. Snell's law of refraction ($n \sin I = n' \sin I'$) then becomes

$$ni = n'i' = A, \tag{2.68}$$

where i and i' are the paraxial angles corresponding to the exact angles, I and I'.

Combining Eqs. (2.67) and (2.68), we therefore have

$$n(hc + u) = n'(hc + u') = A, \tag{2.69}$$

which gives

$$n'u' = nu - hc(n' - n),$$

which is Eq. (2.16).

The quantity A is sometimes known as the **refraction invariant**, for obvious reasons. We will see that it is a very important quantity in aberration theory.

2.8.1 Other expressions for the Lagrange invariant

We have seen, from Eqs. (2.24) and (2.28), that the Lagrange invariant is given by

$$n(u\bar{h} - \bar{u}h) = n'(u'\bar{h} - \bar{u}'h). \tag{2.24}$$

The refraction invariant of a paraxial chief ray, from Eq. (2.67), is

$$\bar{A} = n(\bar{h}c + \bar{u}). \tag{2.70}$$

Therefore,

$$n\bar{u} = \bar{A} - n\bar{h}c, \tag{2.71}$$

and similarly,

$$nu = A - nhc. \tag{2.72}$$

Therefore,

$$\begin{aligned} H &= nu\bar{h} - n\bar{u}h \\ &= (A - nhc)\bar{h} - (\bar{A} - n\bar{h}c)h \\ &= A\bar{h} - \bar{A}h. \end{aligned} \tag{2.73}$$

In Fig. 2.22 we show a chief ray incident at a height $\bar{h}$ at a surface. This ray crosses the axis at a distance $\bar{l}$ from the surface, and the object height is η. The angle of the chief ray is $\bar{u}$. As before, the marginal ray angle is u and the object distance is l.

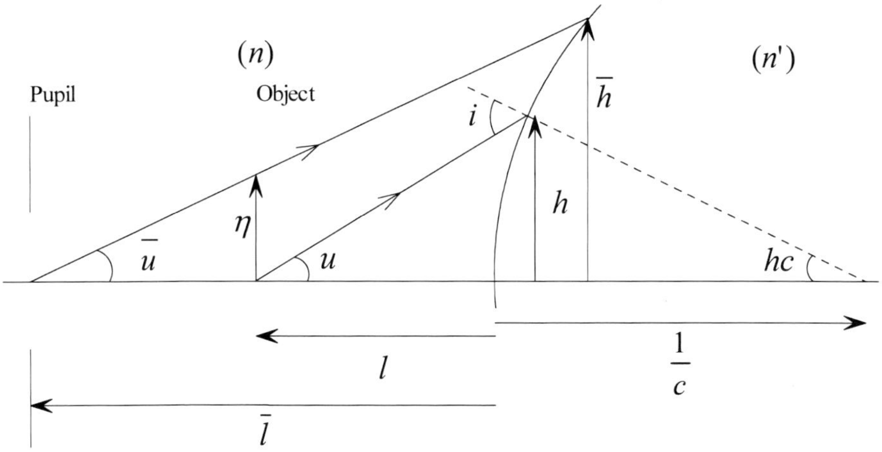

Figure 2.22. Construction for an expression for *H*.

Since

$$\eta = \bar{h} + l\bar{u} \tag{2.74}$$

and

$$\bar{h} = -\bar{l}\,\bar{u}\,, \tag{2.75}$$

it follows that

$$\begin{aligned} H &= nu\eta = nu(\bar{h} + l\bar{u}) \\ &= nu(-\bar{l}\,\bar{u} + l\bar{u}) \\ &= nu\bar{u}(l - l). \end{aligned} \tag{2.76}$$

2.9 The eccentricity, *E*

It is useful to define another quantity that, like the Lagrange invariant, gives us a relationship between chief ray and marginal ray quantities. First, we define, at some surface in an optical system, a quantity E, from the equation

$$HE = \frac{\bar{h}}{h}. \tag{2.77}$$

E is sometimes known as the eccentricity. It is a measure of the distance of a surface from the aperture stop, since the chief ray height $\bar{h}$ is zero at the aperture stop, by definition. Naturally, E will, in general, be different at every surface in the system.

We can then derive an expression for $\bar{A}$, the angle of incidence of a paraxial chief ray, as follows:

$$H = A\bar{h} - \bar{A}h, \tag{2.73}$$

so

$$\begin{aligned}\bar{A} &= \frac{A\bar{h} - H}{h} \\ &= \frac{H}{h}\left(\frac{\bar{h}A}{H} - 1\right) \\ &= \frac{H}{h}(AhE - 1).\end{aligned} \tag{2.78}$$

Similarly,

$$H = n(u\bar{h} - \bar{u}h), \tag{2.24}$$

so

$$\begin{aligned}n\bar{u} &= \frac{nu\bar{h} - H}{h} \\ &= \frac{nuHEh - H}{h} \\ &= \frac{H}{h}(nuEh - 1).\end{aligned} \tag{2.79}$$

We can therefore see that if we know the paraxial marginal ray data (h, u, and A) at a surface, we can completely determine the paraxial chief ray data, if the value of E is also known.

2.9.1 The determination of *E*

Suppose that the heights of a paraxial marginal ray are known at surface i and surface $i + 1$ in an optical system. These heights are h_i and h_{i+1} (Fig. 2.23). Suppose that we also know the heights of a paraxial chief ray at surface i and surface $i + 1$. These are $\bar{h}_i$ and $\bar{h}_{i+1}$.

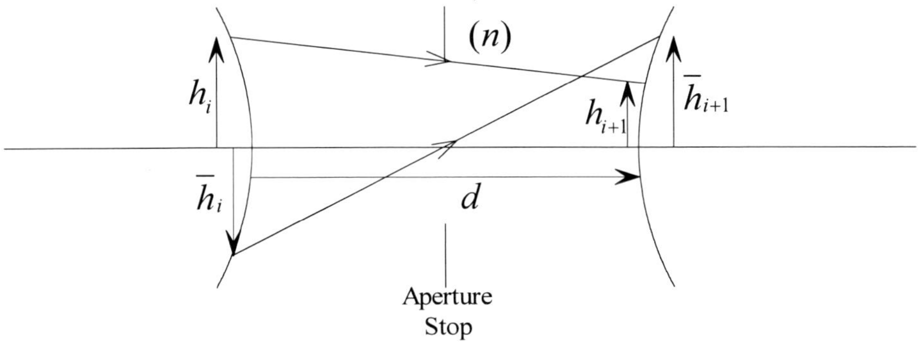

Figure 2.23. Determination of *E*.

Then,

$$HE_{i+1} = \frac{\bar{h}_{i+1}}{h_{i+1}} \quad \text{and} \quad HE_i = \frac{\bar{h}_i}{h_i}. \tag{2.80}$$

Therefore,

$$\begin{aligned} H(E_{i+1} - E_i) &= \frac{\bar{h}_{i+1}}{h_{i+1}} - \frac{\bar{h}_i}{h_i} \\ &= \frac{\bar{h}_{i+1} h_i - \bar{h}_i h_{i+1}}{h_i h_{i+1}}. \end{aligned} \tag{2.81}$$

Therefore,

$$\begin{aligned} E_{i+1} - E_i &= \frac{1}{H} \frac{\bar{h}_{i+1} h_i - \bar{h}_i h_{i+1}}{h_i h_{i+1}} \\ &= \frac{1}{H} \frac{(\bar{h}_i + d\bar{u}) h_i - \bar{h}_i (h_i + du)}{h_i h_{i+1}} \\ &= \frac{d}{H} \frac{\bar{u} h_i - u \bar{h}_i}{h_i h_{i+1}}. \end{aligned} \tag{2.82}$$

Substituting

$$H = n(u\bar{h} - \bar{u}h), \tag{2.24}$$

we arrive at

$$E_{i+1} - E_i = -\frac{d}{nh_i h_{i+1}}. \qquad (2.83)$$

Since E is zero at the stop, by definition, it follows that we can calculate E at the surfaces adjacent to the stop using Eq. (2.83), and we can of course extend this to calculate E at all surfaces in a system. Then, using E in Eqs. (2.77), (2.79), and (2.78), we can calculate all of the chief ray quantities $\bar{h}, \bar{u}$, and $\bar{A}$, *without ever tracing a chief ray.*

In the past, before computers were available, this result gave a saving in the amount of computation. More importantly today, it should be noted that if we move the stop from surface i to surface $i+1$, the value of E changes by an amount given by Eq. (2.83). Furthermore, this change *is the same at all surfaces*. Note that surfaces i and $i+1$ do not have to be real physical surfaces in a lens; they can be dummy surfaces. This result will be used in Chapter 6 to derive equations that give the changes in the Seidel aberrations with stop position.

References

1. W. T. Welford, *Geometrical Optics, Optical Instrumentation*, North-Holland, (1962).
2. W. T. Welford, *Aberrations of Optical Systems*, Adam Hilger (1986).
3. R. Kingslake, *Lens Design Fundamentals*, Academic Press (1978).

CHAPTER 3
RAY TRACING

In the previous chapter, we discussed paraxial optics, also referred to as "first-order" optics, in which we considered the properties of an optical system close to the optical axis. In effect, we have assumed that lenses form perfect images, with no aberrations.

Outside the paraxial region, this assumption is invalid. Real lenses have aberrations, which must usually be reduced. The reduction of aberrations is, of course, the function of the lens designer, and the main subject of this book.

Aberrations are calculated by a process known as ray tracing, probably first used by William Gascoigne, in which the exact form of Snell's law is used, rather than its paraxial approximation. Gascoigne died in 1641, at the Battle of Marston Moor, aged 24, without publishing his method. However, he is thought to have used it to design the telescopic sight and the micrometer eyepiece.

There are at least three methods of looking at such aberrations, but the transverse ray aberration, shown in Fig 3.1 is the most easy to understand, and often the most useful. (The other methods are wavefront aberrations, discussed below, and longitudinal aberrations, which are not considered here. All three methods give essentially the same information but in different ways; in some situations one method is more useful than the others, but this is very often a matter of personal preference.)

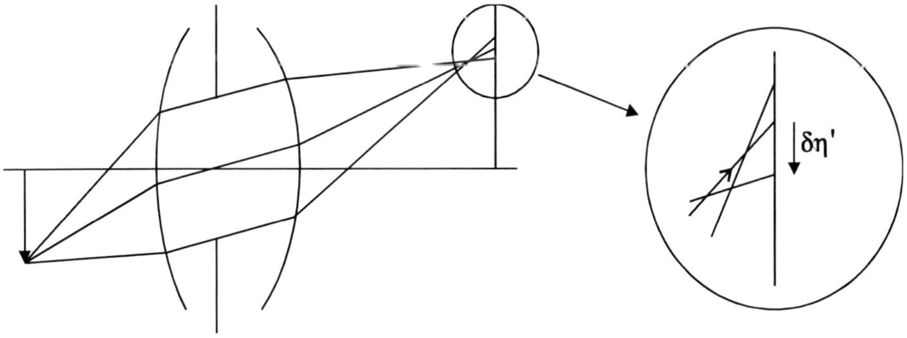

Figure 3.1. Transverse ray aberrations.

3.1 Introduction

In Fig. 3.1, a fan of exact rays is propagated from an off-axis object point. There is a chief ray, passing through the center of the aperture stop and several other rays from the same object point. In general, these rays do not all intersect in the image plane. The differences between the chief ray and any other rays in the image plane are the transverse ray aberrations.

The rays shown in the diagram are meridian rays (in the *y*-*z* plane). In this case the transverse ray aberration is simply the difference between the *y*-coordinates of the rays at the image plane and is the component of transverse aberration, $\delta\eta'$. In the more general case of skew rays, the transverse ray aberration is a vector quantity, with components in the *x*- and *y*-directions, $\delta\xi'$ and $\delta\eta'$. It is, of course, possible for the image surface to be curved, in which case it might be necessary to consider the *z*-component of aberration, but in practice, evaluation is usually carried out over a nominally plane image.

To carry out an exact ray trace, we will always start with a ray whose position and direction is known. We then calculate the intersection point of the ray with the first surface. After this we calculate the direction of the ray after refraction (or reflection, or diffraction) at the surface; and this process is repeated at all surfaces in the system until we find the intersection point of the ray with the image surface, as illustrated in Fig. 3.2.

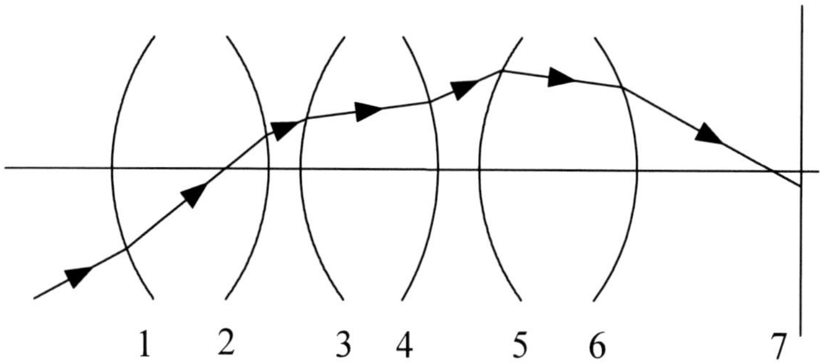

Figure 3.2. Ray tracing through several surfaces.

Therefore, the general problem with ray tracing is that we need to determine the position and direction of a ray after refraction or reflection at a surface. We will assume that we already know the direction of the ray in the space before the surface, and the ray position at some point in the object space.

3.2 A simple trigonometric method of tracing meridian rays

In later sections of this chapter modern ray tracing will be described using algebraic formulae based on coordinate geometry. But, for many decades, most ray tracing was carried out using trigonometric methods.

We begin by tracing a ray through a curved interface of radius *R* separating two mediums with refractive indices, *n* and *n*′, as shown in Fig. 3.3. Suppose we have an incident ray AC that crosses the optical axis at a distance *L* from the vertex of a spherical surface. Let the angle between the ray and the optical axis be U. (Note that U is no longer assumed to be small; we use the uppercase U to distinguish it from the paraxial ray angle, which is denoted by the use of the lowercase u.)

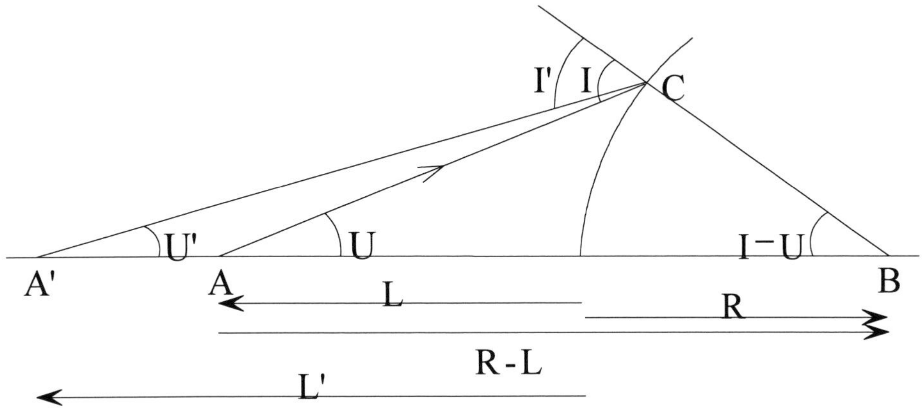

Figure 3.3. Trigonometric ray tracing.

Inspection of the triangle ABC shows that CB = R and AB = R−L. Since sin(π−I) = sin I and we already know U, we can write an expression for I:

$$\frac{\sin I}{R-L}=\frac{\sin U}{R}. \tag{3.1}$$

The use of Snell's law then gives us I':

$$n' \sin I' = n \sin I. \tag{3.2}$$

Also,

$$\text{Angle CBA} = I-U = I' - U', \tag{3.3}$$

and therefore

$$U' = U + I' - I. \tag{3.4}$$

We can now calculate L' by taking Eq. (3.1) and adding a prime (′) to all relevant quantities:

$$\frac{\sin I'}{R-L'}=\frac{\sin U'}{R}, \tag{3.5}$$

which gives

$$L' = R - \frac{R \sin I'}{\sin U'}. \tag{3.6}$$

Finally, we subtract the axial spacing between the surfaces to get the value of L for the next surface.

This method of calculating I, I', U' and L', using Eqs. (3.1), (3.2), (3.4), and (3.6), is one of the simplest methods of ray tracing. This method (and variations of it) has been used for decades as the basis of lens design, but it has several disadvantages:

1. It fails when R is infinite, or L or L' are infinite; in these cases, other equations are used.
2. In this form it is only useful for meridian rays.
3. It requires the calculation of three trigonometric functions at each surface.

For these reasons, the trigonometric method is not suitable for computer calculations. Modern lens design programs use algebraic formulae for ray tracing, which deal very simply with skew rays and avoid any difficulties when the object or image is at infinity, or when the refracting surface is plane. The derivation of the algebraic method begins with a vector treatment of refraction at an interface.

3.3 The vector form of Snell's law

In Fig. 3.4, we have an incident ray defined by a unit vector **r** with an angle of incidence I. After refraction at the surface, the refracted ray is defined by a unit vector **r**′, with an angle of refraction I'. The normal to the surface at the point of intersection of the ray with the surface is denoted by a unit vector **a**.

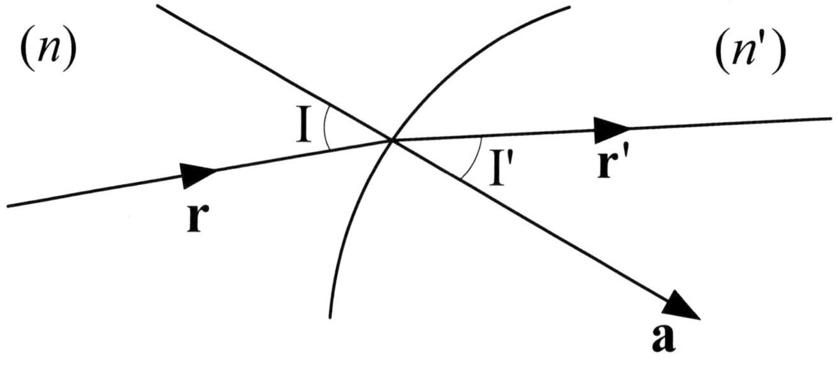

Figure 3.4. The vector form of Snell's law.

The vector product of **r** and **a** is, by definition,

$$\mathbf{r} \times \mathbf{a} = r\, a \sin I\, \mathbf{n}, \tag{3.7}$$

where **n** is a unit vector in the direction orthogonal to **r** and **a**. Since **r** and **a** are, by definition, unit vectors,

$$\mathbf{r} \times \mathbf{a} = \sin I\, \mathbf{n}. \tag{3.8}$$

Snell's law states that **r**, **r′**, and **a** are coplanar, so we can write it in vector form as

$$n\, \mathbf{r} \times \mathbf{a} = n'\, \mathbf{r}' \times \mathbf{a}, \tag{3.9}$$

which is the vector form of Snell's law. This expression is complete in that it not only includes the statement that

$$n \sin I = n' \sin I', \tag{3.2}$$

but also the statement that the incident ray, the refracted ray, and the normal to the surface are coplanar.

However, in this form, Eq. (3.9) is not convenient for ray tracing purposes, since we need an equation that gives us **r′** directly as a function of **r**, **a**, n, and n'.

In order to obtain such an equation, we can define a unit vector in the plane of **r**, **r′**, and **a**, and parallel to the surface at the point of incidence. This unit vector is denoted by **b**.

From Fig. 3.4 we can see that

$$\mathbf{r} = \mathbf{a} \cos I + \mathbf{b} \sin I, \tag{3.10}$$

and similarly

$$\mathbf{r}' = \mathbf{a} \cos I' + \mathbf{b} \sin I'. \tag{3.11}$$

Therefore,

$$\mathbf{b} = \frac{\mathbf{r} - \mathbf{a}\cos I}{\sin I} = \frac{\mathbf{r}' - \mathbf{a}\cos I'}{\sin I'}. \tag{3.12}$$

From this we see that

$$(\mathbf{r} - \mathbf{a} \cos I) \sin I' = (\mathbf{r}' - \mathbf{a} \cos I) \sin I, \tag{3.13}$$

or, using Snell's law again,

$$(\mathbf{r} - \mathbf{a}\cos I)\, n = (\mathbf{r}' - \mathbf{a}\cos I)\, n', \tag{3.14}$$

so that we finally have

$$n'\,\mathbf{r}' = n\,\mathbf{r} + \mathbf{a}\,(n'\cos I' - n\cos I). \tag{3.15}$$

An alternative proof is given by Welford[1] but the proof given above (from R.W. Smith), is simpler.

3.3.1 Definition of direction cosines

Before discussing the algebraic method of ray tracing in any detail, we include a reminder of the definition of direction cosines.

In a Cartesian coordinate system, any vector **r**, of length r, can be resolved into components $x_\mathbf{r}$, $y_\mathbf{r}$, and $z_\mathbf{r}$, such that

$$\mathbf{r} = \mathbf{i}\,x_\mathbf{r} + \mathbf{j}\,y_\mathbf{r} + \mathbf{k}\,z_\mathbf{r}\,, \tag{3.16}$$

where **i**, **j**, and **k** are unit vectors in the directions of the x, y, and z axes, respectively.

The direction cosines are then defined as

$$L = x_\mathbf{r} / r,\;\; M = y_\mathbf{r} / r \text{ and } N = z_\mathbf{r} / r, \tag{3.17}$$

where $r = |\mathbf{r}|$. Therefore we can also write **r** as

$$\mathbf{r} = r\,(\,\mathbf{i}\,L + \mathbf{j}\,M + \mathbf{k}\,N\,). \tag{3.18}$$

The significance of the direction cosines is shown in Fig. 3.5, where we see that L, M, and N are the cosines of the angles between the ray vector and the x-, y-, and z-axes, respectively.

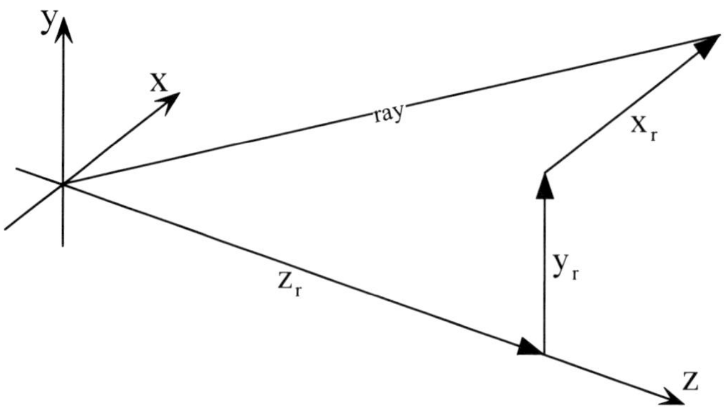

Figure 3.5. Direction cosines.

3.4 Ray tracing (algebraic method)

The general problem in ray tracing is illustrated in Fig. 3.6. A ray with direction cosines L, M, and N, passes through a point A. Let this surface be surface (i–1), and let the coordinates of A with respect to the vertex of surface i be x, y, and z. (Using italics to denote the coordinates at the previous surface.)

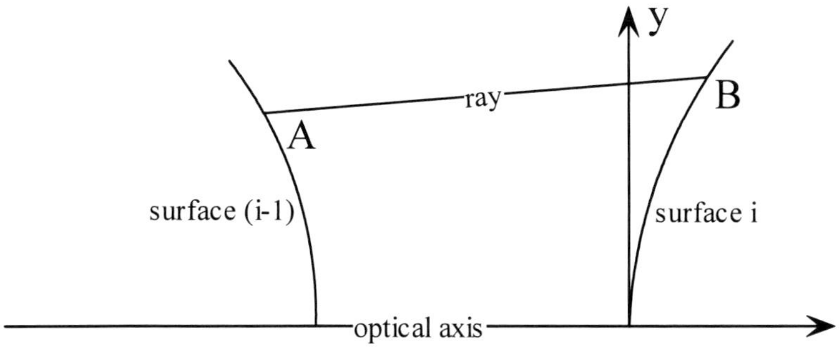

Figure 3.6. Ray tracing—the transfer.

We wish to compute the coordinates of the point of intersection B of the ray at surface i, and then compute the direction cosines after refraction at surface i. First we compute the distance AB **along the ray** from the vertex plane to the surface. Let AB = Δ; then the coordinates of B will be

$$\begin{aligned} \mathrm{x} &= x + L\,\Delta, \\ \mathrm{y} &= y + M\Delta, \\ \mathrm{z} &= z + N\,\Delta. \end{aligned} \tag{3.19}$$

The equation of the spherical surface can be written from Eq. (1.10) (we will omit the subscripts) as

$$\mathrm{z} = \frac{1}{2} c\left(\mathrm{x}^2 + \mathrm{y}^2 + \mathrm{z}^2\right). \tag{3.20}$$

Combining Eqs. (3.19) and (3.20), we have

$$z + N\Delta = \frac{1}{2} c\left[(x + L\Delta)^2 + (y + M\Delta)^2 + (z + N\Delta)^2\right], \tag{3.21}$$

which can be rearranged to give a quadratic equation for Δ, remembering that $L^2 + M^2 + N^2 = 1$:

$$0 = c\Delta^2 - 2\{N - c(Lx + My + Nz)\}\Delta + c(x^2 + y^2 + z^2). \tag{3.22}$$

We write

$$F = c\left(x^2 + y^2 + z^2\right) - 2z, \tag{3.23}$$

and

$$G = N - c(Lx + My + Nz); \tag{3.24}$$

so Eq. (3.22) becomes

$$0 = c\Delta^2 - 2G\Delta + F. \tag{3.25}$$

This can be solved to give

$$\Delta = \frac{G \pm \sqrt{(G^2 - cF)}}{c}. \tag{3.26}$$

The two solutions to Eq. (3.26) are a result of the fact that, mathematically, the ray intersects the spherical surface at two points. We normally need the smaller solution, the solution with the minus sign, because G will normally be positive, as N is normally positive. However, Eq. (3.26) is not suitable for ray tracing because, as c approaches zero (a plane surface, which is quite common), the expression for Δ becomes indeterminate. To get a more suitable result, we use a little-known result for the solution of a quadratic equation in the standard form:

$$ax^2 + bx + c = 0.$$

The most common solution to this equation, which we used to get Eq. (3.26), is

$$x = \frac{-b \pm \sqrt{b^2 - 4ac}}{2a}, \tag{3.27}$$

but this becomes indeterminate as a approaches zero. An alternative form is

$$x = \frac{2c}{-b \pm \sqrt{b^2 - 4ac}}. \tag{3.28}$$

If we use this form of the solution, we obtain

$$\Delta = \frac{F}{G + \sqrt{\left(G^2 - cF\right)}}, \tag{3.29}$$

which correctly becomes zero as c approaches zero (it is not indeterminate). We therefore use Eq. (3.29) to calculate Δ, and the three equations (3.19) to compute x, y, and z.

In order to complete the ray trace, we need to calculate I, the angle of incidence, and I', the angle of refraction, so that we can then calculate the new direction cosines, L', M', and N'.

First we need to have expressions for the direction cosines of the normal to the surface at the point C. By inspection, for a spherical surface, these direction cosines can be seen to be

$$\alpha = -cx$$

and

$$\beta = -cy \tag{3.30}$$

$$\gamma = (1 - cz).$$

Then $\cos I = \mathbf{r} \cdot \mathbf{a} = \alpha L + \beta M + \gamma N$

$$= -cxL - cyM + N(1 - cz)$$

$$= N - c\,(Lx + My + Nz). \tag{3.31}$$

Using Eq. (3.19),

$$\begin{aligned} \cos I &= N - c\,(L(x + L\Delta) + M(y + M\Delta) + N(z + N\Delta) \\ &= N - c\,(Lx + My + Nz) - c\,(L^2 + M^2 + N^2)\,\Delta \\ &= G - c\,\Delta \end{aligned} \tag{3.32}$$

However, from Eq. (3.26),

$$G - c\,\Delta = \sqrt{\left(G^2 - cF\right)}, \tag{3.33}$$

so we see that

$$\cos I = \sqrt{\left(G^2 - cF\right)}. \tag{3.34}$$

To calculate cos I, we square Snell's law and use

$$\sin^2 I = 1 - \cos^2 I \tag{3.35}$$

to get a form of Snell's law in terms of cosines instead of the usual sines:

$$n' \cos I' = \sqrt{n'^2 - n^2 \left(1 - \cos^2 I\right)}; \tag{3.36}$$

or, using $q = n / n'$, as defined in Eq. (1.3),

$$\cos I' = \sqrt{\left\{1 - q^2 \left(1 - \cos^2 I\right)\right\}}. \tag{3.37}$$

We then use Eq. (3.15), which, resolved into its components, becomes

$$\begin{aligned} n' L' &= n L + \alpha \,(n' \cos I' - n \cos I) \\ n' M' &= \text{n} M + \beta \,(n' \cos I' - n \cos I) \\ n' N' &= \text{n} N + \gamma \,(n' \cos I' - \text{n} \cos I) \end{aligned} \tag{3.38}$$

or, using q,

$$\begin{aligned} L' &= q L + \alpha \,(\cos I' - q \cos I) \\ M' &= q M + \beta \,(\cos I' - q \cos I) \\ N' &= q N + \gamma \,(\cos I' - q \cos I). \end{aligned} \tag{3.39}$$

Substituting from Eq. (3.30), we finally get

$$\begin{aligned} L' &= q L - c x \,(\cos I' - q \cos I) \\ M' &= q M - c y \,(\cos I' - q \cos I) \\ N' &= q N + (1 - c z)\,(\cos I' - q \cos I). \end{aligned} \tag{3.40}$$

3.4.1 Precision

It is important to realize that precision in ray tracing equations can be a problem unless the calculations are arranged suitably. Microcomputers may use 4, 8, or more bytes to represent a real floating point number. A number represented by 4 bytes usually has a precision of about 6 decimal places, and a number represented by 8 bytes usually has a precision of about 12 decimal places.

It is quite possible for a lens with overall dimensions of the order of meters to have wavefront aberrations that are less than 0.1 μm. In other words, the aberrations are smaller than the overall dimensions of the system by a factor of 10^7, and it is clearly advisable to use at least an 8-byte representation of real numbers, even though none of the lens parameters are usually measured in the workshop to this precision.

3.5 Calculation of wavefront aberration (optical path difference)

In Fig. 3.7, we have a chief ray passing through the lens along the path OABCO′, and another ray, with a path denoted by ODEFGO′.

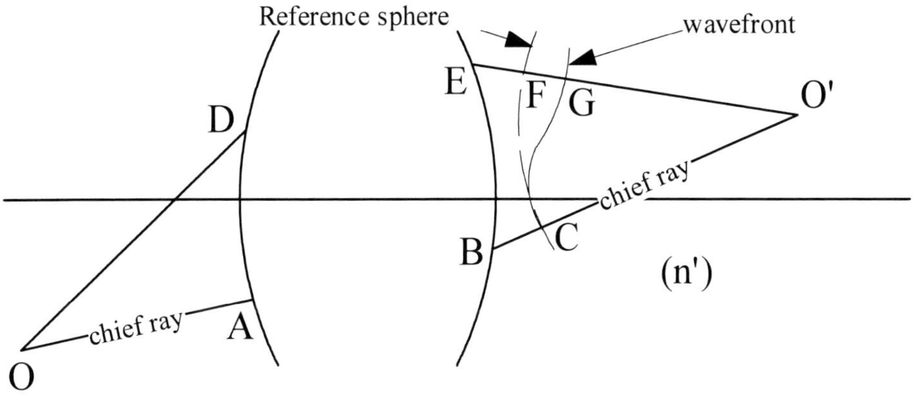

Figure 3.7. Wavefront aberration or optical path difference.

We define the wavefront aberration to be

$$W = [\mathrm{FG}] = n'\,\mathrm{FG}, \tag{3.41}$$

where F is the point of intersection of the ray with the reference sphere that passes through the point C and is centered on the image point O′, and where G is the point of intersection of the ray with a geometrical wavefront that passes through C.

Therefore, the wavefront aberration, as defined above, is an optical path length and is equal to the geometric distance FG multiplied by the refractive index in the image space, n'.

In Fig. 3.8 we have shown the center of the reference sphere to be O′, which is assumed to be the point of intersection of the two rays. However, in the general case of skew rays, it is unlikely that there will be a point of intersection, and we therefore assume, more generally, that O′ is the point of intersection of the chief ray with the image surface.

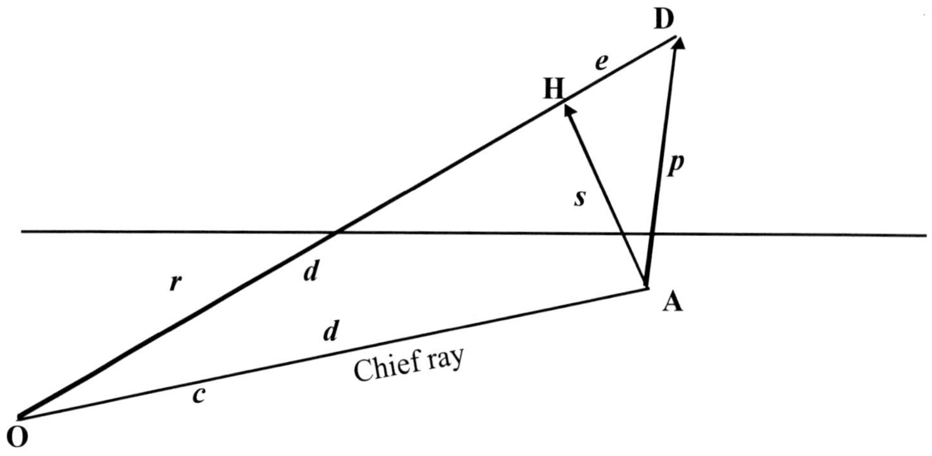

Figure 3.8. Start of the wavefront aberration calculation.

Since the points C and G are on the same wavefront,

$$\begin{aligned}[\text{OABC}] &= [\text{ODEFG}] \\ &= [\text{ODEF}] + [\text{FG}].\end{aligned}$$

Therefore,

$$\begin{aligned}W &= [\text{FG}] = [\text{OABC}] - [\text{ODEF}] \\ &= \{[\text{OA}] - [\text{OD}]\} + \{[\text{ABC}] - [\text{DEF}]\}.\end{aligned} \tag{3.42}$$

We therefore calculate the optical path length from the object to the reference sphere along the two rays, and then take the difference to be the wavefront aberration. This calculation includes the optical paths [OA] and [OD], which might be very large. In the case of an object at infinity, they will be infinite, so an alternative method is needed to calculate [OA]−[OD].

In Fig. 3.8,

O is the object point, which might be at a very large distance from the lens,
A is the point of intersection of the chief ray with the first surface,
d is the distance OA , which is unknown,
c is a unit vector along the chief ray, with components $\bar{L}$, $\bar{M}$ and $\bar{N}$,
p is a vector from A to D, with components $(x-\bar{x})$, $(y-\bar{y})$, and $(z-\bar{z})$,
D is the point of intersection of the other ray with the first surface,
r is a unit vector along the other ray, with components L, M, and N,
H is a point along the other ray, a distance d from O,
s is the vector **AH**,
e = HD = OD − OA, which is the quantity to be determined.

From Fig. 3.8,

$$\mathbf{s} = \mathbf{p} - e\,\mathbf{r}. \tag{3.43}$$

Since the angles OAH and OHA are equal, and since the sum of the angles OAH and AHD = 180 deg, it follows that

$$\cos(\text{OAH}) = (\mathbf{p} - e\,\mathbf{r}) \cdot \mathbf{c} = -\cos(\text{AHD}) = -(\mathbf{p} - e\,\mathbf{r}) \cdot \mathbf{r}.$$

This then gives

$$e\,\{\mathbf{r} \cdot \mathbf{r} + \mathbf{r} \cdot \mathbf{c}\} = \mathbf{r} \cdot \mathbf{p} + \mathbf{c} \cdot \mathbf{p}, \tag{3.44}$$

and therefore

$$e = \frac{(rp + rc)}{(1 + rc)} = \frac{\left[(L + \bar{L})(x - \bar{x}) + (M + \bar{M})(y - \bar{y}) + (N + \bar{N})(z - \bar{z})\right]}{\left(1 + L\bar{L} + M\bar{M} + N\bar{N}\right)}. \tag{3.45}$$

This expression is therefore used to determine [OA] – [OD]. Note that, if the object is at infinity, OA and OD are both infinite, but e becomes

$$e = L\,(x - \bar{x}) + M\,(y - \bar{y}) + N\,(z - \bar{z}), \tag{3.46}$$

which is finite.

Therefore,

$$W = [\text{FG}] = [\text{OABC}] - [\text{ODEF}] = -n \cdot e + \{[\text{ABC}] - [\text{DEF}]\}, \tag{3.47}$$

which is a simple quantity to compute, and is always finite, as long as the image is at a finite distance.

3.6 Ray tracing through aspheric and toroidal surfaces

In the case of aspheric and toroidal surfaces, it is not possible to find the point of intersection of the ray with the refracting surface analytically, and it is necessary to use an iterative method of some type.

We will not discuss the details of this calculation; rather, we will indicate the general principles followed in aspheric and toroidal ray tracing. Full details of the equations are given by Welford.[2]

An aspheric surface is normally described by the following equation:

$$z = \frac{cr^2}{\left[1 + \sqrt{1 - (1 + k)c^2 r^2}\right]} + a_4 r^4 + a_6 r^6 + a_8 r^8 + \ldots. \tag{3.48}$$

The first term in this expression is the standard form of a conic surface, Eq. (1.21), but there are extra terms proportional to r^4, r^6, r^8, r^{10},.... In most cases, the power series is not taken beyond the r^{10} term.

A toroidal surface is described by

$$z = r_x \pm \sqrt{\left\{\left[r_x - r_y \pm \sqrt{\left(r_y^{\,2} - y^2\right)}\right]^2 - x^2\right\}}, \tag{3.49}$$

where r_x is the radius of curvature of the surface in the x-z plane and r_y is the radius of curvature of the surface in the y-z plane.

These equations can be written in the form

$$g(x, y, z) = z - \left\{\frac{cr^2}{\left[1+\sqrt{1-(1+k)c^2r^2}\right]} + a_4r^4 + a_6r^6 + a_8r^8 + \ldots\right\} \text{(asphere)} \tag{3.50}$$

and

$$g(x, y, z) = z - \left(r_x \pm \sqrt{\left\{\left[r_x - r_y \pm \sqrt{\left(r_y^{\,2} - y^2\right)}\right]^2 - x^2\right\}}\right) \text{(toroid).} \tag{3.51}$$

The general method of determining the intersection of a ray with these surfaces, which cannot be handled analytically, is to use an iterative technique, starting at some point that is on the ray but not on the surface.

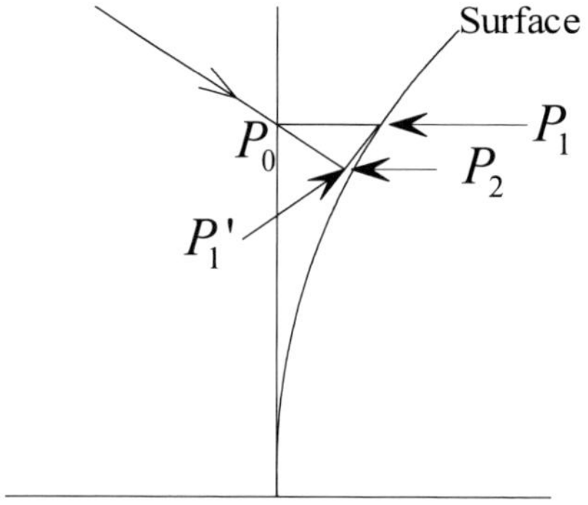

Figure 3.9. Ray tracing through toroids and asphericals.

This point, P_0 in Fig. 3.9, is usually found either by computing the point of intersection with the vertex plane, or by computing the point of intersection of the ray with a spherical or conic surface that is known to be close to the actual surface. Suppose the coordinates of P_0 are x_0, y_0, and z_0. We then calculate the coordinates of a point on the surface, with the same x and y coordinates as P_0, from either Eq. (3.50) or (3.51). This point, P_1, is not on the ray, except in the special case when the ray is parallel to the axis.

Knowing the point P_1, we can find the equation of the tangent plane through P_1, and then we can find the point of intersection of the ray with the tangent plane, P_1'. This point P_1' is generally closer to the surface than the original point P_0. This process is repeated until we have found a point on the ray that is close enough to the surface. For practical purposes, this iterative process can be terminated when the change in ray coordinates between successive approximations is less than about $\lambda/1000$.

In order to find the equation of the tangent plane at P_1, P_2, etc., we need the direction cosines of the normal to the surface, which are found by calculating the partial differentials of $g(x,y,z)$ with respect to x, y, and z, respectively. If these partial differentials are $\partial g/\partial x$, $\partial g/\partial y$, and $\partial g/\partial z$, respectively, the direction cosines will be

$$\alpha = -\frac{\partial g/\partial x}{\sqrt{\left[\left(\partial g/\partial x\right)^2 + \left(\partial g/\partial y\right)^2 + \left(\partial g/\partial z\right)^2\right]}}, \tag{3.52}$$

with similar expressions for β and γ. Since the ray passes through the point P_0, the equation of the ray can be written as

$$\frac{(x - x_0)}{L} = \frac{(y - y_0)}{M} = \frac{(z - z_0)}{N}. \tag{3.53}$$

At the point P_1, with coordinates x_1, y_1 and z_1, the equation of the tangent plane can be written as

$$0 = \alpha(x - x_1) + \beta(y - y_1) + \gamma(z - z_1). \tag{3.54}$$

Solving Eqs. (3.53) and (3.54), we find the coordinates of P_1':

$$z_1' = \frac{N[\alpha(x_1 - x_0) + \beta(y_1 - y_0) + \gamma z_1]}{[\alpha L + \beta M + \gamma N]}, \tag{3.55}$$

and

$$x_1' = x_0 + \frac{L}{N} z_1' \quad \text{and} \quad y_1' = y_0 + \frac{M}{N} z_1'. \tag{3.56}$$

As stated above, this process is repeated until the points P_i and P_{i+1} are sufficiently close.

For refraction, we use the vector form of Snell's law, Eq. (3.38):

$$\begin{aligned} n'L' &= n\,L + \alpha\,(n'\cos I' - n\cos I) \\ n'M' &= n\,M + \beta\,(n'\cos I' - n\cos I) \\ n'N' &= n\,N + \gamma\,(n'\cos I' - n\cos I), \end{aligned} \tag{3.38}$$

where

$$\cos I = \alpha L + \beta M + \gamma N. \tag{3.31}$$

In practice, these equations typically converge with sufficient accuracy in about five iterations.

3.7 Decentered and tilted surfaces

Although this book is mainly concerned with the design of centered systems, we now discuss the problem of ray tracing through decentered systems. The standard method of defining a decentered surface is to define a set of decentrations, dx, dy, and dz, and a set of rotations, α, β, and γ. (These quantities are not at all related to the direction cosines of the normal to a surface used in the previous section, which are also referred to as α, β, and γ; but this notation is almost standard, and we will not change it.)

When a surface is decentered, we first apply decentrations dx, dy, and dz. If the coordinates of a point, referred to as a "global" axis, are x, y, and z, the coordinates referred to an origin at the decentered vertex of the surface will be

$$\begin{aligned} x' &= x - dx \\ y' &= y - dy \\ z' &= z - dz. \end{aligned} \tag{3.57}$$

We then apply the three rotations, in the order α (clockwise about the x-axis), β (clockwise about the new y-axis), and then γ (anticlockwise about the new z-axis). The three rotations are most simply expressed in matrix notation as follows:

$$\begin{pmatrix} x' \\ y' \\ z' \end{pmatrix} = GBA \begin{pmatrix} x \\ y \\ z \end{pmatrix} \qquad \begin{pmatrix} L' \\ M' \\ N' \end{pmatrix} = GBA \begin{pmatrix} L \\ M \\ N \end{pmatrix}, \tag{3.58}$$

where

$$A = \begin{pmatrix} 1 & 0 & 0 \\ 0 & \cos\alpha & -\sin\alpha \\ 0 & \sin\alpha & \cos\alpha \end{pmatrix} \quad B = \begin{pmatrix} \cos\beta & 0 & -\sin\beta \\ 0 & 1 & 0 \\ \sin\beta & 0 & \cos\beta \end{pmatrix} \quad G = \begin{pmatrix} \cos\gamma & \sin\gamma & 0 \\ -\sin\gamma & \cos\gamma & 0 \\ 0 & 0 & 1 \end{pmatrix}$$

and where the matrix products are evaluated from right to left.

These rotations are applied to both the ray coordinates, as described above, and to the ray direction cosines. After the decentrations and rotations, we have ray positions and direction cosines in the local coordinate system of the surface, and we then use the conventional ray tracing equations for transfer and refraction.

After completing the refraction, we then transform back to the global coordinate system by applying the above operations in reverse, or, in some cases, we may wish to continue the ray tracing in the transformed coordinate system.

3.8 Ray tracing at reflecting surfaces

In order to deal with reflecting surfaces, a common procedure is to adopt the convention that we use the same equations as for refraction, apart from setting $q = -1$, as stated in Eq. (1.5). This convention is quite effective, but it must be remembered that, if it is adopted, the unit vector defining the line along which the ray travels **does not** indicate the direction of propagation.

To illustrate this, consider the result when we use Eq. (3.15):

$$n'\,\mathbf{r}' = n\,\mathbf{r} + \mathbf{a}\,(n'\cos I' - n\cos I). \tag{3.15}$$

Writing $q = n / n'$ as before, this becomes

$$\mathbf{r}' = q\,\mathbf{r} + \mathbf{a}\,(\cos I' - q\cos I). \tag{3.59}$$

Now consider the case of a ray traveling along the z-axis and a surface normal to the ray. $\mathbf{r}$ and $\mathbf{a}$ are both equal to $\mathbf{k}$ (the unit vector parallel to the z-axis), and $q = -1$, so we have

$$\begin{aligned} \mathbf{r}' &= -\,\mathbf{k} + \mathbf{k}\,[1 - (-1)(\ 1)] \\ &= \mathbf{k}\,(-1 + 2) = \mathbf{k}. \end{aligned} \tag{3.60}$$

In other words, the vector $\mathbf{r}'$ has exactly the same direction as the original vector $\mathbf{r}$. Clearly, the ray direction has in fact been reversed, so we must be aware of this effect. Freeman[2] has suggested that if the unit vector $\mathbf{r}'$ is multiplied by an integer that equals $+1$ after an even number of reflections, and equals -1 after an odd number of reflections, this problem can be resolved quite simply, but we will not discuss this matter further here.

References

1. W. T. Welford, *Aberrations of Optical Systems*, Adam Hilger (1986).
2. D. E. L. Freeman, "Simple method for finding the valid ray-surface intersection," Proc. SPIE Vol. **1354** (1990).

CHAPTER 4
ABERRATIONS

After the rays have been traced exactly through the optical system, we need to determine how much they depart from the perfection of the paraxial region. As we have seen from Fermat's principle and the law of Malus, we can consider these aberrations either in terms of the geometrical ray intercept errors at the image plane, or as optical path length differences.

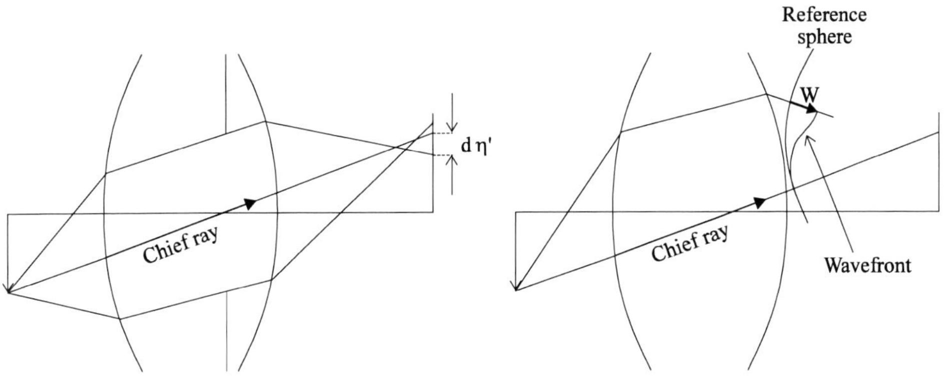

Figure 4.1. Transverse and wavefront aberrations.

4.1 The relationship between transverse and wavefront aberrations

Figure 4.1 shows an aberrated wavefront converging toward an ideal image point defined by the chief ray. We also show a reference sphere centered on the ideal image point, with radius R_{rs}. An aberrated ray intersects the image surface with an intercept error $d\eta'$, and is perpendicular to the wavefront. For the sake of clarity, the diagram illustrates the two-dimensional case, but the results that we will derive are valid for the three-dimensional case.

The ray intersects the reference sphere at a point with coordinates x and y. Because of the aberration, the ray does not follow the path of the normal to the reference sphere. If the angle between the ray and the normal to the reference sphere is denoted by $\delta U'$, we can see that, to a first approximation, the y-component of transverse ray aberration will be given by

$$\delta\eta' = \delta U' R_{rs}. \tag{4.1}$$

Since the ray is normal to the wavefront, it follows that $\delta U'$ equals the angle between the wavefront and the reference sphere, so

$$\delta U' = -\frac{1}{n'}\frac{\partial W}{\partial y}. \tag{4.2}$$

The factor of $1 / n'$ is needed because the wavefront aberration is defined to be an optical path length, and not a geometric length. Therefore, the y-component of transverse ray aberration is given by

$$\delta\eta' = -\frac{R_{rs}}{n'}\frac{\partial W}{\partial y}. \tag{4.3}$$

Similarly, the x-component of transverse aberration is given by

$$\delta\zeta' = -\frac{R_{rs}}{n'}\frac{\partial W}{\partial x}. \tag{4.4}$$

These expressions are derived more rigorously by Welford,[1] and by Born and Wolf.[2]

For practical purposes, these equations are not very useful, as they require a knowledge of the ray coordinates at the reference sphere.

It is conventional in aberration theory to assume that the reference sphere is at the exit pupil position, perhaps on the assumption that diffraction may be considered to take place at the exit pupil. Since we do not consider diffraction in geometrical optics, this convention seems to be unjustified, but, in any case, the position of the reference sphere is usually unknown, and it is more convenient to avoid the use of reference sphere coordinates if possible. This can be achieved if we use the concept of **relative pupil coordinates**, as suggested by Hopkins.[3]

Suppose we define a relative pupil coordinate y_{rel} by the following equation

$$y_{\mathrm{rel}} = y_{\mathrm{rs}} / y_{\mathrm{marg\text{-}rs}}, \tag{4.5}$$

where y_{rs} = y-coordinate of the ray at the reference sphere,
and $y_{\mathrm{marg}\text{-}rs}$ = y-coordinate of a marginal ray at the reference sphere.

Therefore, y_{rel} is a relative coordinate that equals $+1$ for a ray at the extreme top of the pupil, and equals -1 for a ray at the bottom edge of the pupil. Then,

$$\delta\eta' = -\frac{R_{rs}}{n'}\frac{\partial W}{\partial y} \tag{4.3}$$

becomes

$$\delta\eta' = -\frac{R_{rs}}{n'}\frac{\partial W}{\partial y_{rel}}\frac{1}{y_{marg\text{-}rs}}. \tag{4.6}$$

The sine of the slope angle U' in the image space is $y_{marg\text{-}rs}/R_{rs}$, so

$$\delta\eta' = -\frac{-1}{n'\sin U'}\frac{\partial W}{\partial y_{rel}}, \tag{4.7}$$

where n′ sinU' is the numerical aperture of the lens.

Similarly,

$$\delta\xi' = -\frac{-1}{n'\sin U'}\frac{\partial W}{\partial x_{rel}} \tag{4.8}$$

gives the x-component of the transverse ray aberration.

Since Eqs. (4.7) and (4.8) are differential relationships, it is possible to convert them into integral relationships as follows:

$$W = -n' \sin U' R_{rs} \int \delta\eta' \, dy, \tag{4.9}$$

and a similar expression if we integrate with respect to x. Although we would not actually calculate wavefront aberrations this way, the integrand in Eq. (4.9) is simply the area under a sectional plot of transverse ray aberration as a function of relative pupil coordinate; these plots are discussed in the next section.

4.2 Ray aberration plots

One of the most useful aids in the understanding of ray trace results is to plot curves of ray error as a function of aperture. To do this, we trace a number of rays (typically, about 20 rays, but this will depend on the software), and for each ray we calculate the aberration.

In the finite conjugate case, shown in Fig. 4.2, we trace the rays from points on the object surface, and the rays therefore have different angles in the object space. In the infinite conjugate case, the rays from a given object point are naturally parallel to each other in the object space. In both cases the rays are distributed across the entire width of the entrance pupil (AA′), unless there is vignetting caused by obstruction of rays at other apertures in the lens.

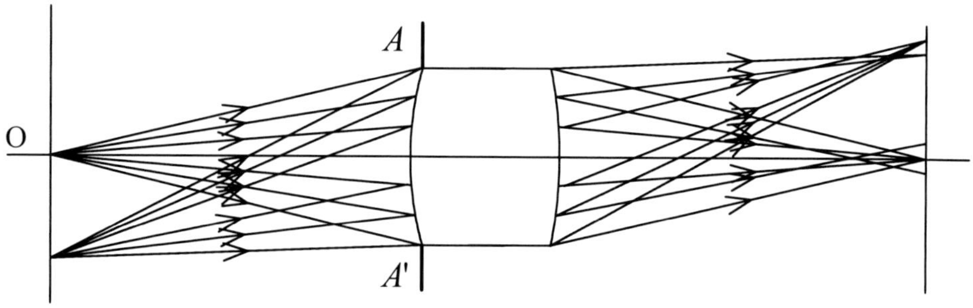

Figure 4.2. Ray tracing for aberration curves.

This process is usually repeated for a small number of object points; two are shown in Fig. 4.2. Typically, it is sufficient to calculate the aberration curves at three or four object points, because the number of lenses in which aberrations depend on high powers of the object coordinate is small. However, in the case of some wide-angle lenses, four object points are sometimes not enough. The only recommendation that we can make is to suggest that the lens designer should always make a final check of the aberrations at more field points than were used for the actual optimization of the design.

The aberration that is plotted is usually either the transverse ray aberration or the wavefront aberration. In either case, we plot the aberrations on the vertical axis as a function of the position of the ray in the entrance pupil on the horizontal axis, between A and A' of Fig. 4.2. We obtain plots of the form shown in Figs. 4.3 and 4.4.

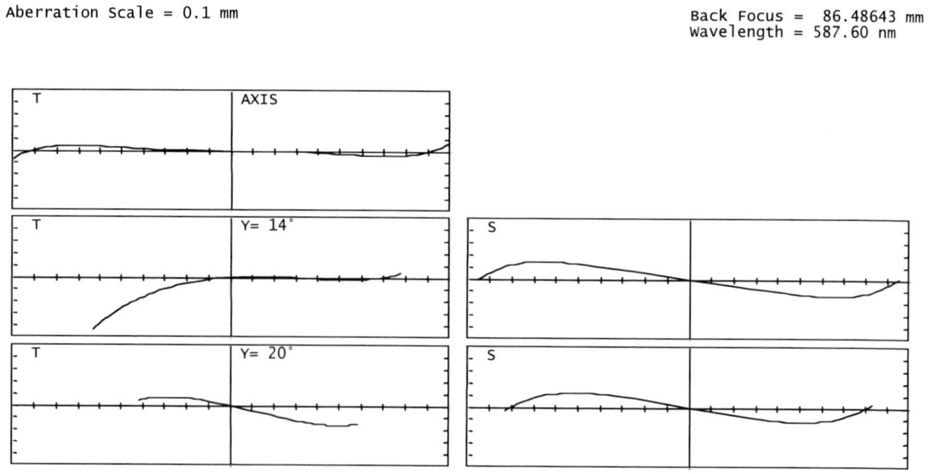

Figure 4.3. A typical set of transverse ray aberration curves.

Consider the ray intercept curves in Fig. 4.3. The top box in Fig. 4.3 shows the transverse ray aberration of rays from an axial object point, similar to O in Fig. 4.2. In this particular case, the scale for the aberrations is set so that 0.10 mm corresponds to half the full height of the box, i.e., one small division corresponds to 0.02 mm. The width of the horizontal axis corresponds to the size of the entrance pupil (AA'). Since we normally wish to trace a set of rays that extends to the full width of the pupil, this curve will usually fill the horizontal axis.

The second set of curves shows the aberrations of rays from an object point that is 14 deg off axis. The curve labelled "T Y=14°" shows the aberrations of meridian rays. Note that the curve does not extend to the full horizontal width of the box, indicating that this lens has some vignetting. The curve labelled "S" shows the aberrations of skew rays. For this curve we plot the x-component of transverse aberration, $d\xi'$, as a function of the x-coordinate of the ray in the entrance pupil. Conventionally, the skew rays are initially in the plane that includes the chief ray, and that is perpendicular to the y-z plane. Since the lens has rotational symmetry, sometimes we do not plot the aberrations of rays with negative values of x in the entrance pupil because the aberrations will be identical—apart from a change of sign—to the aberrations of rays with positive values of x.

Since these are skew rays, there are also y-components of transverse aberration, $d\eta'$, in addition to the x-components, $d\xi'$, but the values of $d\eta'$ are not always displayed because the plots can easily become confusing if too much information is presented.

In these plots we have introduced two new terms, T and S. "T" stands for "tangential section," which refers to the section of the pupil in which the x-coordinates are zero, i.e., in the meridian plane. "S" is the "sagittal section," which refers to the section of the pupil in which only x varies, and in which y is constant and equal to the y-coordinate of the chief ray. In other words, the sagittal section is defined, in each space in the lens, to be the plane that contains the chief ray and that is perpendicular to the tangential section.

The transverse aberration curves can give us direct information concerning the blur of the image of a point object, and therefore of the possible resolution of the lens. For example, in the curve shown at the bottom of Fig. 4.3 and labelled "T Y=20°" we can see that the maximum positive transverse aberration is about 0.02 mm and the maximum negative transverse aberration is about -0.04 mm. Therefore, the total blur is about 0.06 mm, and we can reasonably expect that this figure represents an approximate limit to the resolution of the lens. If we calculate the optical transfer function (OTF) of this lens (described in more detail in Sec. 4.4, below), we can expect that the OTF will drop to a small value at about 16 mm^{-1}, since $0.06 \sim 1/16$. With experience, it is possible to make useful guesses of the maximum spatial frequency resolution from the transverse aberration curves, although these guesses should always be confirmed by direct calculation of the OTF.

The curves in Fig. 4.4 show plots of wavefront aberration as a function of aperture. They are most useful for lenses that are nearly diffraction-limited, that is to say their aberrations are so small that resolution is limited by the finite size of the pupil, rather than by aberrations. In other words, the wavefront aberrations are a small fraction of the wavelength of the light.

We can now comment on some features of the transverse aberration curves and wavefront aberration curves in Figs. 4.3 and 4.4. For instance, the area under the plot of sagittal transverse aberration for $x > 0$ at 14 deg is clearly negative. We could estimate that the area under the curve of transverse aberration is about -0.02 mm. (Remember that the horizontal scale shows **relative** ray coordinates, and is therefore dimensionless). Since this example is an *f*/4.5 lens, the numerical aperture ($n' \sin U'$) is 0.111 and the wavefront aberration or optical path difference (OPD) will be given by

$$W = -n' \sin U' \, R_{rs} \int \delta\eta' \, dy \tag{4.10}$$

$$\approx -\ 0.111\ (-0.020)\ \text{mm}$$

$$= +\ 0.0022\ \text{mm}. \tag{4.11}$$

This result is in good agreement with the computed wavefront aberration, which appears to be about 0.002 mm at the edge of the pupil if we estimate its value from the 14-deg sagittal section wavefront aberration curve shown below.

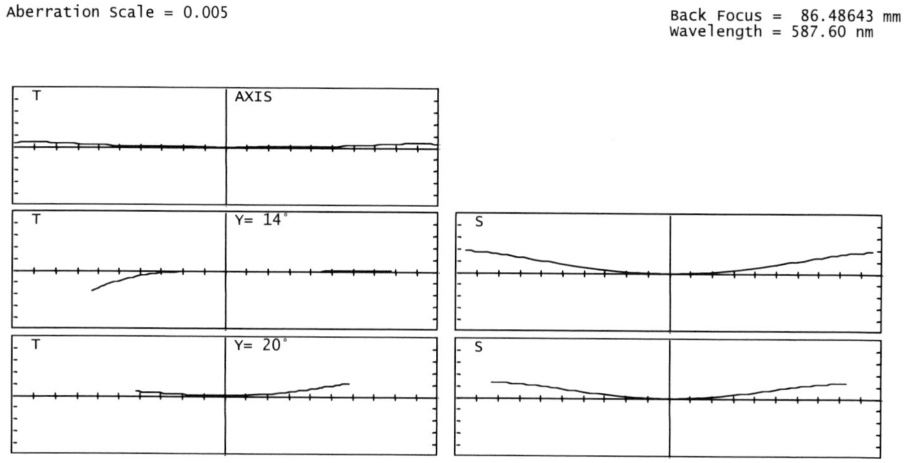

Figure 4.4. A typical set of wavefront aberration curves.

4.3 Spot diagrams

Another method of obtaining information about the aberrations of a lens is to plot spot diagrams. To generate these diagrams we trace a large number of rays, typically of the order of 100, from each field point, and often in several colors. We usually wish each ray to represent an equal amount of energy, and the ray positions are arranged so that the rays are equally spaced in the entrance pupil. The exact form of the ray distribution is not important as long as we trace enough rays. The usual distributions are either a rectangular or a polar grid of rays.

A typical output from a spot diagram calculation is shown in Fig. 4.5. The same lens was used as the example for Figs. 4.3 and 4.4. Each row represents the aberration at one position in the object, so that the first row shows the aberrations of rays from an axial object point, while the second and third rows show the aberrations of rays from points that are 14 and 20 deg off-axis, respectively. The small bar in the top left of the plot shows the scale for the spot diagrams; in this case, the bar indicates 0.100 mm.

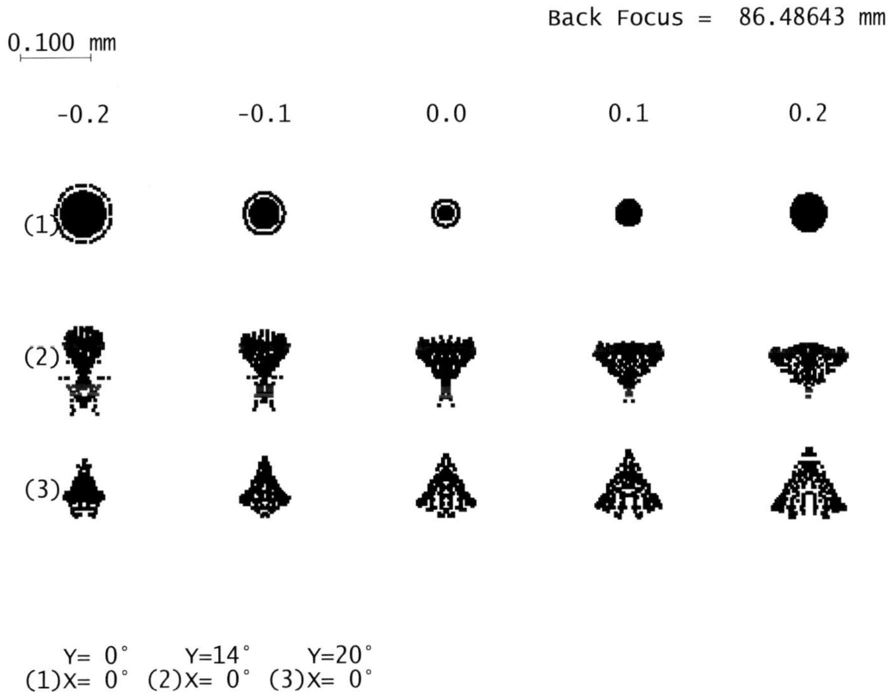

Figure 4.5. A typical set of spot diagrams.

Each column shows the aberrations in one of (in this case) five focus positions. The left column shows the aberrations when we defocus the image surface by −0.2 mm from the paraxial image plane, i.e., we move the image surface 0.2 mm toward the lens. The other columns show the aberrations in four other focus positions; the center column shows the aberration in the paraxial image surface.

Spot diagrams are intended to give a visual representation of the image of a point object, and they do this quite effectively, but it is quite difficult to infer from spot diagrams which type of aberration is present. On the other hand, a lens designer using the information presented in the next section can identify certain types of aberrations. With this information it is sometimes possible to make changes to the lens design that will reduce one or more of the aberrations.

4.4 Aberrations of centered optical systems

So far we have discussed aberrations in a general way, without any discussion of the precise nature of aberrations that we might encounter in the design of centered optical systems. We will now show that only certain types of aberrations are possible in a centered optical system.

In Fig. 4.6, we show a pupil plane, with coordinate axes that we denote as x and y, as before. We also show an image plane with coordinate axes that we denote as ξ' and η'. Note that the ξ' axis in the image plane is parallel to the x-axis in the pupil plane, but when we are referring to the image we will use ξ' to avoid any possible confusion. Similarly, the η' axis in the image plane is parallel to the y-axis.

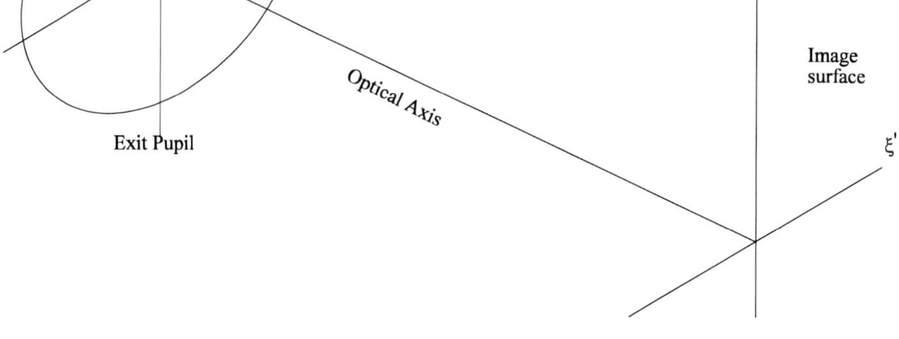

Figure 4.6. Aberration coordinates.

Note that we need not deliberately specify whether the pupil plane is an entrance pupil or an exit pupil, as it is of no significance in the following discussion. Similarly, it is of no significance whether we define the ξ' and η' axes to be in the object plane or the image plane, but it is probably slightly more simple conceptually if we start by considering the two planes to be the image plane and the exit pupil, as shown in Fig. 4.6.

We now have to consider the directions of the ξ' and η' axes. Since we are assuming that the lens has rotational symmetry about the z-axis, the absolute direction of the η' axis is of no significance, and we may, without any loss of generality, assume that all image points lie along the η' axis. This assumption, in fact, is made throughout almost all lens designs with centered optical systems. For example, when we are analyzing a lens by ray tracing or with spot diagrams, we assume that the object points lie in the meridian plane, as long as the lens has rotational symmetry.

We can now see that, for any given lens, any ray passing through the lens can be defined only by the values of η', r and ϕ, so the wavefront aberration of a given lens can only be a function of these quantities:

$$W = W(\eta', r, \phi), \tag{4.12}$$

and we now need to discuss what form this function can take. One way that we can do this is to consider a view toward the object and pupil planes along the direction of the z-axis, looking from the image plane, as shown in Fig. 4.7. B is the off-axis object point, the image of which is shown in Fig. 4.6 at η'. C is the point of intersection of the ray with the pupil, and O is a point on the optical axis of the lens. The point D defines another point in the pupil, equivalent to C, but on the other side of the meridian plane. In other words, D can be thought of as the mirror image of C, if we visualize a reflection in the meridian plane.

We can see from Fig. 4.7 that the triangle OBC defines the ray BC, and the triangle OBD defines a ray BD on the opposite side of the meridian (y) plane. By symmetry, the ray BD has the same wavefront aberration as the ray BC, since the only difference between the two rays is a change in sign of the x-coordinates. Similarly, the triangles OEF and OEG define two rays from an object point E below the optical axis, with a negative η-coordinate. The rays represented by these two triangles must also have, by symmetry, the same wavefront aberration as the original ray BC.

We therefore can say that the aberration depends only on the lengths of the sides of this triangle, which are

$$r^2,\ \eta^2 \text{ and } (r^2 + \eta^2 - 2\,\eta\, r \cos\phi), \tag{4.13}$$

and because r^2 and η^2 form part of the third term in Eq. (4.13), the aberration depends only on

$$r^2, \eta^2 \text{ and } \eta\, r \cos\phi\,. \tag{4.14}$$

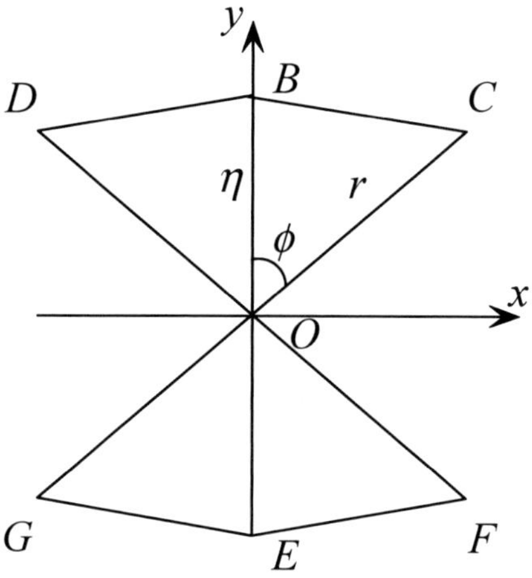

Figure 4.7. Illustration of object and pupil planes, looking along the optical axis from the image.

Assuming that the wave aberration can be expressed as a power series in the above terms, we have

$$\begin{aligned} W &= {}_0w_{20} r^2 \quad \text{Defocus} \\ &+ {}_1w_{11}\, \eta\, r \cos\phi \quad \text{Change in scale} \\ &+ {}_2w_{00}\, \eta^2 \\ &+ {}_0w_{40}\, r^4 \quad \text{Spherical aberration} \\ &+ {}_1w_{31}\, \eta\, r^3 \cos\phi \quad \text{Coma} \\ &+ {}_2w_{22}\, \eta^2 r^2 \cos^2\phi \quad \text{Astigmatism} \\ &+ {}_2w_{20}\, \eta^2 r^2 \quad \text{Field curvature} \\ &+ {}_3w_{11}\, \eta^3 r \cos\phi \quad \text{Distortion} \\ &+ {}_4w_{00}\, \eta^4 \\ &+ {}_0w_{60}\, r^6 \quad \text{Spherical aberration} \\ &+ \text{etc.} \end{aligned} \tag{4.15}$$

In the above expansion, the notation that we have used represents the general terms in this expansion as

$${}_iw_{jk}\, \eta^i r^j \cos^k\phi\,.$$

This notation, which is due to Hopkins,[3] has the advantage that the suffixes indicate the nature of the aberration, and so, if we were to discuss, for example,

${}_1w_{31}$, it should be immediately clear which term we are discussing. However, Welford[1] uses a different notation, and we include for completeness the above expansion in Welford's notation:

$$
\begin{aligned}
W &= a_1 r^2 \quad \text{Defocus} \\
&+ a_2 \eta r \cos\phi \quad \text{Change in scale} \\
&+ a_3 \eta^2 \quad \text{Zero, by definition} \\
&+ b_1 r^4 \quad \text{Spherical aberration} \\
&+ b_2 \eta r^3 \cos\phi \quad \text{Coma} \\
&+ b_3 \eta^2 r^2 \cos^2\phi \quad \text{Astigmatism} \\
&+ b_4 \eta^2 r^2 \quad \text{Field curvature} \\
&+ b_5 \eta^3 r \cos\phi \quad \text{Distortion} \\
&+ b_6 \eta^4 \quad \text{Zero, by definition} \\
&+ \text{etc.}
\end{aligned}
\tag{4.16}
$$

Note in the above equations that $r \cos\phi$ is simply equal to y.

In the above lists we have named the aberrations, as some readers will already be aware of their significance, but we will now discuss the individual aberrations in detail.

4.4.1 First-order aberrations

First we discuss the two term: ${}_2w_{00}\eta^2$ and ${}_4w_{00}\eta^4$. These are constant for any given object point, and since the wavefront aberration at the center of the pupil is zero, by definition, ${}_2w_{00}\eta^2$ and ${}_4w_{00}\eta^4$ must be zero.

4.4.1.1 Defocus — ${}_0w_{20}r^2$

We shall show that this term in the expansion represents a simple longitudinal change in focus. Considering only the meridian plane, for simplicity,

$$W = {}_0w_{20}\, y^2 . \tag{4.17}$$

Differentiating, and substituting into Eq. (4.7),

$$\delta\eta' = \frac{-2}{n' \sin U'}\, {}_0w_{20}\, y. \tag{4.18}$$

A transverse ray aberration that is proportional to the first power of the aperture, y, as indicated in Eq. (4.18), simply represents a situation where the aberrations are computed relative to an out-of-focus image surface. This term is, therefore, zero if we determine the aberrations at the paraxial image surface, but if other aberrations are found to depend on r^2 or y^2, they will indicate that the aberration depends, in some way, on a variation of focus.

4.4.1.2 Lateral image shift — ${}_1w_{11}\eta\, r \cos\phi = {}_1w_{11}\eta\, y$

Since this aberration depends on the first power of y, if we differentiate with respect to y using Eq. (4.7), we will find that the transverse ray aberration is constant for a given value of η. It therefore follows that all of the rays have the same intersection point, but that the actual image point is laterally displaced from the expected image point. Since the displacement is proportional to the object size, η, it follows that we have an image of the wrong size, but if there are no other aberrations the image is otherwise perfect. Therefore, this term corresponds to a change in magnification, which is not usually considered to be an aberration.

4.4.2 The five monochromatic third-order (Seidel) aberrations

In this section we discuss the five aberrations that are dependent on the fourth power in aperture and field, when expressed in the form of wavefront aberrations:

$$\begin{aligned} W = & {}_0w_{40}\, r^4 \text{ Spherical aberration} \\ & + {}_1w_{31}\, \eta\, r^3 \cos\phi \text{ Coma} \\ & + {}_2w_{22}\, \eta^2 r^2 \cos^2\phi \text{ Astigmatism} \\ & + {}_2w_{20}\, \eta^2 r^2 \text{ Field curvature} \\ & + {}_3w_{11}\, \eta^3 r \cos\phi \text{ Distortion} \end{aligned} \tag{4.15}$$

Since the transverse ray aberrations can be obtained by differentiating the wavefront aberrations, it follows that these transverse aberrations are dependent on the third power of aperture and field, and for this reason, they are also known as **third-order** aberrations. (Similarly, paraxial optics is sometimes referred to as **first-order** optics, but this is not a term that we will use). The third-order aberrations are often referred to as **Seidel** aberrations, because they can be computed analytically using equations that were originally derived by Seidel. Another term that is used as a synonym for "third-order aberrations" is "primary aberrations."

4.4.2.1 Spherical aberration — ${}_0w_{40}\, r^4 = {}_0w_{40}\,(x^2 + y^2)^2$

This term is independent of object size η or azimuth angle ϕ and, expressed as a wavefront aberration, it is clearly proportional to r^4. The wavefront aberration associated with third-order spherical aberration is shown in Fig. 4.8. Spherical aberration is the only monochromatic aberration that can occur on-axis. If it is present, it will be constant, to the third-order approximation, over the whole field of view. However, there are higher-order aberrations that can cause spherical aberration to vary over the field.

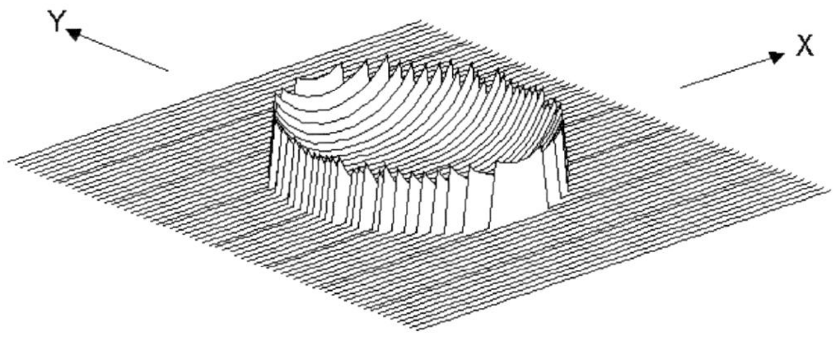

Figure 4.8. Wavefront aberration in the presence of spherical aberration.

Spherical aberration is independent of ϕ, and the aberrations of rays in the *y-z* section (the tangential section) are identical to the aberrations of rays in the *x-z* section (the sagittal section), as shown in Fig. 4.9. Since we are considering plots of transverse ray aberration, the aberrations of the meridian rays are proportional to y^3 and the aberrations of the sagittal rays are proportional to x^3.

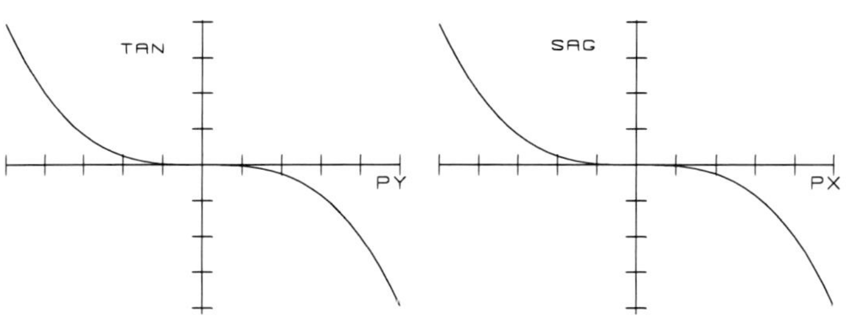

Figure 4.9. Transverse ray aberrations in the presence of spherical aberration.

If we consider through-focus spot diagrams, in the presence of spherical aberration, we find a set of spot diagrams as shown in Fig. 4.10. On one side of the best focus there is a characteristic sharp center to the spot diagram, and on the other side of focus there is a characteristic bright ring around the edge of the spot diagram. This effect can also be seen experimentally in the "star test," as long as the aberrations are sufficiently large that diffraction effects do not dominate.

Figure 4.10. Through-focus spot diagrams when spherical aberration is present.

This effect can be understood if we consider the ray intersection pattern in the presence of spherical aberration as shown in Fig. 4.11, which shows a diagrammatic representation of the ray intersection pattern that produces the spot diagrams of Fig. 4.10. We can see, from Fig. 4.11, that in the extreme right focal position, several rays are close together near the center of the image. This position corresponds to the right-hand spot diagram in Fig. 4.10.

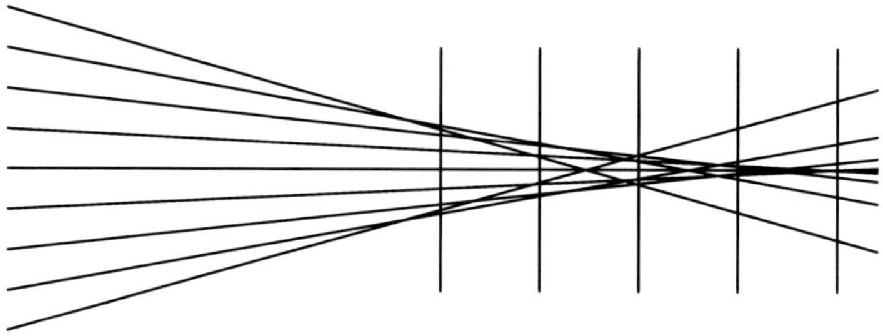

Figure 4.11. Ray intersection pattern in the presence of spherical aberration.

The extreme left focal position in Fig. 4.11 shows several rays close together near the edge of the image; this corresponds to the left-hand spot diagram in Fig. 4.10. The central focal position in Fig. 4.11 and the central spot diagram in Fig. 4.10 show the approximate position of the "disk of least confusion." This is a term which is very popular in elementary text books, but which is almost never used in practical lens design because it does not correspond to any useful criterion of image quality. However, the concept of the disk of least confusion does at least introduce the idea that the best focus is not necessarily the paraxial focus, which is a very important concept; in this sense, if it helps to minimize confusion in the mind of the reader, it has served a useful purpose.

4.4.2.2 Coma — ${}_1W_{31}\eta r^3 \cos\phi = {}_1W_{31}\eta r^2 y$

The wavefront associated with coma is shown in Fig. 4.12. In the meridian section (the y-z section) the wavefront aberration is proportional to y^3. In the x-z section, when $y = 0$ and $\cos\phi = 0$, the wavefront aberration is zero.

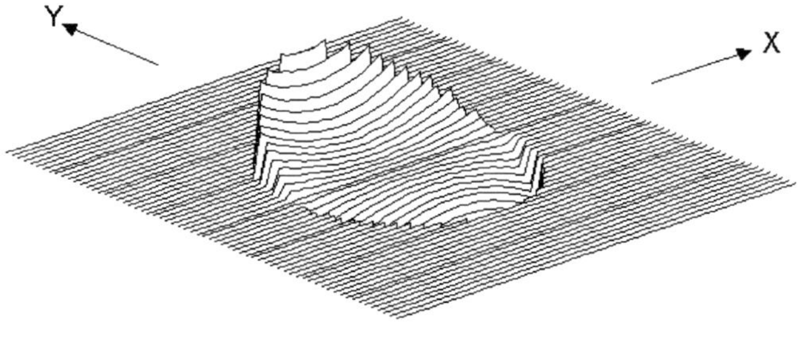

Figure 4.12. Wavefront aberration in the presence of coma.

The transverse aberration ($\delta\eta'$) in the meridian section is proportional to y^2, and this is shown in the left panel of the ray intercept curves in Fig. 4.13. When we consider rays in the *x-z* section (the sagittal section), the *x*-component of aberration ($d\xi'$) of the rays is found to be zero, since $\partial W/\partial x = 0$.

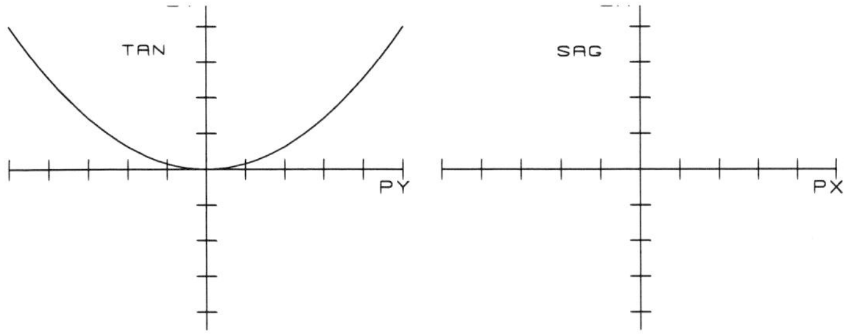

Figure 4.13. Transverse ray aberrations in the presence of coma.

However, even for rays in the sagittal section, where the wavefront aberration is zero, we can see from Fig. 4.12 that the wavefront is tilted, and it follows that the *y*-component of transverse aberration ($\delta\eta'$) is non-zero. Since coma is linearly proportional to η, it is the most important off-axis aberration at small field angles. The geometrical effect of coma is illustrated in a set of through-focus spot diagrams below.

Figure 4.14. Through-focus spot diagrams in the presence of coma.

4.4.2.3 Astigmatism and field curvature — $_2w_{22}\eta^2 r^2 \cos^2\phi$ and $_2w_{20}\eta^2 r^2$, or $_2w_{22}\eta^2 y^2$ and $_2w_{20}\eta^2(x^2 + y^2)$

The wavefront aberrations associated with astigmatism and field curvature are illustrated in Fig. 4.15. Since the wavefront aberration is proportional to r^2, these aberrations will generate a defocus effect of some sort. Specifically, the $_2w_{20}\eta^2 r^2$ term represents a defocus that is proportional to η^2, and it therefore indicates the presence of field curvature. The $_2w_{22}\eta^2 r^2 \cos^2\phi = {}_2w_{22}\eta^2 y^2$ term represents a similar aberration except that it is purely cylindrical, giving only a defocus for the tangential section.

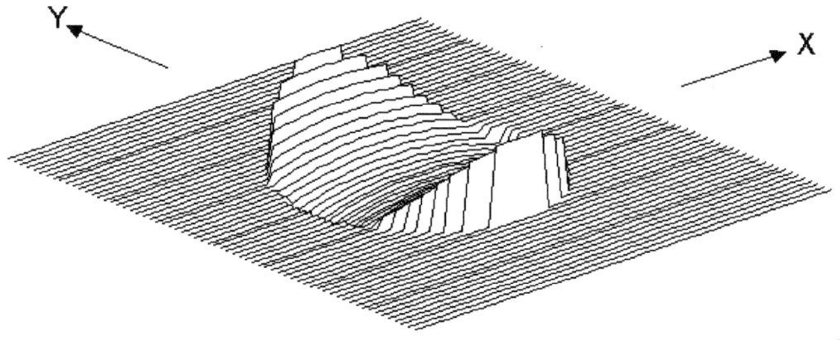

Figure 4.15. Wavefront aberration with astigmatism and field curvature present.

In both sections the transverse aberrations are directly proportional to aperture, but if there is astigmatism, the aberrations of rays in the *y-z* section will differ from the aberrations of rays in the *x-z* section, as shown in Fig. 4.16.

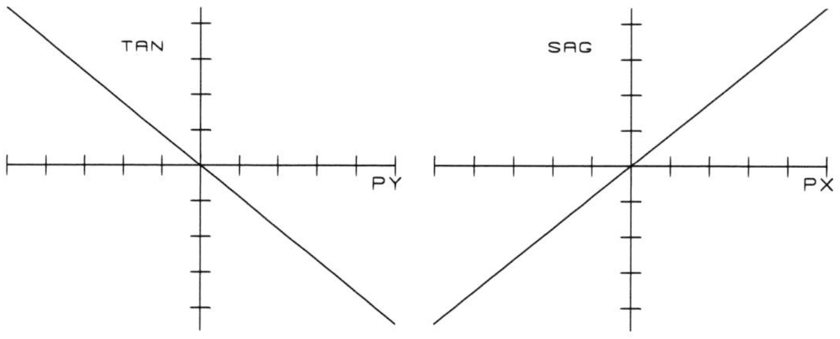

Figure 4.16. Transverse ray aberration with astigmatism and field curvature present.

These two aberrations are proportional to the square of the field angle. The effect of astigmatism and field curvature are shown in a set of through-focus spot diagrams in Fig. 4.17.

We can see from these diagrams that there are two positions in which we find a perfect line focus. The horizontal line focus, shown in the second spot diagram from the left in Fig. 4.17, is the tangential focus, assuming that we are tracing rays from an off-axis point in the meridian plane, with $x = 0$. This is the focus for rays in the tangential section, i.e., the meridian plane. Similarly, the vertical line focus, shown in the second spot diagram from the right in Fig. 4.17, is the sagittal focus for the same object point. This is the focus for rays in the sagittal section, i.e., the section perpendicular to the meridian plane, and containing the chief ray.

Figure 4.17. Through-focus spot diagrams with astigmatism and field curvature present.

Astigmatism and field curvature can also be plotted as longitudinal aberrations as a function of field radius, as shown in Fig. 4.18. For this calculation we compute the position of the sagittal and the tangential foci with respect to the reference image surface, which is usually, but not necessarily, the paraxial image plane.

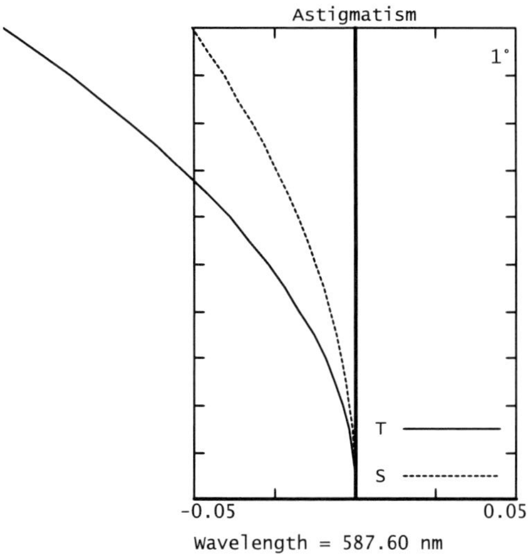

Figure 4.18. The formation of the sagittal and tangential focal lines.

4.4.2.4 Distortion — ${}_3w_{11}\,\eta^3\, r\cos\phi$

Third-order distortion produces a lateral image displacement that is proportional to the third power of the field angle. Often, distortion is described as a fraction, as follows:

$$\text{Distortion} = \frac{\delta\eta'}{\eta'}. \tag{4.19}$$

The effect of this division by η' is to make distortion, as defined in this way, to be a function of the square of the field angle.

Distortion can be calculated from ray trace results, depending on the conjugates at which the lens is to be used, as follows.

4.4.2.4.1 The finite conjugate case

Assuming that the object and image surfaces are plane, the finite conjugate case is the simplest to consider. To calculate distortion, we can trace a real (i.e. exact) chief ray, and also a paraxial chief ray, from the same object point. The distortion can then be computed from

$$\delta\eta' = \eta'_{\text{exact}} - \eta'_{\text{paraxial}}$$

$$\text{Distortion} = \frac{\eta'_{\text{exact}} - \eta'_{\text{paraxial}}}{\eta'_{\text{paraxial}}} \tag{4.20}$$

A lens operating at finite conjugates, if distortion is corrected, will produce an image that is exactly the same shape as the object, but of a different size in general.

4.4.2.4.2 The infinite conjugate case

To visualize this case, consider what happens in the finite conjugate case as the object distance becomes very large.

If the object is a plane surface a distance l from an optical system, as in Fig. 4.19, then

$$\tan\theta = \eta / l. \tag{4.21}$$

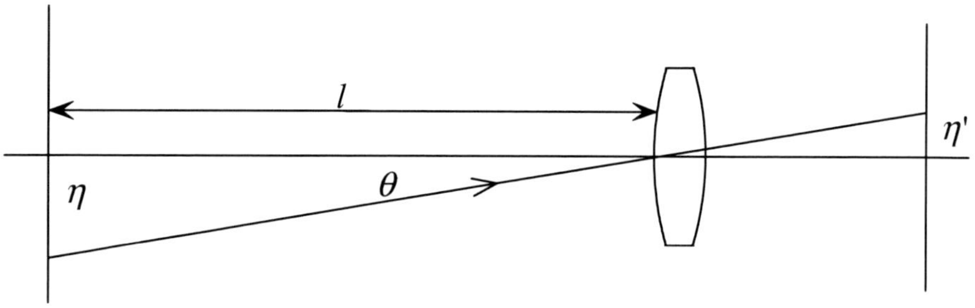

Figure 4.19. Chief ray angle for an object at infinity.

In the case of an object at infinity, it follows that distortion should logically be calculated by tracing an exact chief ray at the correct angle, $U = \theta$, and comparing the image size with the paraxial image size, computed by tracing a paraxial chief ray with angle equal to $u = \tan\theta$. Then distortion can be computed from Eq. (4.20) above. This is the normal definition of distortion for the infinite conjugate case. If the image plane is at the paraxial focus (it quite often is not), then the ideal image size is given by

$$y'_{\text{paraxial}} = f'\tan\theta, \tag{4.22}$$

and a normal lens is sometimes referred to as an f-tan θ lens for this reason. It has the property that, if we form an image of a plane object that is perpendicular to the optical axis of the lens—perhaps a photograph of the facade of a building, which is effectively at infinity—the plane surface will be reproduced in the correct proportions. However, if the object is not perpendicular to the optical axis, the image will show an asymmetric "keystone" distortion even for a distortion-free objective.

4.4.2.4.3 The afocal case

In the afocal case, with an object and image at infinity, we find that the requirements depend on how the system will be used. In the case of typical binoculars, which are often scanned across a scene, it seems to be accepted that the most satisfactory form of distortion correction is one in which angles in the object space are imaged as angles in the image space, with an appropriate magnification. In other words, ideally,

$$\theta' = m\theta, \tag{4.23}$$

where θ is the chief ray angle in the object space, θ' is the chief ray angle in the image space, and m is the angular magnification.

4.4.2.4.4 Effect of pupil aberrations and defocus on distortion

In calculations of distortion for the cases of an image at a finite distance, when the lens is analyzed for an image position which is not at the paraxial focus, y'_{exact} and y'_{paraxial} in Eq. (4.20) will both be affected by the amount of defocus. These effects are determined by the real and paraxial chief ray angles in the image space. Suppose, for some reason, that the real and paraxial chief ray angles were of opposite sign. (This is an extreme case, but not impossible.) As we defocus, the real image size can increase, while the paraxial chief ray height decreases, and the calculated value of distortion will obviously change. This behavior illustrates the point that, when we defocus, the choice of the chief ray becomes significant, and we may get different results depending on whether the paraxial chief ray is aimed at the paraxial entrance pupil position, or at the real entrance pupil position, which is affected by pupil aberrations. In this discussion, we have assumed that the real chief ray is aimed at the real entrance pupil position, which should always be the case.

4.4.2.4.5 F-theta lenses

Many readers will be aware that there is a special class of lenses, usually intended for laser-scanning applications, which is required to satisfy the condition that

$$\eta' = f'\theta, \tag{4.24}$$

and, for obvious reasons, they are known as *f*-theta lenses. The reason for this requirement is that the image size is required to be proportional to the angle of a scanning mirror, which is scanning a laser beam over the image surface.

4.4.2.4.6 Effect of a curved object on distortion

If the object is not plane, the image may not be distortion-free, even if the lens introduces no distortion as defined above. The simplest case to visualize is an oblique image of a sphere. If we image a sphere off-axis with a nondistorting lens, the sphere will effectively be projected onto a plane surface as an ellipse, and this ellipse will be imaged again as an ellipse. Of course, the field angle would have to be quite large before this effect was noticeable, but this example serves to illustrate that even an aberration that is as apparently simple as distortion can produce surprising results.

4.4.3 Higher-order aberrations

It should be remembered that the Seidel aberrations are only the first terms in a power series, and that in almost all real design problems the high-order aberrations can be significant. The next set of terms depends on the sixth power in aperture and field, and just as the Seidel aberrations are often known as third-order aberrations, these terms are known as fifth-order aberrations, or, sometimes, secondary aberrations. In the case of complex lenses, where correction of the Seidel aberrations is possible, the high-order aberrations will be the limiting factor in determining the performance of the system. However, it is still essential for the Seidel aberrations to be corrected, even in these cases.

Since almost all lenses have significant higher-order aberrations, and since the higher-order aberrations are usually very difficult to reduce without making a lens more complex, it is almost always important to adjust the third-order aberrations in such a way as to minimize the effect of the higher-order aberrations. This process, known as "aberration balancing," is very important in the final stages of lens design. In any particular problem, the lens designer will probably be concerned with the maximization of the optical transfer function, or perhaps the Strehl ratio, and many papers have been written on these important but relatively advanced topics (see also Born and Wolf[2]). In this discussion we will look at the topic of aberration balancing in a relatively simple way.

4.4.3.1 Balancing spherical aberration

Suppose that we have a lens with uncorrectable fifth-order spherical aberration. (This is typically the case with medium- to large-aperture systems). In any complex lens design it is possible to reduce the amount of third-order spherical aberration to a reasonable value. When a lens is actually used, the user will usually also have the freedom to adjust the focus of the lens in order to get the best resolution.

Suppose we have a fixed amount of fifth-order spherical aberration. The transverse aberration curve will be as shown in Fig. 4.20(a). In this diagram we have assumed that the wavefront aberration is negative, which is the most common situation. If we add some third-order aberration of the opposite sign, the effect on the transverse aberration curve will be of the form shown in Fig. 4.20(b), assuming that we calculate the aberration at the paraxial image surface.

However the lens is not necessarily used at the paraxial focus. By defocusing the lens, we can effectively add a wavefront aberration that is proportional to r^2. This therefore corresponds to a linear term (that is, a tilt) in the transverse aberration curve. The average transverse ray aberration can been reduced, as shown in Fig. 4.20(c). Although we have shown transverse aberrations in Fig. 4.20, similar reductions would also be seen if we were to show the wavefront aberrations.

Incidentally, if the defocus term is removed and assessment carried out at the paraxial focus, the transverse aberration curve shown in Fig. 4.20(c) is changed by the removal of the linear term, to a curve of the form shown in Fig. 4.20(d). We can see that the effect of the change of focus is simply a tilt in the transverse aberration curve. When analyzing existing lens designs, for example from the patent literature, we sometimes find curves of this type. The obvious conclusion is that the design was optimized for a defocused image surface, but this is rarely stated in the patent.

4.4.3.2 Balancing coma

A similar situation can arise in the case of some lenses with fifth-order coma, whose wave aberration is expressed by a term of the form

$$W = {}_1w_{51}\,\eta\, r^5 \cos\phi. \qquad (4.25)$$

If we differentiate this expression, we find that in the tangential section the transverse aberration depends on y^4, since the Seidel coma gives us a transverse aberration proportional to y^2. By introducing some third-order coma of the opposite sign, the effect of fifth-order coma can be minimized just as was done with fifth-order spherical aberration. Figure 4.21 shows the form of the transverse aberration curve that results.

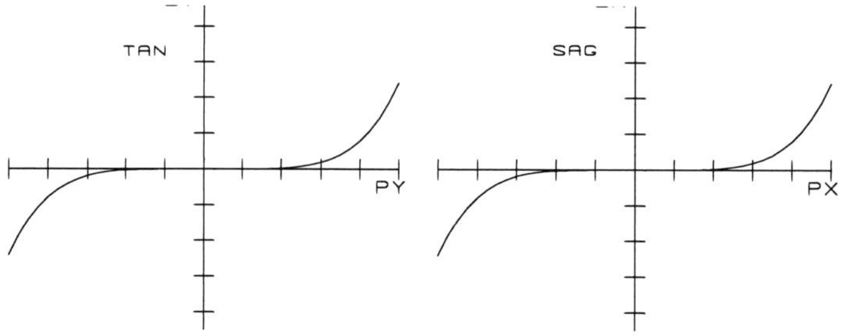

(a)

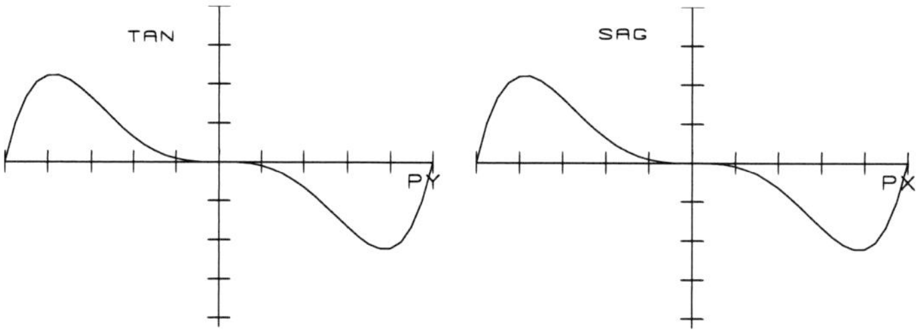

(b)

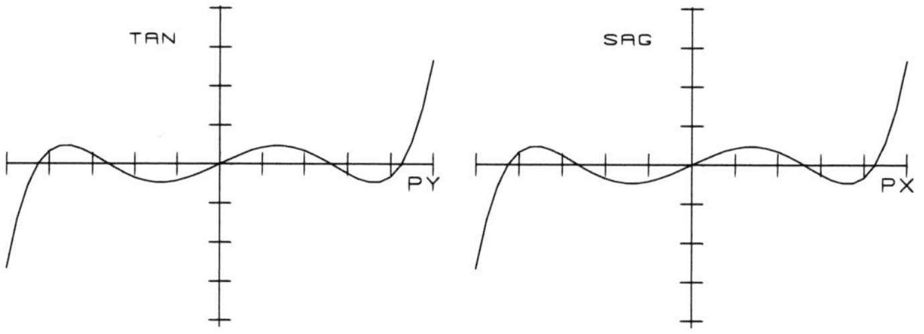

(c)

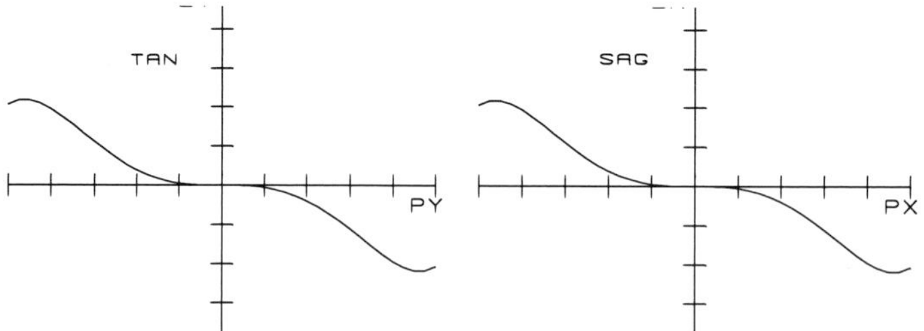

(d)

Figure 4.20. (a) Transverse aberration curves for fifth-order spherical aberration, (b) fifth-order spherical aberration balanced with third-order spherical aberration, (c) fifth-order spherical aberration balanced with third-order spherical aberration and defocus, (d) fifth-order spherical aberration balanced with third-order spherical aberration and defocus, but with defocus removed.

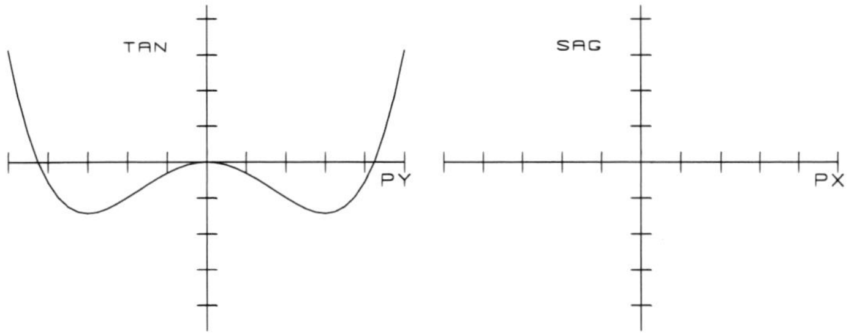

Figure 4.21. Transverse aberration curves with balanced third-order and fifth-order coma.

4.4.3.3 Balancing astigmatism and field curvature

Figure 4.18 showed typical astigmatism curves for a lens with a small field angle. At such small field angles, higher-order astigmatism will be negligible. The sagittal and tangential astigmatism will be quadratic functions of field angle. This example is typical of simple lenses where the tangential astigmatism is significantly larger than the sagittal astigmatism.

Many lenses designed for moderate or large field angles (e.g., camera lenses with semi-fields of about 20 deg or more) have significant amounts of fifth-order astigmatism and field curvature. In addition, they often have significant amounts of fifth-order oblique spherical aberration; these higher-order aberrations must be balanced by appropriate third-order aberrations. Figure 4.22 shows a typical set of astigmatic curves for a photographic objective. Correction of astigmatism should be sufficiently great that the inward curving sagittal surface is bent away from the lens at the same time as the tangent focus at an intermediate field point is moved on the other side of the paraxial image plane. Thus the two astigmatic surfaces cross toward the edge of the field and near the paraxial image plane. This compromise provides reasonably good resolution over a wide field angle.

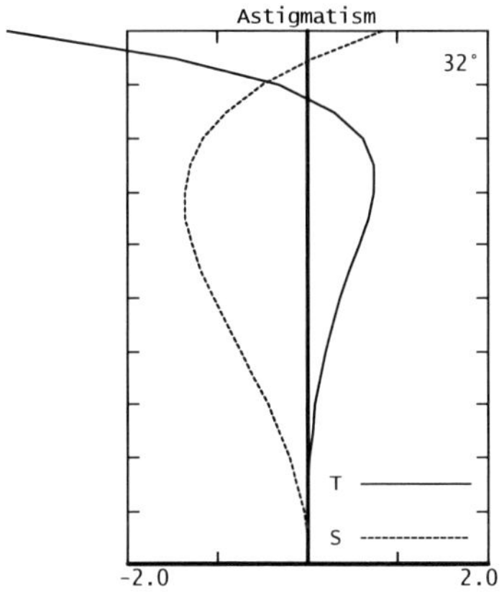

Figure 4.22. Astigmatism of a typical wide-angle photographic objective.

4.4.3.4 Balancing distortion

Third-order distortion produces a lateral image displacement that is proportional to the third power of the field angle. As we have already seen, distortion is often described as a fraction,

$$\text{Distortion} = \frac{\delta\eta'}{\eta'} \tag{4.19}$$

The effect of this division by η' is to make the third-order distortion appear to be a function of the square of the field angle. Similarly, fifth-order distortion, when expressed as a fraction, is proportional to the fourth power of the field angle. In wide-angle lenses these two types of distortion can be balanced in the same way as we have already seen with other aberrations.

4.5 Modulation transfer function (MTF)

While aberration plots give the optical designer a good indication of the quality of a lens, it is widely accepted that a measurable, objective performance criterion is highly desirable, especially for users or evaluators of the lens. MTF is the most widely used criterion, as it applies to common imaging situations with incoherent illumination and where image contrast, or subjective "sharpness," is important. In fact, many modern lenses are specified and measured directly in terms of their MTF performance.

As shown in Fig. 4.23, we consider an optical system that is imaging a grating with a sinusoidal variation of intensity. It can be shown that with incoherent illumination, the image will also have a sinusoidal variation of intensity, but with reduced contrast.

As stated below, the MTF is the ratio of the image contrast to the object contrast; it is of course a function of the spatial frequency, and the use of plots of MTF against frequency (usually in cycles/mm) is very common in lens design. We shall see many examples in later chapters.

The optical transfer function is defined as

$$\text{OTF} = \frac{\text{Image contrast}}{\text{Object contrast}} \tag{4.26}$$

$$\text{Contrast} = \frac{I_{\text{max}} - I_{\text{min}}}{I_{\text{max}} + I_{\text{min}}} \tag{4.27}$$

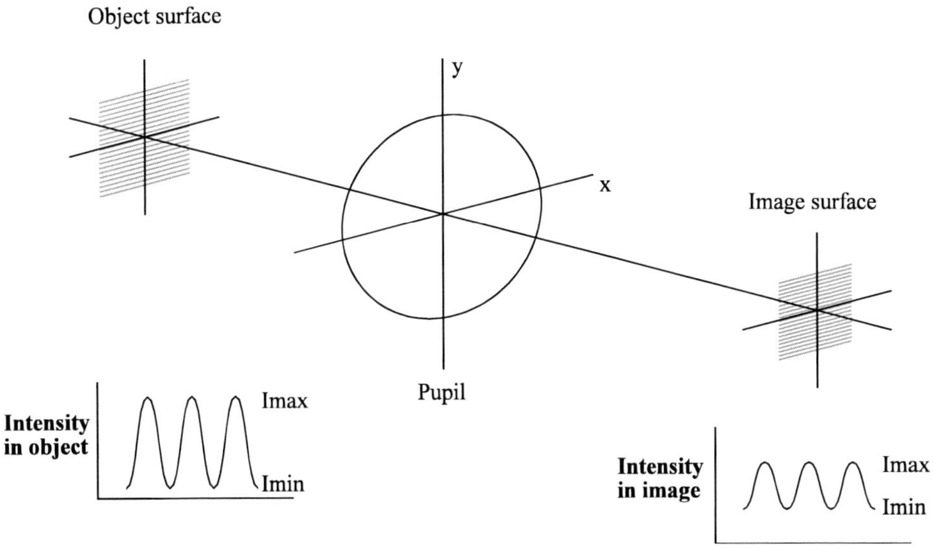

Figure 4.23. Object and image contrast.

Strictly, we should distinguish between the optical transfer function (OTF) and the modulation transfer function (MTF). The OTF is a vector quantity that takes into account any variation of phase in the image whose amplitude is the MTF.

In practice, MTF is what lens designers are most concerned about. The phase term, sometimes called the phase transfer function (PTF), represents the displacement of the sinusoidal image from its ideal position expressed as a phase angle as a function of spatial frequency. If this exceeds 180 deg, then it is possible for the MTF to become effectively negative. This represents a phase reversal, which means that the image has reversed contrast. Many lenses actually show this behaviour at high spatial frequencies.

4.5.1 Theory

For a system that is used with incoherent illumination, the MTF is given by the Fourier transform of the line spread function. MTF is usually computed by this method, but there are other methods that are often faster.

We use the auto-correlation integral to calculate the true, or "diffraction" OTF:

$$D(s)=\frac{1}{A}\iint_S \exp\left\{\left(\frac{2\pi i}{\lambda}\right)(W(x+s/2,y)-W(x-s/2,y))\right\}dx\times dy, \tag{4.28}$$

where $D(s)$ = OTF
A = pupil area
S = area common to two sheared pupils, as shown in Fig. 4.24
W = wavefront aberration
s = reduced spatial frequency, equal to $f\lambda/(NA)$ where NA is the numerical aperture.

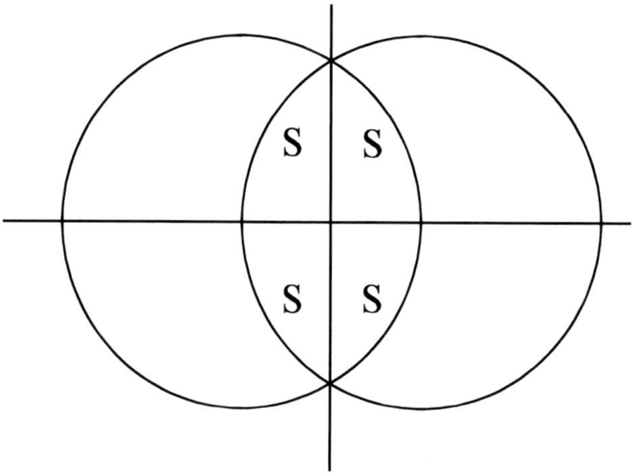

Figure 4.24. S = area of integration for the calculation of MTF.

4.5.2 The geometrical approximation

In the geometrical optics approximation, we assume that λ approaches zero, and it can then be shown that the MTF (in the tangential case) is given by

$$\iint_A \exp\,(-2\pi i \times f \times \delta y) \times dA. \tag{4.29}$$

This is integrated over the whole pupil, and in practice this can be approximated conveniently by a simple summation.

4.5.3 Practical calculation

Since we almost always need to know the MTF in both the sagittal and tangential azimuths, we calculate

$$\text{MTF} = \sum \exp(-2\pi i \times f \times \delta x) \tag{4.30}$$

for the sagittal MTF, and the corresponding expression

$$\text{MTF} = \sum \exp(-2\pi i \times f \times \delta y) \tag{4.31}$$

for the tangential MTF.

It is convenient to simply trace a relatively large number of rays (typically over 100), and of course we include in the summation only the rays that actually pass through the system. We can also trace rays in several different wavelengths, and include them in the summations with appropriate spectral weighting factors.

It is also convenient to apply a correction factor, by multiplying the geometrical MTF by the diffraction-based MTF for a perfect system. The result of this is that the geometrical results will be correct when we have large aberrations and also when we have very small aberrations. In between, the geometrical MTF will usually give a pessimistic result.

We have to choose how many rings of rays to use. The ray pattern is chosen as shown in Fig. 4.25, where it can be seen that each segment of the semi-circle has an equal area. It should be clear that, with n rings of rays, we will have n^2 rays. In order to save computing time, when the optical system has plane symmetry (about the meridian, or tangential, y-plane), the program need not trace rays with negative values of x.

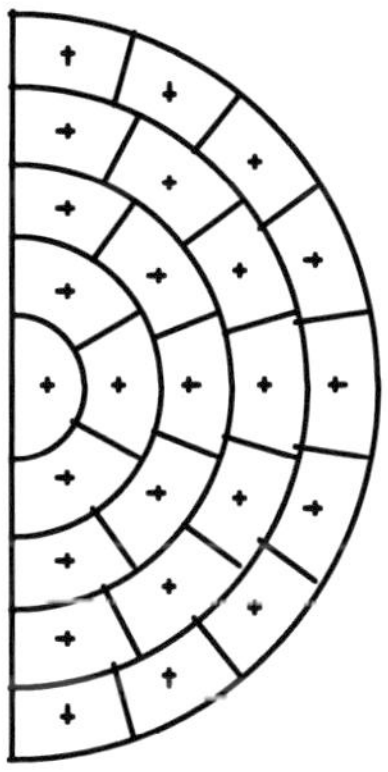

Figure 4.25. Ray pattern for calculating geometric MTF and spot diagrams.

4.5.4 The diffraction limit

From Eq. (4.28) it can be shown that the highest limiting spatial frequency allowed by diffraction of light from the grating object into the finite-sized lens pupil is given by the following equation:

$$f_{\text{lim}} = 2\frac{NA}{\lambda}. \tag{4.32}$$

For example, with an $f/2$ camera lens, the numerical aperture, *NA*, is 0.25, and for a wavelength of $\lambda = 0.0005$ mm, the cut-off frequency will be:

$$f_{\text{lim}} = 2 \times 0.25 / 0.0005 = 1000 \text{ cycles / mm}.$$

This corresponds to a grating with a spacing of 0.001 mm = 1 μm, but note that at this frequency the MTF is actually zero; at 500 cycles/mm (2-μm spacing) the MTF of a perfect lens would be about 40%, so this might be a more realistic indication of the usefulness of an $f/2$ lens. However, you should note that very few detectors can resolve 500 or 1000 cycles/mm, so it should be clear that many very useful lenses are not diffraction-limited.

References

1. W. T. Welford, *Aberrations of Optical Systems*, Adam Hilger (1986).
2. M. Born and E. Wolf, *Principles of Optics*, Cambridge University Press (1999).
3. H. H. Hopkins, *Wave Theory of Aberrations*, Oxford University Press (1950).
4. M. J. Kidger, "The calculation of the optical transfer function using Gaussian quadrature," *Optica Acta*, Vol. **25**, No. 8, p. 665-680 (1978).

CHAPTER 5
CHROMATIC ABERRATION

5.1 Variation of refractive index—dispersion

We have already seen, in the previous chapters, that the optical properties of a lens depend on the refractive index of the glass; this refractive index is a function of the wavelength of light and, for all transmitting materials, the refractive index varies in the manner indicated in Fig. 5.1.

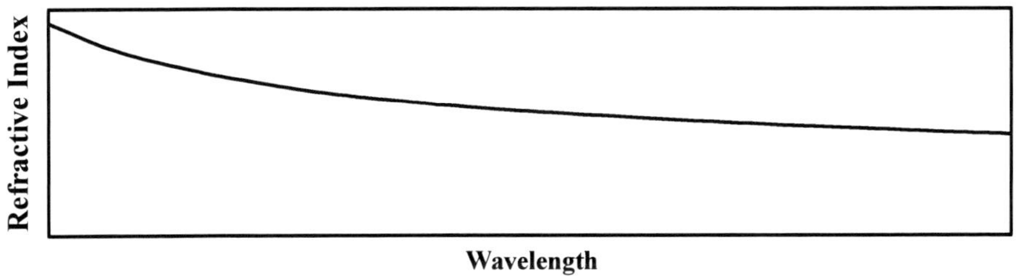

Figure 5.1. The variation of refractive index with wavelength.

This curve can be fitted to a number of polynomials. One of the most commonly used is that due to Schott, first given in their 1967 glass catalog[1]:

$$n^2 = A_0 + A_1\lambda^2 + a_2\lambda^{-2} + A_3\lambda^{-4} + A_4\lambda^{-6} + A_5\lambda^{8}. \quad (5.1)$$

The effect of this variation of refractive index, known as dispersion, is to produce a wavelength-dependent variation of optical properties, known collectively as "chromatic aberration." Sir Isaac Newton invented the reflecting telescope because he considered this aberration impossible to remove since all glasses have dispersion. Fortunately for the development of modern optical instruments, in 1729 Chester Moore Hall, a barrister, proved him wrong. He "made" an achromatic doublet by combining a positive-powered low-dispersion (crown) glass and a negative-powered high-dispersion (flint) glass. He realized that this would work as required by experimenting with prisms until he found a pair of prisms, using different glasses, which produced deviation without dispersion. He reasoned, correctly, that the same principle would work with lenses. In fact he asked one optician to make the crown component and another to make the flint, but unfortunately for Hall, they both subcontracted to George Bass, who discovered the purpose of these two lenses. Hall never published his invention, but Dolland, who did patent the achromatic doublet, developed the idea. Not surprisingly in the circumstances, a legal dispute followed—an early illustration of the risk of trying to keep a trade secret, rather than patenting!

As we shall see, achromatization is not perfect, so residual chromatic aberration remains significant for almost all lenses that are required to work over the whole visible waveband, from wavelengths of about 400 to 700 nm; and depending on the performance of the lens, it can even be significant for wavebands as small as a few picometers. Since all of the optical properties depend on wavelength, we can usefully distinguish between chromatic variations in first-order (paraxial) quantities and chromatic variations in third-order and higher-order aberrations.

5.1.1 Longitudinal chromatic aberration (axial color) of a thin lens

The power of a thin lens can be derived from the power of the thick lens given in Eq. (2.58) by setting d equal to zero:

$$K = (n-1)\,(c_1 - c_2). \tag{5.2}$$

Clearly, the power will be larger at short wavelengths, and for a single converging lens such as that shown in Fig. 5.2, the focal point for red light will be farther from the lens than that for blue light.

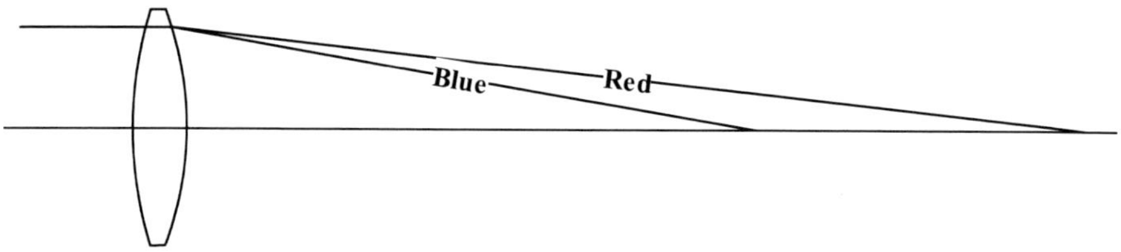

Figure 5.2. Longitudinal chromatic aberration (axial color).

If δn is the dispersion of the glass, i.e., the chromatic variation of n, the change of power will be

$$\begin{aligned}\delta K &= \delta n\,(c_1 - c_2)\\ &= K\frac{\delta n}{(n-1)}\,.\end{aligned} \tag{5.3}$$

This is a first-order chromatic aberration, since it is simply a chromatic change of focus.

5.1.2 The Abbe *V*-value

The reciprocal of the quantity $\delta n/(n-1)$ determines the variation of focal length. This quantity is known as the Abbe *V*-value or, more often, simply as *V*:

$$V = \frac{n-1}{\delta n}. \tag{5.4}$$

In the visible spectrum, the value of refractive index that is used is measured at the helium *d* line, while the dispersion is conventionally taken to be the difference between the refractive indices at the *F*- and *C*- wavelengths, so that

$$V_d = \frac{n_d - 1}{n_F - n_C}. \tag{5.5}$$

The wavelengths at which refractive indices are traditionally measured are as follows:

Spectral Line	Wavelength (nm)	Source and color
h	404.65	Mercury violet
g	435.83	Mercury blue
F′	479.99	Cadmium blue
F	486.13	Hydrogen blue
e	546.07	Mercury green
d	587.56	Helium yellow
C′	643.85	Cadmium red
C	656.27	Hydrogen red
r	706.52	Helium red

Figure 5.3 shows the range of values of n_d and V_d that is obtainable from one of the optical glass manufacturers, in this case, Schott.[1] Note that the V_d axis is traditionally reversed.

When designing lenses for the normal visible waveband, it is possible to use the optical glass data for the standard set of three wavelengths, (*d*, *C*, *F*), although some designers prefer to use the mercury green line at 546.07 nm (*e*, *C*′, *F*′) in place of the helium yellow line because it is closer to the peak of the visual response of the eye. However, for lenses that are required to operate at a different waveband from the visible, the designer is advised to generate a glass diagram for an appropriate set of wavelengths. These wavelengths will be determined primarily by the spectral emission of the source and by the spectral sensitivity of the detector, although other factors such as the transmissivity of the lens may also affect the choice of an appropriate set of wavelengths.

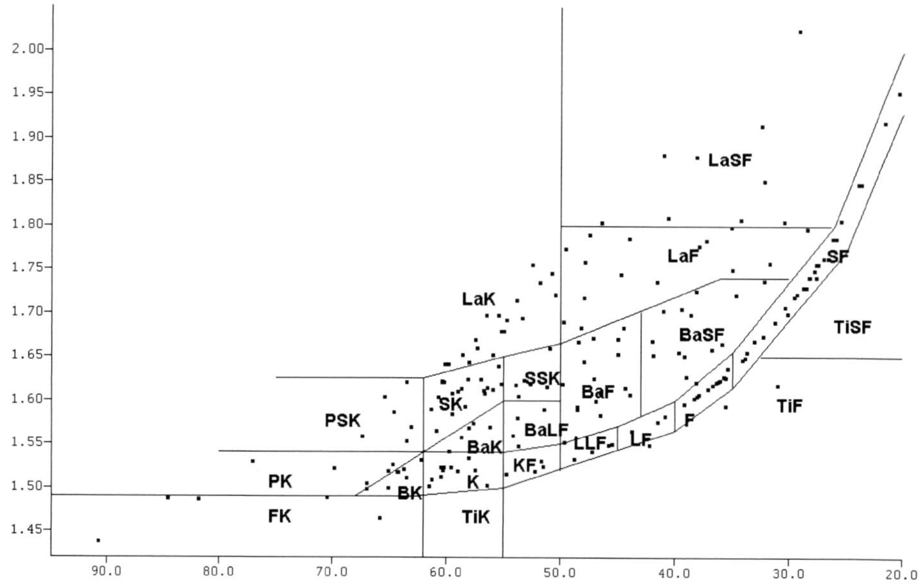

Figure 5.3. The Schott optical glass n_d vs. V_d diagram.

5.1.3 Secondary spectrum

Suppose we wish to correct the chromatic variation in power δK in a thin-lens positive doublet. We will then find that the focal length is not constant with wavelength, but varies, as shown in Fig. 5.4. If the C and F wavelength foci are brought together, then the focal length for the d-line will be shorter. This effect is known as secondary spectrum, and is difficult to correct.

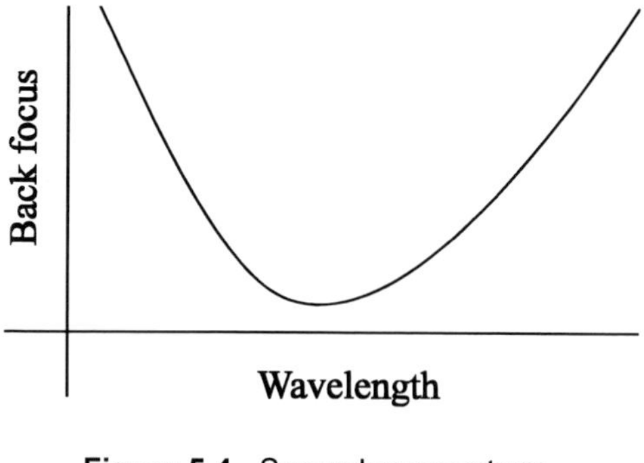

Figure 5.4. Secondary spectrum.

If we are to correct δK for the C-F waveband, we must make

$$\delta K_{C-F} = 0,$$

and, therefore, applying Eqs. (5.3) and (5.4) we require

$$\frac{K_1}{V_1} + \frac{K_2}{V_2} = 0, \tag{5.6}$$

where V_1 and V_2 refer to the two elements of the doublet over the C-F waveband.

The sum of the individual lens element powers equals the overall power of the doublet K:

$$K = K_1 + K_2. \tag{5.7}$$

So the values of K_1 and K_2 that satisfy these two conditions for an achromatic doublet are

$$K_1 = \frac{KV_1}{V_1 - V_2} \quad \text{and} \quad K_2 = \frac{-KV_2}{V_1 - V_2}. \tag{5.8}$$

That is, the second element power must be of the opposite sign to that of the doublet—this was the innovation that Newton had not anticipated.

However, since the curve of refractive index versus wavelength shown in Fig. 5.1 is nonlinear, its gradient, δn, varies with wavelength. Therefore the variation of power over the d-F waveband,

$$\delta K_{d-F} = \left(\frac{K_1}{V_{1_{d-F}}} + \frac{K_2}{V_{2_{d-F}}} \right), \tag{5.9}$$

where $V_{1d\text{-}F}$ and $V_{2d\text{-}F}$ refer to the d–F waveband, i.e.,

$$V_{d-F} = \frac{n_d - 1}{n_F - n_d}. \tag{5.10}$$

The relative partial dispersion, p, is defined as

$$p = \frac{n_F - n_d}{n_F - n_C} \tag{5.11}$$

so

$$p = \frac{V_{C-F}}{V_{d-F}}. \tag{5.12}$$

Then, from Eqs. (5.9) and (5.12),

$$\delta K_{d-F} = \left(\frac{K_1 p_1}{V_{1_{C-F}}} + \frac{K_2 p_2}{V_{2_{C-F}}} \right). \tag{5.13}$$

Using the *C-F* values in Eq. (5.8) for the two element powers, it follows that

$$\frac{K_1 p_1}{V_{1_{C-F}}} = \frac{KV_1}{(V_1 - V_2)} \cdot \frac{p_1}{V_1} = \frac{Kp_1}{(V_1 - V_2)} \text{ and } \frac{K_2 p_2}{V_{2_{C-F}}} = \frac{KV_2}{(V_2 - V_1)} \cdot \frac{p_2}{V_2} = \frac{Kp_2}{(V_2 - V_1)}, \tag{5.14}$$

and therefore

$$\delta K_{d-F} = K \frac{(p_1 - p_2)}{(V_1 - V_2)}. \tag{5.15}$$

In other words, if we want to correct secondary spectrum we need to find two glasses with different V_1 and V_2, but with $p_1 = p_2$. From plots of p versus V, which may be found in the glass manufacturer's catalog (Fig. 5.5 is again from Schott[1]) we find that few pairs of glasses satisfy this condition and that for most glasses (the so-called normal glasses) $(p_1 - p_2)/(V_1 - V_2) = -0.0005$ for this set of wavelengths.

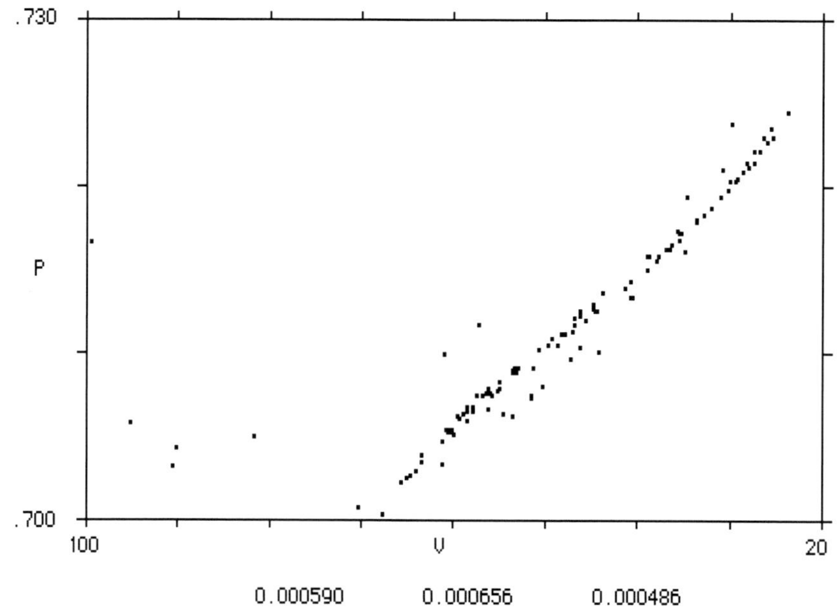

Figure 5.5. Relative partial dispersion diagram for Schott glasses.

For example, an achromatic doublet with normal glasses and a focal length of 5 m will have a difference in focus between the *d* wavelength and the *C* and *F* wavelengths of 2.5 mm. Since a lens like this might be a diffraction-limited telescope objective, this amount of aberration is very significant. It is for this reason (as well as the practical difficulties of obtaining and mounting large, homogenous pieces of glass) that derivatives of Newton's reflecting objective continue to be used for large aperture telescopes. We shall explore some of these in Chapter 13.

5.1.4 Transverse chromatic aberration (lateral color)

If we have a single lens, separated from the aperture stop, a chief ray will be deviated as it passes through the lens. The effect of the chromatic variation of refractive index is to cause dispersion of the chief ray. Therefore the images in the red and blue parts of the visible spectrum will be separated from each other, as shown in Fig. 5.6.

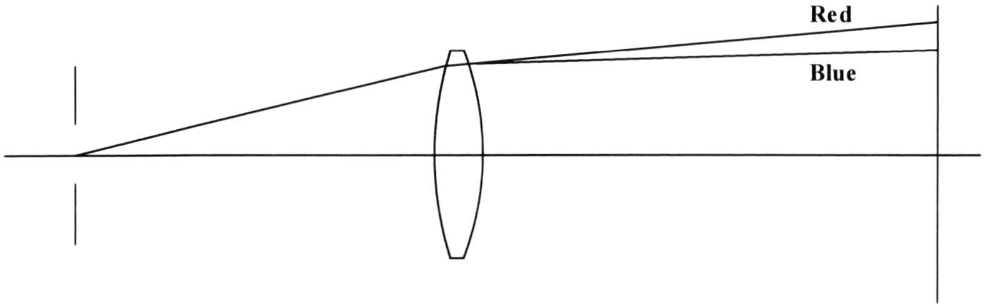

Figure 5.6. Transverse chromatic aberration (lateral color).

In addition to the lateral color a simple system such as this thin lens will also have axial color, superimposed on the lateral color. However, in the interest of clarity, we have not shown the axial color in Fig. 5.6. Lateral color is also a first-order aberration, since it is simply a chromatic lateral focal shift.

5.2 The Conrady method for calculation of chromatic aberration

Naturally, chromatic aberrations can be computed by repeating ray trace calculations in as many different wavelengths as required. Very often, it is sufficient to carry out the ray tracing in three wavelengths. This will increase the ray tracing time by a factor of three, which might be significant in some cases.

There is, however, a very useful result, due to Conrady, that permits calculation of the chromatic aberration, within certain limitations, by tracing an exact ray at a single wavelength.

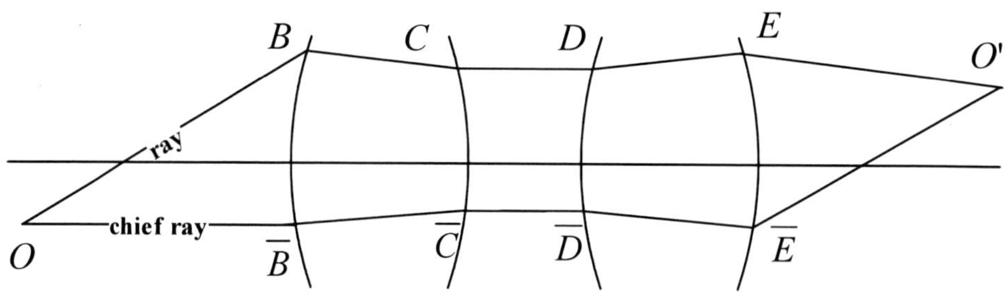

Figure 5.7. The Conrady chromatic aberration formula.

If we trace two rays from O to O′, at a wavelength λ, the optical path difference is given by, from Eq. (3.57),

$$W_\lambda = \left[\overline{\mathrm{O\bar{B}\bar{C}\bar{D}\bar{E}O'}}\right] - \left[\overline{\mathrm{OBCDEO'}}\right]$$

$$W_\lambda = \sum_i n_i D_{Ci} - \sum_i n_i D_i$$

$$W_\lambda = \sum_i n_i (D_{Ci} - D_i), \tag{5.16}$$

where D_{Ci} is the length along the chief ray from one surface to the next, and D_i is the corresponding length along any other ray from the same object point in the same space. The summation extends over all spaces in the system from the object space to the image space.

If we trace the same rays in a different wavelength, $\lambda + \delta\lambda$, the refractive index will change from n to $n + \delta n$, and the paths of the rays will therefore change so that the optical path difference becomes:

$$\begin{aligned} W_{\lambda+\delta\lambda} &= W_\lambda + \delta W_\lambda \\ &= \sum_i (n+\delta n)(D_{Ci} + \delta D_{Ci}) - \sum_i (n+\delta n)(D_i + \delta D_i) \qquad (5.17) \\ &= \sum_i n(D_{Ci} - D_i) + \sum_i n(\delta D_{Ci} - \delta D_i) + \sum_i \delta n(D_{Ci} - D_i) + \sum_i \delta n(\delta D_{Ci} - \delta D_i). \end{aligned}$$

The first term is the same as Eq. (5.16), so

$$\begin{aligned} \delta W_\lambda &= W_{\lambda+\delta\lambda} - W_\lambda \\ &= \sum_i n(\delta D_{Ci} - \delta D_i) + \sum_i \delta n(D_{Ci} - D_i) + \sum_i \delta n(\delta D_{Ci} - \delta D_i). \qquad (5.18) \end{aligned}$$

We will assume that the changes in D_{Ci} and D_i are small. If this is the case, we can ignore the second-order term $\sum_i \delta n(\delta D_{C_i} - \delta D_i)$ and Eq. (5.18) becomes

$$\delta W_\lambda = \sum_i n(\delta D_{Ci} - \delta D_i) + \sum_i \delta n(D_{Ci} - D_i). \tag{5.19}$$

We now consider the term $\sum_i n\delta D_i$. This represents the change in optical path length in wavelength λ when, instead of the physically possible ray, we go to the neighboring path which is not a physical ray (it is physically possible at $\lambda + \delta\lambda$, but not at λ). By Fermat's principle the optical path length along neighboring rays is stationary, which implies that $\sum_i n(\delta D_{C_i} - \delta D_i)$ is a second-order small quantity and can be ignored.

We are finally left with Conrady's so-called "*D* minus *D*" expression:

$$\delta W_\lambda = \sum_i \delta n(D_{Ci} - D_i). \tag{5.20}$$

In order to calculate δW_λ, therefore, we use the values of D_C for the chief ray and D for the "other" ray, which are obtained as a part of the ray tracing process at wavelength λ. Typically, for a lens designed for the visual waveband, we use values of δn given by $\delta n = n_F - n_C$ but, naturally, other wavebands must be used in other cases.

This result is very useful, but lenses that have been optimized using the Conrady formula must be checked by ray tracing or by spot diagram calculation in several wavelengths. The reasons for this caution are

1. The terms that we have ignored, $\sum_i \delta n(\delta D_{Ci} - \delta D_i)$ and $\sum_i n(\delta D_{Ci} - \delta D_i)$, can be significant in the rare case that one part of a lens has a large amount of chromatic aberration that is corrected in another part of the lens, with a large separation between the two parts. In general, this is not a desirable situation, although there are some novel designs that deliberately make use of this condition to correct secondary spectrum with normal glasses. But the effect of this arrangement will be that the rays of different wavelengths separate so much that the ignored terms become significant.

2. The Conrady formula gives us, directly, the change in wavefront aberration. But often we wish to know the chromatic aberration expressed in the form of transverse aberrations. Although it is possible to

differentiate the wavefront aberration to determine the transverse aberration, in practice it is much simpler and safer to trace the ray in several wavelengths.

5.3 Chromatic variation of aberrations

Since almost all of the aberrations of refracting systems are dependent on refractive index, we find that all aberrations produced by refraction, as opposed to reflection, vary with wavelength. For example, in large aperture refracting systems, spherical aberration can vary with wavelength. This variation is referred to as spherochromatism.

For some reason, which I have never understood, there is a widespread, but **incorrect** belief, that the Conrady formula only includes axial and lateral color. However, since the Conrady formula gives the change of wavefront aberration, this includes any chromatic variation in aberration. Thus the Conrady formula includes the effect of spherochromatism as well as all other chromatic variations of aberration.

References

1. Schott Glas, Mainz, Germany.

CHAPTER 6
SEIDEL ABERRATIONS

6.1 Introduction

The previous chapters have been concerned with various aspects of geometrical optics. From this discussion, the reader should be able to see how, by ray tracing, the aberrations of a lens are calculated and spot diagrams are generated. Furthermore, the wavefront aberrations can be calculated as a part of the ray trace, and used to analyze systems whose performance is close to the diffraction limit. This information can be used to compute the MTF, which itself is a measure of closeness to diffraction-limited performance. However, ray tracing gives us virtually no diagnostic information that is helpful in understanding **why** a lens has the aberrations that it does. Nor does it tell us what parameters should be changed in order to reduce the aberrations.

Although this book is written on the assumption that the reader will be doing lens design with modern lens design software, and that only a very small amount of calculation will be carried out manually, it is still essential for the designer to understand the basic principles of aberration correction. This chapter and the next are concerned with the derivation and discussion of analytical formulae, derived originally by Seidel, which do help us to understand to a third-order approximation why aberrations arise, and what can be done to reduce them. This material is of great importance in understanding lens design.

6.2 Seidel surface contributions

We have already seen in Chapter 3 that the wavefront aberration can be written as the difference between the optical path lengths along a chief ray and another ray from the same object point.

Figure 6.1 shows a lens with three surfaces, an object at O and an image at O′. For simplicity the diagram is drawn so that there is also a real intermediate images formed in each space between the lens surfaces, at B and at D. Using the same approach as in Chapter 3, the wavefront aberration can be written as

$$W = [\mathrm{O\bar{A}B\bar{C}D\bar{E}O'}] \quad [\mathrm{O\Lambda BCDEO'}] \tag{6.1}$$

$$= \{[\mathrm{O\bar{A}B}] - [\mathrm{OAB}]\} + \{[\mathrm{B\bar{C}D}] - [\mathrm{BCD}]\} + \{[\mathrm{D\bar{E}O'}] - \mathrm{DEO'}]\}. \tag{6.2}$$

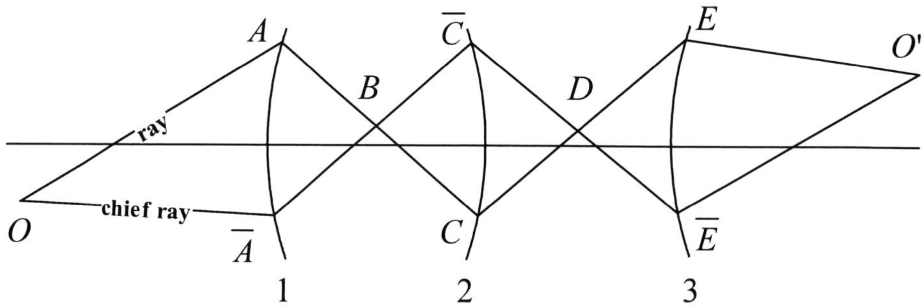

Figure 6.1. The total aberration of a system.

The first term, $\{[O\bar{A}B]-[OAB]\}$ is simply the wavefront aberration due to the first surface. Similarly the second and third terms, $[B\bar{C}D]-[BCD]$ and $[D\bar{E}O']-[DEO']$, are the wavefront aberrations introduced by surfaces 2 and 3, respectively. We therefore see that the wavefront aberration can be expressed as the sum of the wavefront aberration contributions of the individual surfaces in the lens. These contributions can be evaluated independently. In evaluating aberrations of the fifth and higher orders, the contributions cannot all be obtained in this way, since the aberrations of one surface will affect the fifth- and higher-order aberrations of all those that follow. Third-order aberrations, however, can be evaluated using only paraxial ray data, namely the paraxial heights and angles of the axial marginal ray and of the chief ray for the edge of the field. This is the basis of the Seidel analysis. By simplifying to the point where we can use the third-order approximation, we will find that we can derive analytical expressions for the contributions of individual surfaces to spherical aberration, coma, and the other third-order aberrations.

6.2.1 Spherical aberration

In this section we will derive a simple equation for the spherical aberration of a single spherical surface. There are many proofs of this result in the literature. This is one of the simplest. Other proofs are given in the references at the end of the chapter.

Consider a single spherical refracting surface of radius of curvature R, curvature $c = 1/R$, with refractive indices n and n' as shown in Fig. 6.2.

The wavefront aberration of this surface, relative to a reference sphere with a center at O′, can be written as

$$W = [OAO'] - [OBO'] \tag{6.3}$$

$$= n\,(OA - OB) + n'\,(AO' - BO'). \tag{6.4}$$

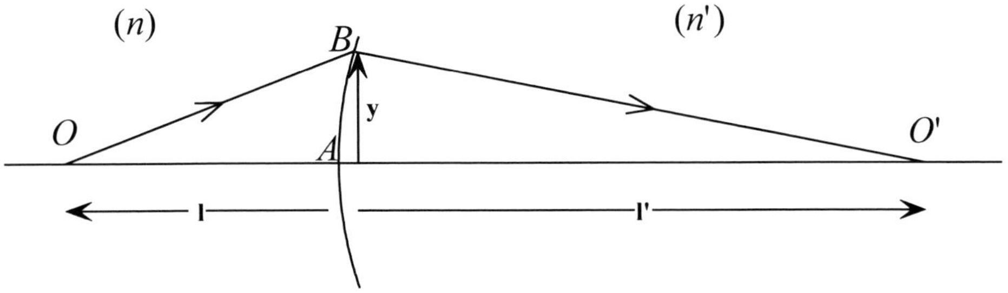

Figure 6.2. Axial image formation—spherical aberration of a single surface.

OA is simply the distance $-l$ to the surface vertex A, and AO′ is the distance l' from the vertex to the ray-axis intersection. So the problem of finding the third-order wavefront aberration reduces to the problem of obtaining suitable approximations for the distances OB and BO′. Clearly,

$$OB^2 = (z - l)^2 + y^2 \tag{6.5}$$

where z is the z-coordinate of B with respect to A. This z-coordinate is given exactly by

$$z = \frac{1 - \sqrt{1 - c^2 y^2}}{c}, \tag{1.13}$$

and can be expanded to give

$$z = \frac{1}{2} c y^2 + \frac{1}{8} c^3 y^4 + \frac{1}{16} c^5 y^6 + \ldots. \tag{1.16}$$

In the Seidel approximation we ignore the terms proportional to y^6 and above, and we then have

$$z \approx \frac{1}{2} c y^2 + \frac{1}{8} c^3 y^4. \tag{6.6}$$

If we substitute Eq. (6.6) in Eq. (6.5), we obtain, with the same approximation,

$$OB^2 = \left(\frac{1}{2} c y^2 + \frac{1}{8} c^3 y^4 - l \right)^2 + y^2 \tag{6.7}$$

$$\mathrm{OB}^2 = l^2 - lcy^2 - \frac{1}{4}lc^3y^4 + \frac{1}{4}c^2y^2 + y^2 \tag{6.8}$$

$$\mathrm{OB}^2 = l^2\left[1 - \left(c - \frac{1}{l}\right)\frac{y^2}{l} - \left(c - \frac{1}{l}\right)\frac{c^2y^4}{4l}\right]. \tag{6.9}$$

Taking the square root and using the paraxial approximation ($y << l$)

$$\mathrm{OB} = -l + \frac{1}{2}y^2\left(c - \frac{1}{l}\right) + \frac{1}{8}c^2y^4\left(c - \frac{1}{l}\right) + \frac{1}{8l}y^4\left(c - \frac{1}{l}\right)^2. \tag{6.10}$$

Then

$$\mathrm{OA} - \mathrm{OB} = -\frac{1}{2}y^2\left(c - \frac{1}{l}\right) - \frac{1}{8}c^2y^4\left(c - \frac{1}{l}\right) - \frac{1}{8l}y^4\left(c - \frac{1}{l}\right)^2. \tag{6.11}$$

Similarly,

$$\mathrm{AO}' - \mathrm{BO}' = -\frac{1}{2}y^2\left(c - \frac{1}{l'}\right) - \frac{1}{8}c^2y^4\left(c - \frac{1}{l'}\right) - \frac{1}{8l'}y^4\left(c - \frac{1}{l'}\right)^2. \tag{6.12}$$

Substituting these into Eq. (6.4), the wavefront aberration is

$$W = n\,(\mathrm{OA} - \mathrm{OB}) + n'\,(\mathrm{AO}' - \mathrm{BO}') \tag{6.4}$$

$$\begin{aligned} W &= \frac{1}{2}y^2\left[n'\left(c - \frac{1}{l'}\right) - n\left(c - \frac{1}{l}\right)\right] \\ &+ \frac{1}{8}c^2y^4\left[n'\left(c - \frac{1}{l'}\right) - n\left(c - \frac{1}{l}\right)\right] \\ &+ \frac{1}{8}y^4\left[\frac{n'}{l'}\left(c - \frac{1}{l'}\right)^2 - \frac{n}{l}\left(c - \frac{1}{l}\right)^2\right]. \end{aligned} \tag{6.13}$$

So far we have made no assumption about the position of O′, but we now choose O′ to be the paraxial image point. The effect of this is that we will derive an expression for the wavefront aberration with respect to a reference sphere centered on the paraxial image point.

If O′ is the paraxial image point, Eq. (2.12) gives us

$$\frac{n'}{l'} = \frac{n}{l} + (n' - n)c, \tag{2.12}$$

or

$$n'\left(c-\frac{1}{l'}\right)=n\left(c-\frac{1}{l}\right) \tag{6.14}$$

and Eq. (6.13) then simplifies to

$$W=\frac{1}{8}y^4\left[\frac{n'}{l'}\left(c-\frac{1}{l'}\right)^2-\frac{n}{l}\left(c-\frac{1}{l}\right)^2\right]. \tag{6.15}$$

Suppose that, instead of tracing a real ray at a height of y at the surface, we trace a paraxial ray, at a paraxial height h, corresponding to the real ray height y. Then, recalling Eq. (2.70), the angle of incidence, $i = hc + u$, and Snell's law expressed as $A = ni = n'i'$,

$$A = n\,(hc + u) = nh\,(c - 1/l) = n'(hc + u') = n'h(c - 1/l') \tag{6.16}$$

so we can use

$$(c - 1/l) = A/(nh) \quad \text{and} \quad (c - 1/l') = A/(n'h) \tag{6.17}$$

in Eq. (6.15), giving

$$W=\frac{1}{8}h^4\left[\frac{n'}{l'}\left(\frac{A}{n'h}\right)^2-\frac{n}{l}\left(\frac{A}{nh}\right)^2\right] \tag{6.18}$$

$$W=\frac{1}{8}h^2\left(\frac{A^2}{n'l'}-\frac{A^2}{nl}\right) \tag{6.19}$$

$$W=\frac{1}{8}A^2h\left(\frac{h}{n'l'}-\frac{h}{nl}\right) \tag{6.20}$$

$$W=\frac{1}{8}A^2h\left(\frac{u}{n}-\frac{u'}{n'}\right), \tag{6.21}$$

since, from Eq. (2.15)

$$u=-\frac{h}{l} \quad \text{and} \quad u'=-\frac{h'}{l'}. \tag{2.15}$$

If we write

$$\delta\left(\frac{u}{n}\right)=\frac{u'}{n'}-\frac{u}{n}, \tag{6.22}$$

we obtain

$$W=-\frac{1}{8}A^2h\delta\left(\frac{u}{n}\right). \tag{6.23}$$

It will be noticed that we have used paraxial ray data to calculate the Seidel aberration coefficient. This is a topic that causes considerable confusion to the newcomer to lens design, since it seems strange that aberration information can be obtained from a ray that is, by definition, aberration-free.

Suppose that we trace the same ray through a lens, using both the paraxial equations and the exact ray trace equations. If the ray is very close to the axis (in the paraxial region, in fact) the heights of the ray that are found from the two calculations will be essentially the same. But as the angle of the ray is increased, the results of the two calculations will differ. If the paraxial ray height is denoted by h, and the real ray height is denoted by y, we can write

$$y=h+O(h^3), \tag{6.24}$$

where $O(h_3)$ refers to terms cubic in h and above. We have already stated that third-order spherical aberration is proportional to y^4, and that all of the third-order aberrations are proportional to the fourth power in aperture and field when expressed as wavefront aberrations. It can be seen that

$$y^4=[\,h+O(h^3)]^4, \tag{6.25}$$

which upon expansion produces a leading term of h^4 and the next higher order terms are $h^3O(h^3)$; that is,

$$y^4=h^4+O(h^6). \tag{6.26}$$

Therefore, the error that is introduced by using paraxial ray data in place of real ray data is of the same order as terms that were ignored in the derivation of Eq. (6.23), and we can therefore use paraxial ray data in the calculation of the Seidel coefficients.

An alternative way of looking at the same question is to suppose that we calculate the wavefront aberration for two different cases. One case is for a ray with a height h, and the other case is for a ray with a height of $h/2$. If we use paraxial ray data for the calculation of W, the paraxial ray heights and angles will be exactly in the ratio of 2:1 in the two cases, and it follows that the calculated values of W will be exactly in the ratio of 2^4:1 = 16:1. This result is clearly correct in view of the fact that third-order spherical aberration is proportional to h^4, by definition.

On the other hand, if we use real ray data for the calculations, the ray heights and angles will not be exactly in the ratio of 2:1, and the calculated values of W will therefore not be in the ratio of 16:1 as expected.

In practice, we remove the factor of 1/8 and define the Seidel spherical aberration coefficient, which is a factor of eight times the wavefront value, as:

$$S_1 = -A^2 h\delta\left(\frac{u}{n}\right). \tag{6.27}$$

6.2.2 Off-axis Seidel aberrations

While Eq. (6.27) is quite simple since the object O is on-axis, its derivation is rather tedious. Since they apply to off-axis objects, the derivation of the expressions for the other Seidel coefficients is significantly more complex. These derivations are given in the references at the end of this chapter.[2–4] The formulae for the off-axis Seidel aberrations of a single spherical surface are

Coma $$S_2 = -A\bar{A}h\delta\left(\frac{u}{n}\right) \tag{6.28}$$

Astigmatism $$S_3 = -\bar{A}^2 h\delta\left(\frac{u}{n}\right) \tag{6.29}$$

Field Curvature $$S_4 = -H^2 c\delta\left(\frac{1}{n}\right) \tag{6.30}$$

Distortion $$S_5 = \frac{\bar{A}}{A}(S_3 + S_4) \tag{6.31}$$

In these equations, A is ni, where i is the angle of incidence of the paraxial marginal ray, $\bar{A}$ is $n\bar{i}$, where $\bar{i}$ is the angle of incidence of the paraxial chief ray, h is the paraxial marginal ray height

$$\delta\left(\frac{u}{n}\right) = \frac{u'}{n'} - \frac{u}{n},$$

H is the Lagrange invariant, $n\,u\,\eta$.

As in the case of spherical aberration, the off-axis Seidel aberrations of the complete lens are found by simply adding the individual surface contributions. Note again that all of the quantities needed to compute the Seidel aberrations are based on the heights and angles of the paraxial marginal ray and paraxial chief ray.

6.2.3 Alternative formula for distortion

All of Eqs. (6.27) through (6.30) are well behaved, and can never become indeterminate apart from the case where $n = 0$, which is physically impossible. However, if we attempt to calculate distortion from the expression given in Eq. (6.31):

$$S_5 = \frac{\bar{A}}{A}(S_3 + S_4).$$

When A=0, we will expect to find a division by zero error. The case when A=0 is a very common case. It occurs, for example, when a collimated marginal ray (with $u = 0$) is incident on a plane surface, or, more generally, it occurs when any surface is normal to the marginal ray. Since S_5 is a wavefront aberration, we can perhaps foresee that S_5 can not be infinite, and we must use an alternative expression for S_5.

In order to derive an alternative expression, we first note from Eq. (2.68) that

$$-u = hc - i, \tag{6.32}$$

and that

$$H = \bar{h}A - h\bar{A}. \tag{2.74}$$

Then, using Eqs. (6.29) and (6.30),

$$\begin{aligned} S_5 &= \frac{\bar{A}}{\mathrm{A}}\left[-\bar{A}^2 h\delta\left(\frac{u}{n}\right) - H^2 c\delta\left(\frac{1}{n}\right)\right] \\ &= \frac{\bar{A}}{A}\left[\bar{A}^2 h\delta\left(\frac{hc - i}{n}\right) - (\bar{h}A - h\bar{A})^2 c\delta\left(\frac{1}{n}\right)\right] \end{aligned} \tag{6.33}$$

$$\begin{aligned} S_5 = \frac{\bar{A}}{A}\Bigg[&\bar{A}^2 h^2 c\delta\left(\frac{1}{n}\right) - \bar{A}^2 Ah\delta\left(\frac{1}{n^2}\right) \\ &- (\bar{h}^2 A^2 - 2h\bar{h}A\bar{A} + h^2\bar{A}^2)c\delta\left(\frac{1}{n}\right)\Bigg] \end{aligned} \tag{6.34}$$

$$S_5 = \frac{\bar{A}}{A}\left[-\bar{A}^2 Ah\delta\left(\frac{1}{n^2}\right) + (2h\bar{h}A\bar{A} - \bar{h}^2 A^2)c\delta\left(\frac{1}{n}\right)\right] \tag{6.35}$$

$$S_5 = -\bar{A}^3 h\delta\left(\frac{1}{n^2}\right) + (2h\bar{h}\bar{A}^2 - \bar{h}^2 A\bar{A})c\delta\left(\frac{1}{n}\right) \tag{6.36}$$

$$S_5 = -\bar{A}^3 h\delta\left(\frac{1}{n^2}\right) + \bar{h}\bar{A}(2h\bar{A} - \bar{h}A)c\delta\left(\frac{1}{n}\right). \tag{6.37}$$

This formula can be evaluated without any possibility of a division-by-zero error.

6.2.4 Aberrations of a plano-convex singlet

We are now in a position to show how some of the formulae for Seidel aberration coefficients can be used to deduce, from first principles, some characteristics of a simple plano-convex singlet made of a glass of refractive index n.

The spherical aberration of a surface is given by

$$S_1 = -A^2 h\delta\left(\frac{u}{n}\right) \tag{6.27}$$

From Fig. 6.3, we can see that the marginal ray heights, h_1 and h_2, are positive at both surfaces. In addition we can see that for an object at infinity, the angle of incidence of the marginal ray at both surfaces is non-zero, and therefore $A\ (= ni\ )$ is also non-zero at both surfaces.

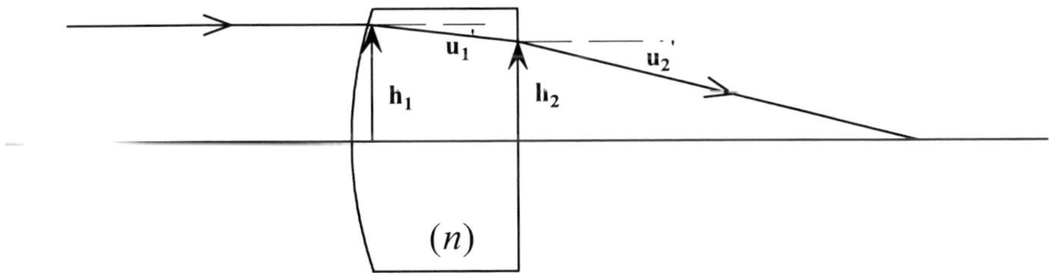

Figure 6.3. Plano-convex singlet.

Therefore, the sign of S_1 depends on the sign of

$$-\delta\left(\frac{u}{n}\right) = -\left(\frac{u'}{n'} - \frac{u}{n}\right) = \frac{u}{n} - \frac{u'}{n'}. \tag{6.38}$$

At the first surface, u is zero, if we assume that the object is at infinity, so that the marginal ray is initially parallel to the axis. Clearly, from the diagram, the angle after refraction, u_1' is negative, so the spherical aberration depends on

$$\frac{u}{n}-\frac{u'}{n'}=-\frac{u'}{n'}>0. \tag{6.39}$$

Therefore, the spherical aberration of the first surface is positive, remembering that this is a wavefront aberration or optical path difference.

If we consider the second surface, which we are assuming is plane, we can see from Snell's law that

$$u_2' = nu_2,$$

since $n'_2 = 1$ and $n_2 = n$. Therefore,

$$\frac{u_2}{n_2}-\frac{u_2'}{n_2'}=\frac{u_2}{n}-nu_2=u_2\left(\frac{1}{n}-n\right). \tag{6.40}$$

Since u_2 and $(1/n - n)$ is negative, the spherical aberration of the second surface is also positive, so we can see that the complete lens has positive spherical aberration.

The significance of this is that we can, using the Seidel formulae, determine properties of the lens that would have been difficult, if not impossible, to deduce by ray tracing. For example, the result deduced in Eqs. (6.39) and (6.40) shows that the spherical aberration will be positive for all values of the refractive index, n.

We can also deduce a simple result for the field curvature coefficient, S_4, which is given by

$$S_4=-H^2c\delta\left(\frac{1}{n}\right). \tag{6.30}$$

At the second surface we have chosen the curvature to be zero, so S_4 is automatically zero at this surface. Clearly H^2 is positive, unless H is zero, which is a trivial case. At the first surface, c is positive and $\delta(1/n) = 1/n - 1$. Therefore the total S_4 for the plano-convex lens is given by

$$\begin{aligned}S_4&=-H^2c\left(\frac{1}{n}-1\right)\\&=-H^2c\left(\frac{1-n}{n}\right)\\&=H^2\frac{K}{n},\end{aligned} \tag{6.41}$$

since $K = c(n - 1)$, from Eq. (2.58), with $c_2 = 0$ because of the plane surface.

This result shows that the S_4 coefficient depends only on the power of the lens, since n is always positive. We will see in Chapter 7 that the result given by Eq. (6.41) is actually a very general result, and applies to any thin lens, although we have derived it for the specific case of a thick plano-convex lens.

6.2.5 First-order axial color and lateral color

Referring to the derivation of the expression for S_1, and to Fig. 6.2, we stated that the monochromatic wavefront aberration for an axial object point could be written as

$$W = n\,(\mathrm{OA} - \mathrm{OB}) + n'(\mathrm{AO}' - \mathrm{BO}'). \tag{6.4}$$

From the Conrady formula [Eq. (5.20)] we can see that the chromatic aberration is given by an expression which is similar to Eq. (6.4), namely,

$$\delta W_\lambda = \delta n\,(\mathrm{OA} - \mathrm{OB}) + \delta n'\,(\mathrm{AO}' - \mathrm{BO}'). \tag{6.42}$$

In the derivation of the spherical aberration result, we used a result

$$\mathrm{OA} - \mathrm{OB} = -\frac{1}{2}y^2\left(c - \frac{1}{l}\right) - \frac{1}{8}c^2y^4\left(c - \frac{1}{l}\right) - \frac{1}{8l}y^4\left(c - \frac{1}{l}\right)^2. \tag{6.11}$$

However in the derivation of the result for axial color we only need to consider the first term, since we are interested in the lowest order aberration, including only terms in y^2. Therefore, we can write, replacing y with the paraxial ray height h,

$$\delta W_\lambda = \frac{1}{2}h^2\left[\delta n'\left(c - \frac{1}{l'}\right) - \delta n\left(c - \frac{1}{l}\right)\right] \tag{6.43}$$

$$\delta W_\lambda = \frac{1}{2}h\left[\left(c - \frac{1}{l'}\right)n'h\frac{\delta n'}{n'} - \left(c - \frac{1}{l}\right)nh\frac{\delta n}{n}\right] \tag{6.44}$$

$$\delta W_\lambda = \frac{1}{2}Ah\left(\frac{\delta n'}{n'} - \frac{\delta n}{n}\right). \tag{6.45}$$

Since, from Eqs. (6.16), (6.17)

$$A = n'h\left(c - \frac{1}{l'}\right) = nh\left(c - \frac{1}{l}\right), \tag{6.46}$$

therefore,

$$\delta W_{\lambda} = \frac{1}{2} A h \delta\left(\frac{\delta n}{n}\right), \tag{6.47}$$

where

$$\delta\left(\frac{\delta n}{n}\right) = \frac{\delta n'}{n'} - \frac{\delta n}{n}. \tag{6.48}$$

The coefficient of axial color is twice the wavefront aberration, or

$$C_1 = A h \delta\left(\frac{\delta n}{n}\right). \tag{6.49}$$

Using a similar approach for the chief ray, Welford[3] shows the result for lateral color to be

$$C_2 = \overline{A} h \delta\left(\frac{\delta n}{n}\right). \tag{6.50}$$

6.2.6 Summary of the Seidel surface coefficients

We can now collect together the equations for the five monochromatic third-order aberrations, and the two chromatic first-order aberrations:

Spherical aberration $$S_1 = -A^2 h \delta\left(\frac{u}{n}\right) \tag{6.27}$$

Coma $$S_2 = -A\overline{A} h \delta\left(\frac{u}{n}\right) \tag{6.28}$$

Astigmatism $$S_3 = -\overline{A}^2 h \delta\left(\frac{u}{n}\right) \tag{6.29}$$

Field curvature $$S_4 = -H^2 c \delta\left(\frac{1}{n}\right) \tag{6.30}$$

Distortion $$S_5 = \frac{\overline{A}}{A}(S_3 + S_4) \tag{6.31}$$

Axial color $$C_1 = Ah\delta\left(\frac{\delta n}{n}\right) \tag{6.49}$$

Lateral color $$C_2 = \bar{A}h\delta\left(\frac{\delta n}{n}\right) \tag{6.50}$$

It is impossible to overstate the importance of these seven equations in lens design. They are not only reasonably simple so that they can be easily understood, but they enable a lens designer to sort out the contribution of each surface to a particular aberration. Although, as we will see, higher-order aberrations are often the factor that limits the resolution that a lens can give, correction of the above seven aberrations is a **necessary condition** for good aberration correction.

6.2.7 A numerical example

In order to illustrate the use of these coefficients, we now give a numerical example, which the reader is encouraged to study in detail, to ensure that the use of the equations has been understood.

The example that we give is a triplet objective with prescription as given below, taken from Kingslake.[4] We have already seen the aberration curves of this lens in Chapter 4.

EFL = 99.996
WAVELENGTHS [nm]
587.60 656.30 486.10
SURFACES

# SURF SPACE	RADIUS	SEPN	INDEX1	V	CLR RAD	GLASS
0		0.00000	1.000000			
1 S	42.98790	4.00000	1.620408	60.30	11.500	S SK16
2 S	-248.07740	10.51018	1.000000		11.500	
3 S	-38.21035	2.50000	1.616589	36.61	9.852	S-F4
4#S	43.95894	9.86946	1.000000		8.885	
5 S	656.66349	4.50000	1.620408	60.30	11.000	S-SK16
6 S	-33.50754	86.48643	1.000000		11.000	
7 S	Plane				37.166	

A drawing of the lens is shown in Fig. 6.4. It has a focal length of 100 mm, and an aperture of *f*/4.5. Therefore the diameter of the entrance pupil is 100/4.5 mm = 22.22 mm, and its radius is 11.11 mm. This is the initial height of the marginal ray. If the semi-field angle is 20 deg, the image size should be 100 tan (20) = 36.4 mm, so the Lagrange invariant is

$$H = n'\ u'\ h' = (-0.111)\cdot(36.4)\ \text{mm} = -4.04\ \text{mm}.$$

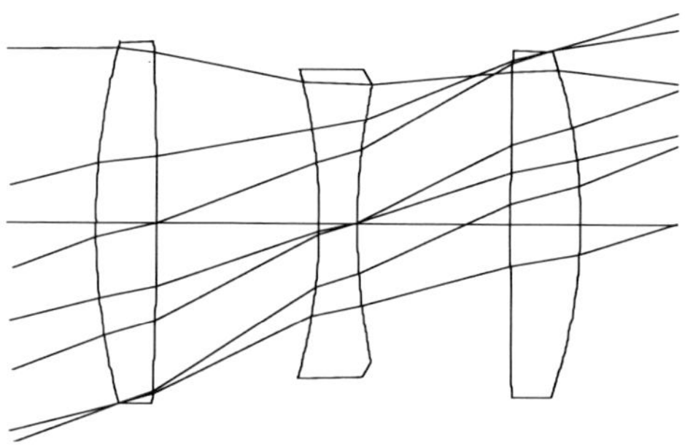

Figure 6.4. The Cooke triplet used for the Seidel calculation example.

First we trace a paraxial marginal ray using Eqs. (2.16) and (2.19) in turn:

$$n'u' = nu - hc(n' - n), \tag{2.16}$$

and

$$h_{i+1} = h_i + u_i'd. \tag{2.19}$$

The initial value of h will be 11.11 mm, as stated above, and the values of h and u at successive surfaces are shown in the table below.

```
EFL =     99.996

LAGRANGE INVARIANT =      -4.0439
```

#	H	U	HBAR	UBAR	D(U/N)	A	ABAR
1	11.11052	0.00000	-7.45218	0.36397	-0.06107	0.25846	0.19061
2	10.71470	-0.09896	-6.28822	0.29099	-0.12608	-0.23034	0.51260
3	8.74777	-0.18714	-1.16717	0.48725	0.16955	-0.41608	0.51779
4	8.67666	-0.02845	-0.44278	0.28975	0.09331	0.27310	0.45213
5	9.42396	0.07572	4.11891	0.46220	-0.05027	0.09007	0.46847
6	9.60951	0.04123	5.39167	0.28284	-0.13656	-0.39790	0.19757
		-0.11111		0.35848			

We also calculate the refraction invariant for the paraxial marginal ray:

$$A = n(hc + u), \tag{2.70}$$

at each surface, and these quantities are enough for us to be able to calculate the spherical aberration coefficient, S_1, and the axial color, C_1. The calculation of S_4 is very simple, using Eq. (6.30). In order to calculate the remaining aberrations, S_2, S_3, S_5 and C_2, we must calculate the refraction invariant for the chief ray $\overline{A}$ at each surface. We may use Eqs. (2.83) and (2.78) for this:

$$E_{i+1} - E_i \quad = \quad \frac{-d}{nh_i \cdot h_{i+1}} \tag{2.83}$$

$$\bar{A} = \frac{H}{h}(AhE - 1). \tag{2.78}$$

First, note that since surface 4 is the stop surface, E is zero by definition at this surface. Taking $i = 4$, we can calculate from Eq. (2.83):

$$E_5 - E_4 = \frac{-d_5}{n_5 h_4 h_5}, \tag{6.51}$$

and since E_4 is zero,

$$E_5 = \frac{-d_5}{n_5 h_4 h_5}, \tag{6.52}$$

and similarly for $i = 5$, to give E_6.

For the surfaces in front of the stop, we have to work forward from the stop calculating E_3, then E_2 and finally E_1.

Having computed the values of E at each surface we can now calculate $\bar{A}$ at every surface; these are tabulated under the heading ABAR in the table above. Also the values of S_2, S_3, S_5 and C_2 are evaluated. Finally we sum the surface contributions to obtain the Seidel coefficients for the complete lens:

#	S1	S2	S3	S4	S5	C1	C2
1	0.045324	0.033427	0.024653	0.145649	0.125599	0.018233	0.013447
2	0.071670	-0.159496	0.354946	0.025239	-0.846073	0.015670	-0.034873
3	-0.256773	0.319542	-0.397653	-0.163236	0.697998	-0.037917	0.047186
4	-0.060387	-0.099973	-0.165511	-0.141889	-0.508917	-0.024685	-0.040867
5	0.003843	0.019991	0.103976	0.009535	0.590398	0.005390	0.028032
6	0.207755	-0.103159	0.051223	0.186858	0.118217	0.024278	-0.012055
Sum	0.011432	0.010331	0.028367	0.062155	-0.059211	0.000969	0.000870

6.3 Stop-shift effects

Because the off-axis aberrations are computed using chief ray values, they will be affected by the location of the aperture stop. If the position of the aperture stop is changed, the path of an off-axis beam through the lens will change, and the off-axis aberrations will be affected. This can be shown graphically in Fig. 6.5 below. Consider a lens that is capable of transmitting the rays 1, 2, 3, and 4. The wavefront emerging from the lens, if neither aperture "a" or "b" is in place, is shown to the right of the lens. If we now locate a stop at position "a," rays between 1 and 3 will form the image.

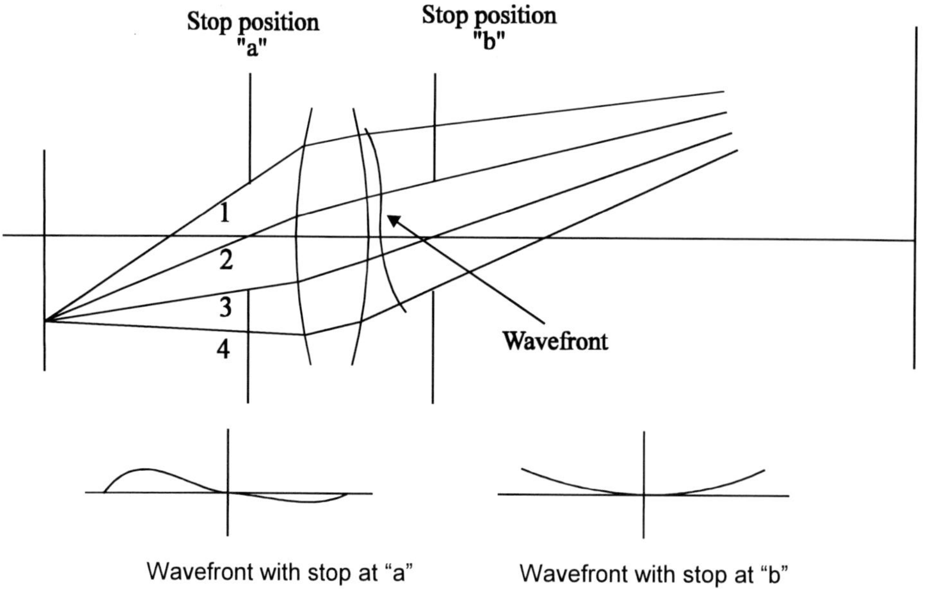

Figure 6.5. Illustration of the effect of a stop-shift along the optical axis.

Now consider what happens if the stop is moved to position "b"; clearly, the rays 2, 3, and 4 are transmitted, but ray 1 is not. The part of the wavefront that forms the image is then limited by rays 2 and 4. Note that in moving the stop along the axis we have also changed the diameter of the aperture stop; what we are actually doing is maintaining a constant numerical aperture in the image space. Note that unless the stop is in a space where the marginal ray is parallel to the axis, we always have to change its diameter in order to maintain the system numerical aperture.

We can see from the wavefront curves at the bottom of Fig. 6.5 that the aberration of the system changes, although we have not made a change to any of the refracting surfaces. We can also see that the change of aberration must depend in some way on the existing aberrations, and that if the initial system is aberration-free a change in stop position will not introduce any aberration.

6.3.1 Derivation of the Seidel stop-shift equations

Suppose, for example, that we have a system consisting of two lenses with a central air space, and that the stop is moved from position "a" to position "b" in Fig. 6.6, and that the numerical aperture of the lens is maintained so the marginal paraxial ray quantities are unaffected.

For the purposes of this section we will consider the possible stop position at "a" to be surface 3, and we will consider the possible stop position at "b" to be surface 4. In the ray tracing program each will be represented by a dummy surface.

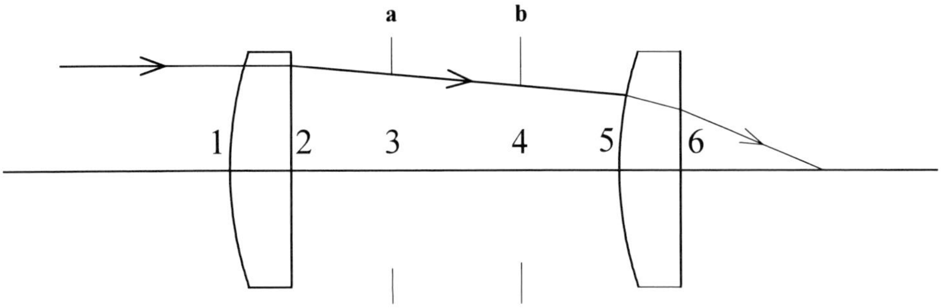

Figure 6.6. Derivation of the stop-shift equations.

If we have traced the paraxial marginal ray we know the values of h_1, h_2, h_3, h_4, h_5, and h_6. From these values we can calculate, using Eq. (2.83),

$$(E_2 - E_1), (E_3 - E_2), (E_4 - E_3), (E_5 - E_4) \text{ and } (E_6 - E_5). \tag{6.53}$$

We will refer to these differences as

$$e_2, e_3, e_4, e_5, \text{ and } e_6. \tag{6.54}$$

So far, all of these quantities are unaffected by the choice of stop position.
If the stop is at position "a" (surface 3), the values of E at the six surfaces will be

$$\begin{aligned} E_1 &= -e_3 - e_2 \\ E_2 &= -e_3 \\ E_3 &= 0 \\ E_4 &= e_4 \\ E_5 &= e_4 + e_5 \\ E_6 &= e_4 + e_5 + e_6. \end{aligned} \tag{6.55}$$

On the other hand, if the stop is at position "b" (surface 4), the values of E at the six surfaces will be

$$\begin{aligned} E_1 &= -e_4 - e_3 - e_2 \\ E_2 &= -e_4 - e_3 \\ E_3 &= -e_4 \\ E_4 &= 0 \\ E_5 &= e_5 \\ E_6 &= e_5 + e_6. \end{aligned} \tag{6.56}$$

Comparing these results in Eqs. (6.66) and (6.67), we see that there is a difference of e_4 between the two cases, and **this difference is the same at all surfaces**. Or, to put it slightly differently, the effect of moving the stop is to change all the values of E by $-e_4$ at all surfaces, where e_4 is $(E_4 - E_3)$ and is given by Eq. (2.84) as

$$E_4 - E_3 = \frac{-d_4}{n_4 h_4 \cdot h_3}.$$

We now use this result to find the effect on the Seidel aberrations. First note that

$$\bar{A} = \frac{H}{h}(AhE - 1) \tag{2.79}$$

or

$$\frac{\bar{A}}{A} = H\left(E - \frac{1}{Ah}\right). \tag{6.57}$$

Note also that this ratio of $\bar{A}/A$ occurs in several relationships:

$$S_2 = \frac{\bar{A}}{A} S_1, \tag{6.58}$$

$$S_3 = \left(\frac{\bar{A}}{A}\right)^2 S_1, \tag{6.59}$$

$$S_5 = \frac{\bar{A}}{A}(S_3 + S_4), \tag{6.31}$$

and

$$C_2 = \frac{\bar{A}}{A} C_1. \tag{6.60}$$

Suppose that at any surface, E changes to $E + \Delta E$. Note, as stated above that ΔE is the same at all surfaces (we will not assume that ΔE is small). Then, $\bar{A}/A$ changes from

$$\frac{\bar{A}}{A} = H\left(E - \frac{1}{Ah}\right) \tag{6.57}$$

to

$$\frac{\bar{A}}{A} = H\left(E + \Delta E - \frac{1}{Ah}\right). \tag{6.61}$$

If we denote, in this section, aberrations **after** a stop-shift with an asterisk, we find that

$$S_2^* = S_1 H\left(E + \Delta E - \frac{1}{Ah}\right) \tag{6.62}$$

or

$$S_2^* = S_2 + S_1 H \Delta E. \tag{6.63}$$

In other words, the effect of the stop-shift is to introduce a change in coma, given by

$$\Delta S_2 = H \Delta E S_1. \tag{6.64}$$

Similarly, the new value of astigmatism is given by

$$S_3^* = S_1 H^2 \left(E + \Delta E - \frac{1}{Ah}\right)^2, \tag{6.65}$$

or,

$$S_3^* = S_1 H^2 \left(E - \frac{1}{Ah}\right)^2 + 2H^2 \Delta E S_1 \left(E - \frac{1}{Ah}\right) + (H\Delta E)^2 S_1, \tag{6.66}$$

$$S_3^* = S_3 + 2H\Delta E S_2 + (H\Delta E)^2 S_1. \tag{6.67}$$

In other words, the effect of the stop-shift is to introduce a change in astigmatism, given by

$$\Delta S_3 = 2H\Delta E S_2 + (H\Delta E)^2 S_1. \tag{6.68}$$

Similarly, we can also show that the effect of the stop-shift is to introduce a change in distortion, given by

$$\Delta S_5 = H\Delta E(3S_3 + S_4) + 3(H\Delta E)^2 S_2 + (H\Delta E)^3 S_1. \tag{6.69}$$

and a change in lateral color, given by

$$\Delta C_2 = H\Delta E C_1. \tag{6.70}$$

Remember that these equations are not limited to small changes in E. They apply to single surfaces and to complete systems, and are independent of the nature of the surface, as long as the system has rotational symmetry. Note, in particular, that these stop-shift results also apply to aspheric surfaces.

6.4 Dependence of the Seidel aberrations on surface curvature

It is instructive to consider the way in which the Seidel surface coefficients depend on the curvature of a surface. For example, consider the Seidel coma coefficient:

$$\text{Coma} \qquad S_2 = -A\bar{A}h\delta\left(\frac{u}{n}\right). \tag{6.28}$$

Since

$$A = n\,(hc + u) \tag{2.70}$$

$$\bar{A} = n\,(\bar{h}c + \bar{u}), \tag{2.71}$$

and

$$n'u' = nu - hc(n' - n), \tag{2.16}$$

then

$$\delta\left(\frac{u}{n}\right) = \frac{u'}{n'} - \frac{u}{n}, \tag{6.22}$$

or

$$\delta\left(\frac{u}{n}\right) = \frac{nu - hc(n' - n)}{n'} - \frac{u}{n}. \tag{6.71}$$

We can see that three of the terms in Eq. (6.28), A, $\bar{A}$ and $\delta(u/n)$ are linear functions of the curvature, c. Therefore, S_2 must be a cubic function of c, as shown schematically in Fig. 6.7. There are zeros at the points where A=0, $\bar{A} = 0$ and $\delta(u/n)$ =0. These points can appear in any sequence, and are not necessarily in the sequence shown in Fig. 6.7. Also, S_2 is not necessarily positive for large positive values of c; the reverse of this is equally likely. However, the general point is that S_2 depends on c^3.

Similarly, we can show that, since S_1 depends on A^2 and on $\delta(u/n)$, S_1 must also depend on c^3; similarly for S_3 and S_5.

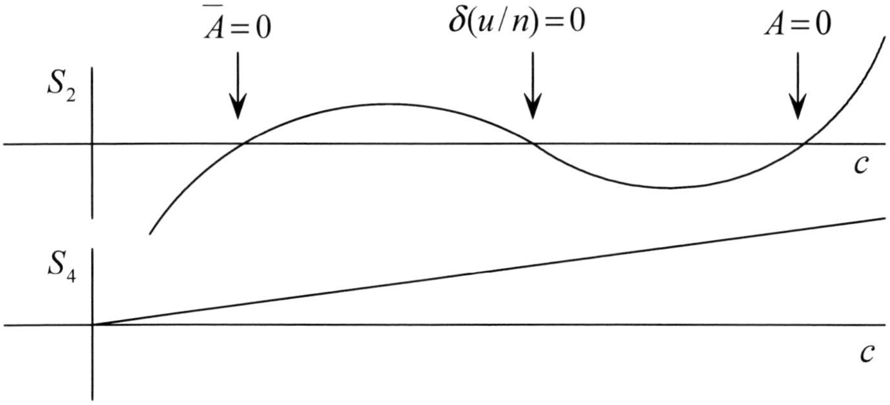

Figure 6.7. The dependence of Seidel coma and field curvature on surface curvature.

These results show that even these four Seidel aberrations are quite complex functions of surface curvature, which is perhaps the design parameter that is the easiest to vary in lens design. In addition, if we make a simple change in one curve, the paraxial ray path at later surfaces will be changed, and therefore the aberration contributions at those surfaces will also be changed.

The situation concerning field curvature is easier. The field curvature coefficient is given by

$$S_4 = -H^2 c\delta\left(\frac{1}{n}\right), \tag{6.30}$$

and this is simply a linear function of c. Further, since Eq. (6.30) does not depend on either of the paraxial ray paths, a change in one surface curvature does not affect the S_4 contributions of other surfaces. This is perhaps the only case in which any aberration is a linear function of surface curvature.

The fact that the relationship between S_4 and c is so simple leads us to the conclusion that field curvature is a particularly fundamental type of aberration, and that it can be corrected in a few very simple ways. On the other hand, it is easy to think of configurations in which field curvature is uncorrectable. The simplest case is perhaps a single thin lens with non-zero power.

It should be stated here that there is another situation where aberrations depend linearly on a design parameter, and this concerns axial and lateral color. Axial color is given by

$$C_1 = Ah\delta\left(\frac{\delta n}{n}\right), \tag{6.49}$$

and lateral color is

$$C_2 = \bar{A}h\delta\left(\frac{\delta n}{n}\right). \tag{6.50}$$

Both of these aberrations are linear functions of the dispersions δn of the glasses. Further, if we change the dispersion of one glass in a lens, we affect the C_1 and C_2 coefficients at both surfaces of that lens, but the C_1 and C_2 coefficients at other surfaces are unaffected. Therefore, first-order axial and lateral color are linear functions of the dispersions of all of the glasses in a lens.

Here, as in the case of field curvature, the relationship is so simple that correction of first-order axial and lateral color is either easy or impossible, because there is no complexity in the relationship that would yield unexpected solutions.

6.5 The aplanatic surface

An optical system which has zero spherical aberration and zero coma is known as an **aplanatic** system. An aplanatic system can be either a complete lens, or a part of a lens, or a single surface. It is the case of a single surface that we shall now discuss.

If we satisfy the condition

$$\delta\left(\frac{u}{n}\right) = 0, \tag{6.72}$$

it will ensure that for the surface in question

$$S_1 = S_2 = S_3 = 0. \tag{6.73}$$

Combining Eq. (6.72) and the paraxial ray trace equation,

$$n'u' = nu - hc(n' - n), \tag{2.16}$$

we find that

$$-hcn = u(n+n). \tag{6.74}$$

If we substitute

$$u' = -\frac{h}{l'} \quad \text{and} \quad u = -\frac{h}{l}, \tag{2.15}$$

we can see that the object distance is given by

$$l = R\frac{n'+n}{n}, \tag{6.75}$$

and the image distance is

$$l' = R\frac{n'+n}{n'}. \tag{6.76}$$

Note that l and l' must always have the same sign. This tells us that an aplanatic surface, or succession of such surfaces, cannot form a real image of a real object. In other words, either the object or image must be virtual, as shown in Fig. 6.8.

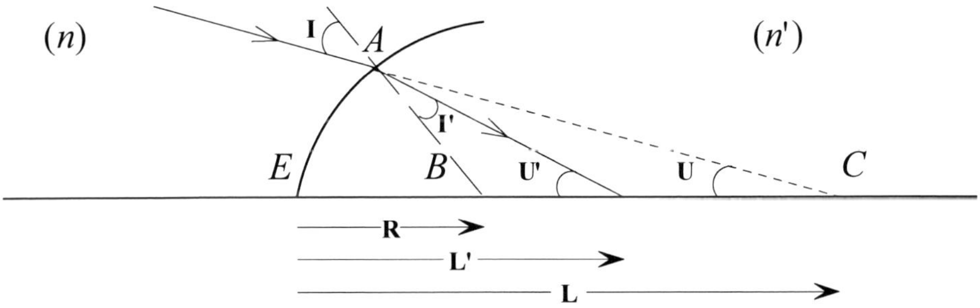

Figure 6.8. The aplanatic condition.

Usually, the term "aplanatic surface" is taken to mean a surface satisfying the above condition, although it has $S_3 = 0$ in addition to $S_1 = S_2 = 0$. A surface that satisfies the condition $A = 0$ or $h = 0$ is also aplanatic in the sense that $S_1 = S_2 = 0$. For some reason, perhaps because these cases are relatively trivial, they are usually not included when the term "aplanatic surface" is used.

Although we have derived Eqs. (6.75) and (6.76) to satisfy the condition that $S_1 = S_2 = S_3 = 0$, we will now show that **spherical aberration is completely corrected** at such a surface, to all orders, not just to the Seidel approximation of

S_1. We will use upper case L, L', U, and U' to denote exact distances and angles, equivalent to the paraxial quantities l, l', u, and u'. We now assume an object distance, L, is given by

$$L = R\frac{n' + n}{n}. \tag{6.77}$$

Therefore,

$$\frac{n'}{n} = \frac{L - R}{R}. \tag{6.78}$$

The triangle ABC in Fig. 6.8 has sides of length R and $(L - R)$, and angles I and U, so we can use the sine rule to write

$$\frac{L - R}{R} = -\frac{\sin I}{\sin U}, \tag{6.79}$$

and so

$$\frac{n'}{n} = -\frac{\sin I}{\sin U}. \tag{6.80}$$

Therefore

and, by Snell's law

$$\begin{aligned} n \sin I &= -n' \sin U, \\ &= n' \sin I', \end{aligned} \tag{6.81}$$

so

$$U = -I'. \tag{6.82}$$

In Fig. 6.8, allowing for the signs of the angles,

$$\text{angle } ABE = I - U = I' - U', \tag{6.83}$$

so, from Eqs. (6.82) and (6.83),

$$U' = -I. \tag{6.84}$$

Then,

$$\frac{L'-R}{R}=-\frac{\sin I'}{\sin U'}$$

$$=\frac{\sin I'}{\sin I}=\frac{n}{n'}. \tag{6.85}$$

Therefore,

$$L'=R\frac{n'+n}{n'}. \tag{6.86}$$

Therefore, we have shown that any ray, incident according to Eq. (6.74), is refracted according to Eq. (6.86), and therefore the intersection point where the ray crosses the optical axis is independent of the height of the ray at the surface. This means that spherical aberration is zero at this surface, to all orders.

Equations (6.82) and (6.84) ($U = -I$ and $U' = -I$) are also expressions of an aplanatic surface.

6.5.1 An example—the classical oil-immersion microscope objective

The aplanatic condition is used in classical oil-immersion microscope objectives, in which the object is immersed in an oil of refractive index n, matching that of the first lens element. An example is shown in Fig. 6.9.

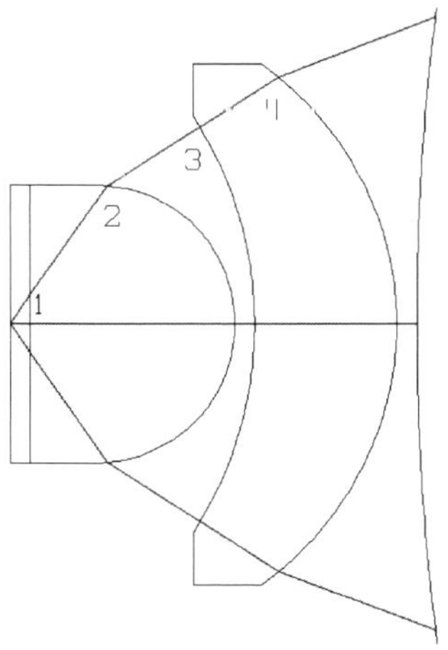

Figure 6.9. The first four surfaces of a classical oil-immersion microscope objective.

Considering the first four surfaces, special conditions are satisfied at each surface:

	Condition	S_1	S_2	S_3	S_4
Surface 1	$n' = n$	0	0	0	0
Surface 2	$\delta(u/n) = 0$	0	0	0	>0
Surface 3	$A = 0$	0	0	>0	<0
Surface 4	$\delta(u/n) = 0$	0	0	0	>0

Because of the aplanatic condition at surfaces 2 and 4, the numerical aperture of the beam is reduced by a factor of n^3. Typically, if the object-space numerical aperture is 1.3, and if $n = 1.5$, the numerical aperture after surface 4 will be $1.3 / 1.5^3$ or 0.38. After surface 4, the beam still diverges with an angle of 0.38, and the remaining components in the design must make the beam converge to form a real image, as shown in Fig. 6.10.

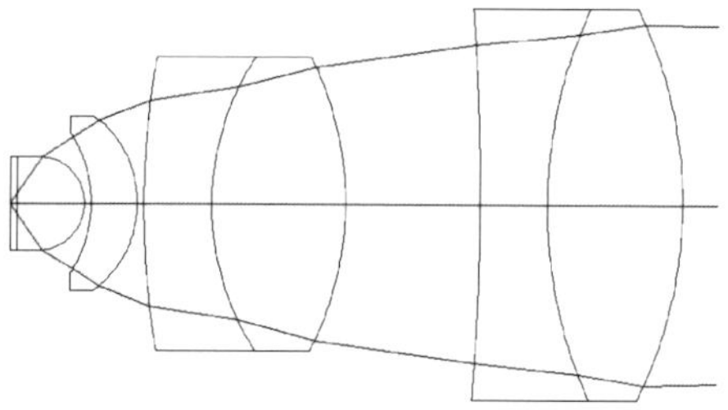

Figure 6.10. Complete microscope objective.

6.6 Zero Seidel conditions

In this section we consider another lens that has been deliberately set up in order to make several of the Seidel aberrations zero at one or more surfaces. This has been done purely as an illustration of the conditions under which these aberrations can be eliminated. The reader should not assume that the design in Fig. 6.11 is a good design, or that it is generally possible to design a lens with all of the Seidel surface contributions zero.

At the first surface, the curvature has been chosen so that the chief ray is perpendicular to the surface. In other words, the entrance pupil position must coincide with the center of curvature of the first surface. Therefore, $\bar{A} = 0$ and $S_2 = S_3 = S_5 = C_2 = 0$. The other three aberrations, S_1, S_4, and C_1 are non-zero.

At the second surface, the curvature has been chosen so that the paraxial marginal ray is perpendicular to the second surface. Therefore $A_2 = 0$ and $S_1 = S_2 = C_1 = 0$. The other four aberrations S_3, S_4, S_5, and C_2 are non-zero.

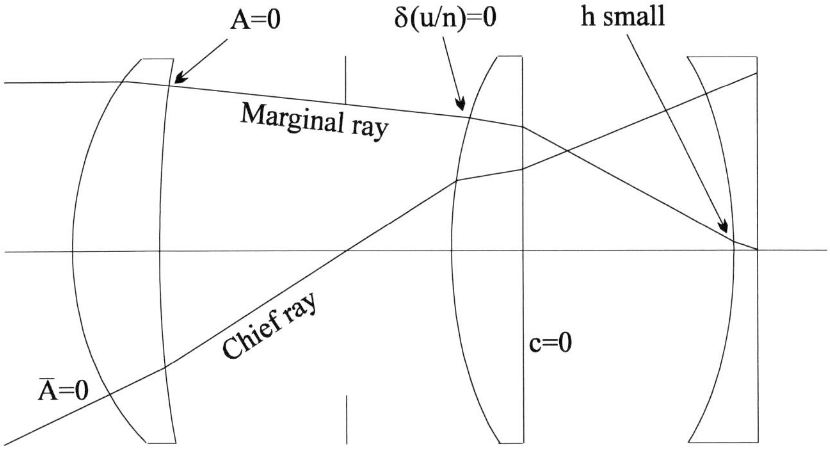

Figure 6.11. Conditions for zero aberration.

We have chosen surface 3 to satisfy the aplanatic condition, $\delta(u/n) = 0$. Therefore, $S_1 = S_2 = S_3 = 0$, but the other four aberrations S_4, S_5, C_1, and C_2 are non-zero.

Surface 4 has been chosen to be plane, so it has zero Petzval sum, or $S_4 = 0$. However, all of the other Seidel aberrations are non-zero.

Surface 5 is chosen to be concave and to be close to the image position, so that $h \cong 0$. Thus, all of the aberrations that depend on the marginal ray height, h, are small. Therefore, $S_1 = S_2 = S_3 = C_1 = C_2 \cong 0$. However, the field curvature, S_4, is negative which allows us to correct the positive field curvature of the rest of the system. Concave surfaces (or negative lenses) used in this way are known as field flatteners, and are quite common. Finally we consider distortion. In order to ascertain whether the distortion is zero, it is simplest to consider the alternative equation for distortion:

$$S_5 = -\bar{A}^3 h\delta\left(\frac{1}{n^2}\right) + \overline{hA}(2h\bar{A} - \bar{h}A)c\delta\left(\frac{1}{n}\right). \tag{6.37}$$

If surface 5 is in contact with the image, h is zero, so we have

$$S_5 = -\bar{h}^2 \bar{A}Ac\delta\left(\frac{1}{n}\right).$$

This will not be zero unless $\bar{A}$ is zero (which is possible), so in general we can say that the distortion will be non-zero. This result is perhaps surprising; we might have intuitively expected that a surface in contact with the image surface would introduce zero distortion, but in general, this is clearly not so.

The last surface, 6, has zero curvature, and $h = 0$. Therefore all of the seven aberrations are zero at this surface. In practice, we would not choose to put surfaces in contact with an object or image, because any imperfections (e.g., scratches) on the surface are in focus, and a real field flattener would be moved slightly away from the image. In this case, the aberrations which were zero in the case discussed above will be slightly different from zero, and these differences must naturally be allowed for, in the design of the lens. All of this is summarized in Table 6.1, below.

Table 6.1. Summary of aberrations for the lens in Fig. 6.13.

		S_1	S_2	S_3	S_4	S_5	C_1	C_2
Surface 1	$\bar{A} = 0$		0	0		0		0
Surface 2	$A = 0$	0	0				0	
Surface 3	$\delta(u/n)=0$	0	0	0				
Surface 4	$c = 0$				0			
Surface 5	h small	small	small	small		small	small	small
Surface 6	$c=0$, h small	0	0	0	0	0	0	0

This illustrates that it is not possible for the seven Seidel aberrations to be corrected at all surfaces of a lens, and therefore aberration correction is almost always achieved by balancing positive aberrations against negative aberrations. There are simple cases, such as the classical astronomical telescope, where both surfaces are free of spherical aberration, and where no other aberrations are corrected, but these are exceptions.

Note that this example is not intended to illustrate any real system of practical use, but in fact there are lenses of a similar form, but using aspheric surfaces. These lenses are used as television projection lenses, where a high aperture and medium field are required.

6.7 Undercorrected and overcorrected aberrations

In some literature on optical design the terms "undercorrected" and "overcorrected" are used to describe aberrations. Normally a single uncorrected positive lens will have positive S_1, S_3, S_4, and C_1 aberrations. Taking S_1 as an example,

Undercorrected spherical aberration means that S_1 is positive.

Corrected spherical aberration means that S_1 is zero.

Overcorrected spherical aberration means that S_1 is negative.

The same applies to the other aberrations listed above.

Since coma, distortion and lateral color may be zero for a single lens in certain situations, we do not find the terms "undercorrected" and "overcorrected" used to describe these three aberrations. Instead, coma is sometimes described as either "outwards" or "inwards" depending on whether the transverse ray aberrations are positive or negative respectively, assuming that the image size, h', is positive. Distortion is described as either "pincushion" or "barrel," depending on whether the magnification of the image increases or decreases with field angle.

6.8 Seidel aberrations of spherical mirrors

When considering spherical mirrors, it is usual to adopt the convention that the refractive index changes sign after reflection at a surface. If we do this, we can use all of the results that we have discussed for the case of refraction in the case of reflection.

A word of warning. This convention can get very confusing when we are considering decentered systems, particularly if the systems have three-dimensional decentrations and tilts. However, in these notes we are considering only centered systems, and in any case, Seidel theory, at least in the form that we are discussing, is only valid for centered systems.

Let us consider the case of a spherical mirror with light incident from the left, as shown in Fig. 6.12. We will assume that the refractive index in the incident medium, n, is +1, and after reflection the refractive index, n', is −1. We will restrict this discussion to the case of an object at infinity, so that $u = 0$.

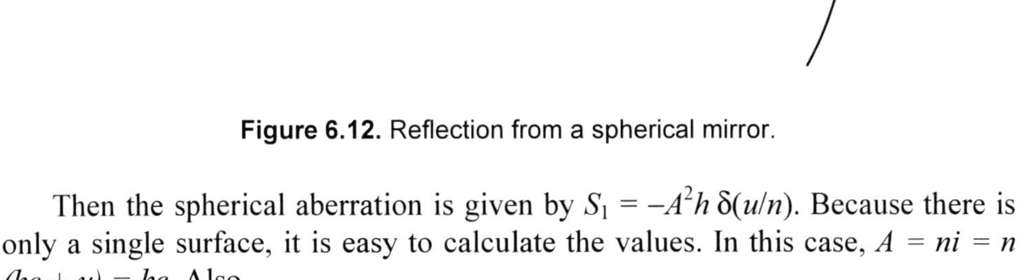

Figure 6.12. Reflection from a spherical mirror.

Then the spherical aberration is given by $S_1 = -A^2 h\,\delta(u/n)$. Because there is only a single surface, it is easy to calculate the values. In this case, $A = ni = n(hc + u) = hc$. Also,

$$n'u' = nu - hc(n' - n) = -hc(-1 - 1) = 2hc.$$

So

$u' = -2hc$ or $hc = -u'/2$. Also, $\delta(u/n) = u'/n' - u/n = -u'$.

Therefore, the spherical aberration, S_1, is given by

$$-u'^2 h(-u')/4 = u'^3 h/4. \tag{3.36}$$

Similarly, the Petzval sum is given by

$$P = -c\left(\frac{1}{n'} - \frac{1}{n}\right) = -c(-1-1) = 2c.$$

Since the power is given by $K = c(n' - n) = -2c$, the Petzval sum is $P = -K$. Therefore, the Petzval sum of a reflecting surface with positive power is negative, which is the opposite of the case of a lens with positive power. As we shall see in Chapter 13, this is a very useful property that is used in catadioptric systems, where lenses and mirrors are combined to correct P.

6.9 Seidel aberration relationships

We have seen that the Siedel coefficients are fairly simply expressed and computed. We shall now look at how they relate to wavefront, transverse, and longitudinal aberrations.

6.9.1 Wavefront aberrations

The first five terms in the aberration polynomial can be written in the following way, using the Seidel coefficients:

$$\begin{aligned} W &= \frac{1}{8}S_1(x_r^2 + y_r^2)^2 \\ &+ \frac{1}{2}S_2 y_r(x_r^2 + y_r^2)h_r \\ &+ \frac{1}{2}S_3 y_r^2 h_r^2 \\ &+ \frac{1}{4}(S_3 + S_4)(x_r^2 + y_r^2)h_r^2 \\ &+ \frac{1}{2}S_5 y_r h_r^3, \end{aligned} \tag{6.87}$$

where x_r and y_r are relative pupil coordinates, so that $(x_r^2 + y_r^2) = 1$ at full aperture and h_r is a relative object size (or image size), so that $h_r = 1$ at full field.

On axis the wavefront aberration at full aperture is 1/8 S_1; the coma at full aperture and full field is 1/2 S_2, and the distortion, expressed as a wavefront aberration, is 1/2 S_5. However, comparing the above equations with Eq. (4.14), ${}_2w_{20}$ (field curvature) corresponds to 1/4 (S_3 + S_4), but ${}_2w_{22}$ (astigmatism) corresponds to 1/2 S_3.

We should point out, here, that we would hardly ever calculate the aberration coefficients ${}_iw_k$ from the Seidel coefficients. If we need to calculate the aberration coefficients ${}_iw_{jk}$ we would do this by ray tracing. The importance of the above relationships is more a matter of understanding. We should also re-state the fact that the third-order (Seidel) aberrations alone are rarely enough to completely determine the image-forming properties of a system.

6.9.2 Transverse ray aberrations

Since the Seidel coefficients are related to the wavefront aberrations by some simple equations, we can calculate the expected transverse ray aberrations for a given amount of Seidel aberration, by differentiating the wavefront aberration as described in Eq. (4.7):

Spherical Aberration

$$\delta\eta' = \frac{S_1}{2n'u'} \tag{6.88}$$

Coma

$$\delta\eta' = \frac{3S_2}{2n'u'} \tag{6.89}$$

Astigmatism and field curvature

$$\delta\eta' = \frac{3S_3 + S_4}{2n'u'} \tag{6.90}$$

$$\delta\xi' = \frac{S_3 + S_4}{2n'u'}. \tag{6.91}$$

Note that $\delta\eta'$ is the y-component of aberration (for rays in the y-z plane, the tangential plane), and $\delta\xi'$ is the x-component of aberration (for rays in the x-z plane, the sagittal plane).

Distortion

$$\delta\eta' = \frac{S_5}{2n'u'} \tag{6.92}$$

Therefore

$$\frac{\delta\eta'}{\eta'} = \frac{S_5}{2n'u'h'} = \frac{S_5}{2H}. \tag{6.93}$$

In each case these aberrations are at the edge of the aperture and at the edge of the field.

6.9.3 The Petzval sum and the Petzval surface

The **Petzval sum** is defined by the following equation:

$$P = \frac{S_4}{H^2}, \tag{6.94}$$

where S_4 is the summation of the contributions from all the surfaces in the system. The **Petzval surface** is defined to be the image surface produced by a given optical system, if the astigmatism S_3 is zero. We will write the curvature of the Petzval surface as c_P.

From Eqs. (6.90) and (6.91) we can see that if $S_3 = 0$, the transverse aberration will be

$$\delta\eta' = \frac{S_4}{2n'u'}, \tag{6.95}$$

and assuming that image space is air ($n' = 1$), this obviously becomes

$$\delta\eta' = \frac{S_4}{2u'}. \tag{6.96}$$

By simple geometry, the longitudinal aberration is therefore

$$\delta z = \frac{\delta\eta}{-u'} = -\frac{S_4}{2u'^2} = -\frac{h'^2 P}{2}, \tag{6.97}$$

since $H = u'\eta'$ in the image space in air.

Since the "sag" of a spherical surface of curvature c, is given in Eq. (1.16), for small distances r from the optical axis, by

$$z = \frac{cr^2}{2}; \tag{1.16}$$

the longitudinal aberration equivalent to an image surface of curvature c_P will be

$$\frac{h'^2 c_P}{2}. \tag{6.98}$$

Comparing Eqs. (3.32) and (3.33), it is clear that the curvature of the Petzval surface is equal in magnitude to the Petzval sum, but opposite in sign, i.e.,

$$P = -c_P. \tag{6.99}$$

Therefore, this equation gives us a very simple means of estimating the significance of the Petzval sum, for moderately small field angles.

6.9.4 The Petzval surface and astigmatic image surfaces

In many lens design problems there is freedom to vary the third-order astigmatism. Occasionally we can also vary the Petzval sum. The following diagrams illustrate some of the possible forms of the sagittal and tangential image surfaces. The Petzval image surface is denoted by "*P*," the tangential image surface is denoted by "*T*" and the sagittal image surface is denoted by "*S*".

First we note that Eq. (6.90) contains a factor of 3 that is absent from Eq. (6.91):

$$\delta\eta' = \frac{3S_3 + S_4}{2n'u'} \tag{6.90}$$

$$\delta\xi' = \frac{S_3 + S_4}{2n'u'}. \tag{6.91}$$

The result of this factor of 3 is that the distance from the Petzval image surface to the tangential image surface is always three times the distance from the Petzval image surface to the sagittal image surface. In other words,

$$PT = 3\ PS \tag{6.100}$$

in all cases.

First, in Fig. 6.13, we show a situation where the astigmatism is zero and the Petzval sum is varied. (This is not a common case, but it is the simplest case to consider first.) If there is no astigmatism the Petzval surface, the tangential surface, and the sagittal surface are coincident. If $S_4 > 0$, (i.e., $P > 0$), the image surfaces are concave and bent toward the lens as shown on the left diagram. If $S_4 = 0$, (i.e., $P = 0$), the image surfaces are naturally plane, and if $S_4 < 0$, (i.e., $P < 0$), the image surfaces are convex and bent away from the lens as shown on the right diagram.

Figure 6.14 shows the situation where S_4=0, (i.e., $P = 0$) and S_3 is variable. As we have already stated, $PT = 3\ PS$, and this can be seen in the diagrams.

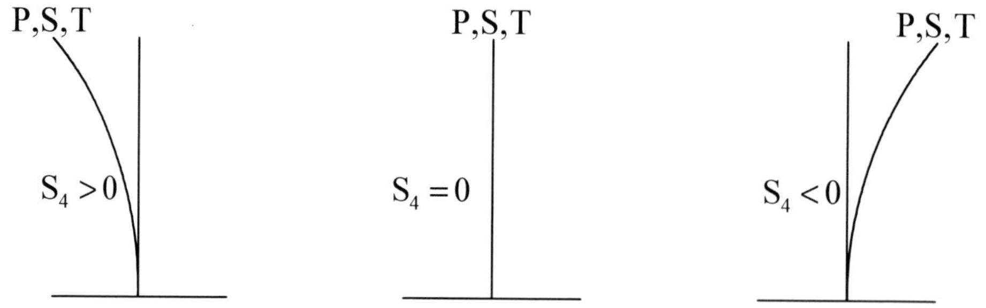

Figure 6.13. Image surfaces when astigmatism, S_3, is zero and the Petzval sum, S_4, is varied.

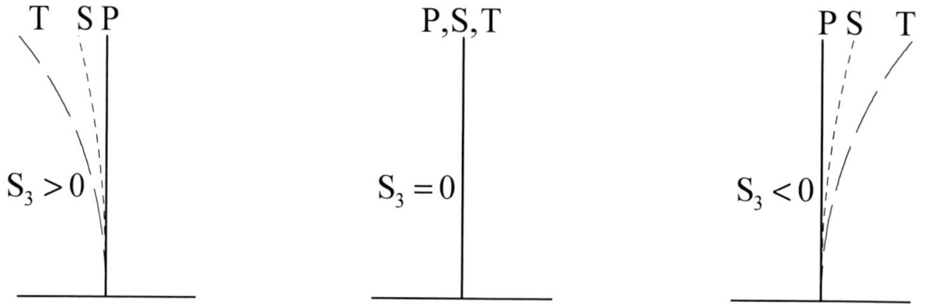

Figure 6.14. Image surfaces when Petzval sum, S_4, is zero and astigmatism, S_3, is varied.

Finally, Fig. 6.15 shows the usual case when $S_4 > 0$ (i.e., $P > 0$) and S_3 can be varied to a certain extent. If S_3 is positive (as it is in the case of a simple singlet or doublet), and S_4 is also positive but essentially fixed, we have image surfaces as shown in the left diagram below. If S_3 is made zero, *P*, *S*, and *T* will coincide, as shown in the second diagram. If S_3 is made equal to $-S_4/3$, the tangential image surface will be plane, as shown in the third diagram. As always, $PT = 3PS$, so we can see that the sagittal image surface is still concave and bent towards the lens. Last, if $S_3 = -S_4$, the sagittal image surface is plane, but the tangential image surface becomes very strongly curved, as shown in the right diagram.

It is clearly possible to choose other cases; in particular, if $S_3 = -S_4/2$, Eqs. (6.90) and (6.91) show that the *S* and *T* surfaces are equally spaced on each side of the paraxial image plane.

We have already seen in Chapter 4 (Fig. 4.22) that in lens systems of some complexity it is possible to reverse the curvature of the *S* and *T* curves at higher field angles, where astigmatism and field curvature of higher orders and opposite signs become comparable or larger than the Seidel values. In such cases the lens can be designed so that the *S* and *T* curves cross each other and the Petzval surface. At that field angle (often, but not always, chosen to be 70% or 85% of full field), the astigmatism is zero.

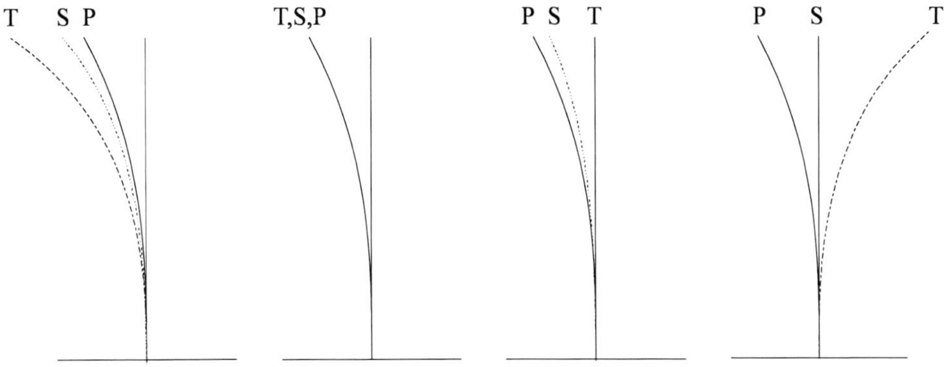

Figure 6.15. Image surfaces when $S_4 > 0$ and S_3 is variable.

6.10 Pupil aberrations

As we saw in Chapter 1 (Fig. 1.6), the paraxial marginal ray describes imaging of the object, while the chief ray describes imaging of the entrance pupil to the exit pupil. So far we have considered the Seidel aberrations of the object's image, but there is also an analogous set of aberrations for pupil imaging.

The pupil Seidel aberrations may be derived simply by interchanging the marginal and chief ray quantities, A and $\bar{A}$, h and $\bar{h}$, u and $\bar{u}$, since the roles of pupil and image planes are reversed.

Pupil spherical aberration $$\bar{S}_1 = -\bar{A}^2 \bar{h}\delta\left(\frac{\bar{u}}{n}\right) \quad (6.101)$$

Pupil coma $$\bar{S}_2 = -\bar{A}A\bar{h}\delta\left(\frac{\bar{u}}{n}\right) = \frac{A}{\bar{A}}\bar{S}_1 \quad (6.102)$$

Pupil astigmatism $$\bar{S}_3 = -A^2\bar{h}\delta\left(\frac{\bar{u}}{n}\right) = \frac{A}{\bar{A}}\bar{S}_2 \quad (6.103)$$

Pupil field curvature $$\bar{S}_4 = -H^2 c\delta\left(\frac{1}{n}\right) = S_4 \quad (6.104)$$

Pupil distortion $$\bar{S}_5 = \frac{A}{\bar{A}}\left(\bar{S}_3 + \bar{S}_4\right). \quad (6.105)$$

Wynne[6] has derived a set of simple relationships between the pupil Seidel and object-image Seidel aberrations. As with $\delta(u/n)$, $\delta(u)$ and $\delta(\overline{u})$ refer to the change of the marginal and chief ray angles as they are refracted at an interface, and the surface contributions are summed over all surfaces of the system.

$$\overline{S}_1 = HE\left[S_5 + \overline{A}\overline{h}Hc\delta\left(\frac{1}{n}\right)\right] \tag{6.106}$$

$$\overline{S}_2 = S_5 - H\delta(\overline{u}^2) \tag{6.107}$$

$$\overline{S}_3 = S_3 - H\delta(u\overline{u}) \tag{6.108}$$

$$\overline{S}_4 = S_4 \tag{6.109}$$

$$\overline{S}_5 = S_2 - H\delta(u^2). \tag{6.110}$$

While pupil aberrations do not have to be as small as object-image aberrations, there are certain lenses where pupil imaging is required to be more accurate, such as Fourier transform lenses, and other lenses where pupil aberrations are deliberately introduced to achieve some effect such as an increase in pupil size off-axis to compensate for a loss of image brightness relative to on-axis.

6.11 Conjugate-shift effects

From the pupil Seidel aberrations, we can use the same methodology that we saw in Sec. 6.3 for stop-shifts, but this time with all surfaces seeing the same change of $\Delta\overline{E}$. Looking back at the marginal and chief rays in Fig 1.6, it is apparent that the "stop" for pupil imagery is the same as the "object" for the normal object-imaging function of the lens. From the stop-shift changes of the pupil aberrations, Wynne used his relationships between pupil and object-image aberrations to derive a set of conjugate-shift changes to the Seidel aberrations[6]. Here, we only summarize the results, with $\vec{S}$ denoting the aberration after such a shift.

$$\begin{aligned}\vec{S}_1 = S_1 &+ \overline{H}\Delta\overline{E}\left[4S_2 - H\delta(u^2)\right] + (\overline{H}\Delta\overline{E})^2\left[6S_3 - 3H\delta(u\overline{u}) + 2S_4\right] \\ &+ (\overline{H}\Delta\overline{E})^3\left[4S_5 - 3H\delta(\overline{u}^2)\right] + (\overline{H}\Delta\overline{E})^4\,\overline{S}_1,\end{aligned} \tag{6.111}$$

$$\vec{S}_2 = S_2 + \overline{H\Delta E}\left[3S_3 + S_4 - H\delta(u\bar{u})\right] + (\overline{H\Delta E})^2\left[3S_5 - 2H\delta(\bar{u}^2)\right] + (\overline{H\Delta E})^3\bar{S}_1, \tag{6.112}$$

$$\vec{S}_3 = S_3 + \overline{H\Delta E}\left[2S_5 - H\delta(\bar{u}^2)\right] + (\overline{H\Delta E})^2\bar{S}_1, \tag{6.113}$$

$$\vec{S}_4 = S_4, \tag{6.114}$$

$$\vec{S}_5 = S_5 + \overline{H\Delta E}\bar{S}_1, \tag{6.115}$$

$$\vec{\bar{S}}_1 = \bar{S}_1. \tag{6.116}$$

These are similar to the stop-shift equations that we looked at earlier in the chapter, but more complicated because they include $\bar{S}_1$. In fact, they indicate that pupil spherical aberration is an important factor in the sensitivity of Seidel aberrations to a change in object distance from the lens, or magnification.

For example, it is easy to see that if $\bar{S}_1$ is zero, S_5 is invariant with conjugate shift. If, in addition, $\left[2S_5 - H\delta(\bar{u}^2)\right] = 0$, then S_3 is also invariant, and so on for conditions of increasing complexity for S_2 and S_1. Wynne also found specific conditions in which combinations of some Seidel aberrations are invariant at two different conjugates.

In general, however, Wynne confirmed a result originally given by J. Clerk Maxwell, that it is impossible for a non-trivial imaging system to correct **all** aberrations over a range of conjugates, or even at two different conjugates. It is for this reason that some photographic lenses change airspaces between one or more lens elements or groups, in an attempt to minimize aberration changes as the lens is refocused over a wide range of object distances.

References

1. H. H. Hopkins, *Wave Theory of Aberrations*, Oxford University Press (1950).
2. A. E. Conrady, *Applied Optics and Optical Design*, Oxford University Press (1929).
3. W. T. Welford, *Aberrations of Optical Systems*, Adam Hilger (1986).
4. R. Kingslake, *Lens Design Fundamentals*, Academic Press (1978).
5. H. A. Buchdahl, *Optical Aberration Coefficients*, Oxford University Press (1954).
6. C. G. Wynne, *Proc. Phys. Soc.* 65B, p. 429-437 (1952).

CHAPTER 7
PRINCIPLES OF LENS DESIGN

In this chapter we will use the concept of thin lenses to understand some special strategies and techniques used in designing lenses. While it is sometimes possible to rely solely on a lens design program to improve a lens, this strategy usually leads to unwieldy components that are impossible to fabricate. By using the insights derived from a study of simple lens arrangements, the lens designer can see which variations in lens parameters will yield the greatest improvements in system performance, and still give designs for lenses that can be made and mounted easily and economically.

7.1 Thin lenses

In many situations it is extremely useful to consider lenses in terms of thin lenses, rather than as a sequence of single surfaces. When we do this, we assume that the lens is of infinitesimal thickness, so that a paraxial ray will intersect the two surfaces of the lens at the same height. Naturally, real lenses are never thin in this sense, but the results derived using this assumption are valid for many real lenses. We will see from many of the lenses discussed in this text that thin lens Seidel theory provides insights into the possibilities of a new lens type, with little or no calculation.

In the thin lens, the only relevant quantities are the two curvatures c_1, c_2 and its refractive index n (Fig. 7.1). We define two new quantities, for a thin lens. The shape of a lens is specified by the shape factor, as

$$B = \frac{c_1 + c_2}{c_1 - c_2}. \tag{7.1}$$

This is a very convenient parameter, as we show in Fig. 7.2. A plano-convex lens, with the first surface curved, has a shape factor of unity. An equi-convex lens (or an equi-concave lens) has a shape factor of zero. A plano-convex lens, with the first surface plane, has a shape factor of –1.0.

Since in almost all lens design problems the shape is a design parameter that the designer can choose quite freely, it is a very powerful design parameter. In addition, within the limitations of the paraxial approximation, as the shape of a thin lens is varied, provided the power K is held constant, then the paraxial marginal and paraxial chief ray paths through the rest of the lens are unaffected.

This means that the Seidel aberrations of other sections of the lens will be unchanged. We will also see, later, that some aberrations are dependent on the shape of a lens, while others are independent. This is a very important distinction in lens design.

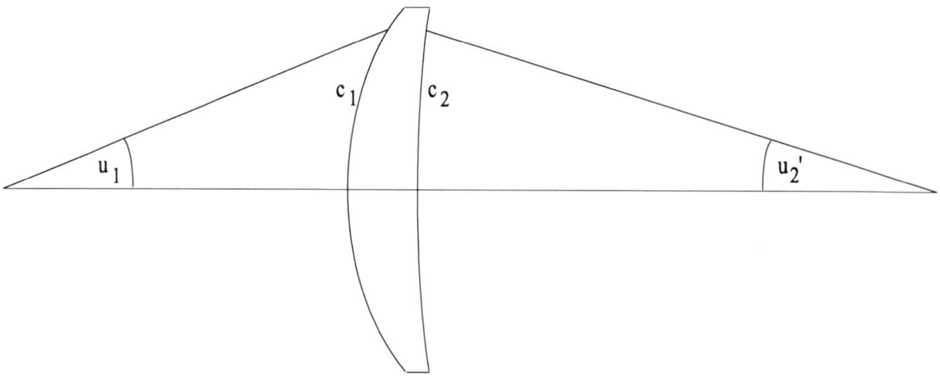

Figure 7.1. Thin lens shape and conjugate factors.

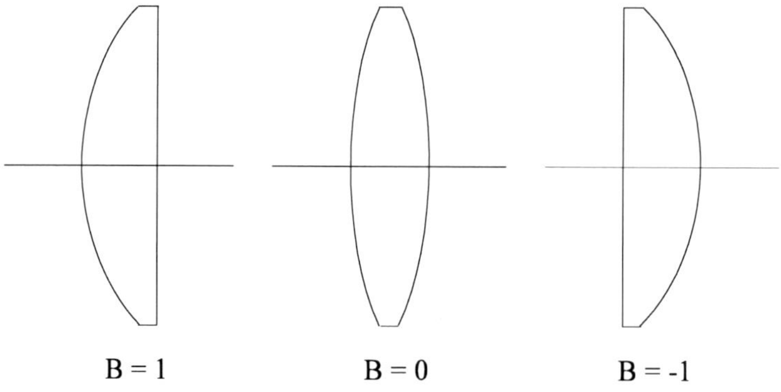

Figure 7.2. The shape factors of different lenses.

The second quantity relevant to the description of the thin lens is the conjugate factor, defined as

$$C = \frac{u_1 + u_2'}{u_1 - u_2'}. \tag{7.2}$$

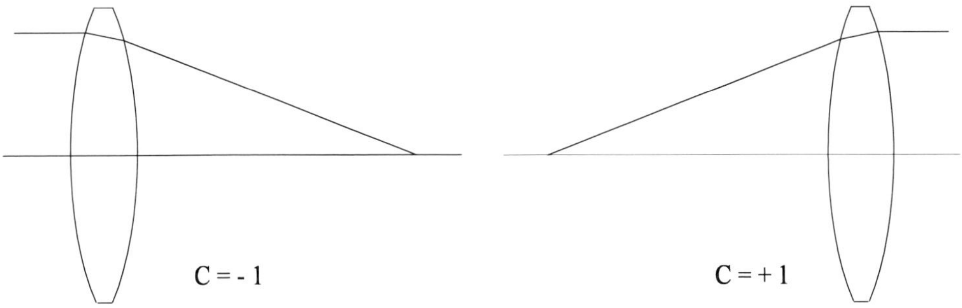

Figure 7.3. The conjugate factor, *C*.

The conjugate factor is related to the transverse magnification of a lens. Specifically, from the Lagrange invariant, the transverse magnification m is given by

$$m = \frac{n_1 u_1}{n_2 u_2'}. \tag{2.33}$$

We will assume that the lens is in air, so that the object space and the image space have a refractive index of unity. Therefore the conjugate factor,

$$\begin{aligned} C &= \frac{u_1 + u_2'}{u_1 - u_2'} \\ &= \frac{m u_2' + u_2'}{m u_2' - u_2'} \\ &= \frac{m+1}{m-1} \end{aligned} \tag{7.3}$$

and the magnification

$$m = \frac{C+1}{C-1}. \tag{7.4}$$

The conjugate factor cannot be as powerful a design parameter as the shape factor, because the magnification is usually determined by the basic layout of a lens. Therefore, it is not often freely variable. However, it can be used to orient the designer to those situations that will provide more or less of a problem because of the magnification requirement.

7.2 Thin lens at the stop

We have seen from Chapter 6 that the position of the aperture stop affects the Seidel aberrations of any lens, unless its aberrations are already zero. This statement is, of course, also true for thin lenses, so this detailed discussion of the aberrations of thin lenses will begin with the special case of a thin lens at the stop. Later we will extend the discussion to thin lenses that are not at the stop, by using the stop-shift Eqs. (6.61) to (6.66).

Because the derivations of the equations for the shape-dependent Seidel aberrations of thin lenses are rather long, we give here only the simpler derivations for the formulae for S_3, S_4, S_5, C_1 and C_2, which are all independent of shape factor. See Welford (Ref. 3 of Chapter 6) for an outline of the derivations of the formulae for S_1 and S_2, which are quoted below.

7.2.1 Spherical aberration

The Seidel coefficient of spherical aberration for a thin lens at the stop is given in terms of the shape factor B and the conjugate factor C by the following equation:

$$S_1 = \frac{h^4K^3}{4}\left\{\frac{n^2}{(n-1)^2} + \frac{n+2}{n(n-1)^2}\left(B + \frac{2(n^2-1)C}{n+2}\right)^2 - \frac{nC^2}{n+2}\right\}. \tag{7.5}$$

7.2.2 Coma

The Seidel coefficient of coma for a thin lens at the stop can be shown to be

$$S_2 = -\frac{h^2K^2H}{2}\left\{\frac{(n+1)B}{n(n-1)} + \frac{(2n+1)C}{n}\right\}. \tag{7.6}$$

7.2.3 Astigmatism

The Seidel coefficient for the astigmatism of a single surface is given by

$$S_3 = -\bar{A}^2 h\delta\left(\frac{u}{n}\right), \tag{6.29}$$

and $\bar{A}$ can be found from the formula given in Chapter 2:

$$\bar{A} = \frac{H}{h}(AhE - 1). \tag{2.79}$$

For the case of a surface at the stop, E is zero, and therefore,

$$\bar{A} = -\frac{H}{h}. \tag{7.7}$$

Substituting Eq. (7.7) in Eq. (6.29) we get, for a surface at the stop,

$$S_3 = -\frac{H^2}{h}\delta\left(\frac{u}{n}\right). \tag{7.8}$$

Considering both surfaces of the thin lens, which are assumed to be both at the stop,

$$S_3 = -\frac{H^2}{h}\left[\left(\frac{u_1'}{n_1'} - \frac{u_1}{n_1}\right) + \left(\frac{u_2'}{n_2'} - \frac{u_2}{n_2}\right)\right]. \tag{7.9}$$

In the case of a thin lens in air, $u_1' = u_2$ and $n_1' = n_2 = 1$.
Then,

$$\begin{aligned} S_3 &= -\frac{H^2}{h}(u_2' - u_1) \\ &= -\frac{H^2}{h}(-hK) \\ &= H^2 K, \end{aligned} \tag{7.10}$$

since $u_2' - u_1 = -hK$ for a thin lens, where K is the power of the lens.

7.2.4 Field curvature

The Seidel coefficient for the field curvature of a single surface is given by

$$S_4 = -H^2 c\delta\left(\frac{1}{n}\right) \tag{6.30}$$

or,

$$S_4 = H^2 c\left(\frac{1}{n} - \frac{1}{n'}\right). \tag{7.11}$$

Considering both surfaces of a lens, where $n_1' = n_2 = n$ is the refractive index of the lens,

$$S_4 = H^2\left[c_1\left(1-\frac{1}{n}\right)+c_2\left(\frac{1}{n}-1\right)\right]$$
$$= H^2(c_1 - c_2)\frac{n-1}{n} \qquad (7.12)$$
$$= H^2\frac{K}{n}.$$

7.2.5 Distortion

The Seidel coefficient of distortion of a single surface is given by

$$S_5 = \frac{\bar{A}}{A}(S_3 + S_4), \qquad (6.31)$$

and as stated above, for a surface at the stop,

$$\bar{A} = -\frac{H}{h}. \qquad (7.7)$$

Adding Eqs. (7.8) and (6.30) for a single surface,

$$S_3 + S_4 = -H^2\left[\frac{\delta\left(\frac{u}{n}\right)}{h} + c\delta\left(\frac{1}{n}\right)\right]$$
$$= -\frac{H^2}{h}\delta\left(\frac{u+hc}{n}\right) \qquad (7.13)$$
$$= \frac{AH^2}{h}\left(\frac{1}{n^2}-\frac{1}{n'^2}\right).$$

Therefore, combining both surfaces,

$$S_5 = \frac{H^3}{h^2}\left(1-\frac{1}{n^2}+\frac{1}{n^2}-1\right) = 0. \qquad (7.14)$$

Perhaps this result should have been obvious by inspection; a ray passing through a zero thickness plane parallel plate is undeviated. If a ray passes through the center of a zero-thickness lens, the two surfaces are parallel at the point of incidence of the ray, and the result is the same.

7.2.6 Axial color

From Eq. (6.49), the Seidel coefficient of axial color of a thin lens in air, made of glass of index n and dispersion δn, is given by

$$C_1 = A_1 h \frac{\delta n}{n} - A_2 h \frac{\delta n}{n}. \tag{7.15}$$

At the first surface,

$$A_1 = n_1 (hc_1 + u_1) = (hc_1 + u_1), \tag{7.16}$$

and at the second,

$$A_2 = n_2' \left(hc_2 + u_2' \right) = \left(hc_2 + u_2' \right). \tag{7.17}$$

Therefore, for the complete lens

$$\begin{aligned} C_1 &= A_1 h \frac{\delta n}{n} - A_2 h \frac{\delta n}{n} \\ &= h \frac{\delta n}{n} [h(c_1 - c_2) + (u_1 - u_2')] \\ &= h \frac{\delta n}{n} \left(\frac{hK}{n-1} + hK \right) \\ &= h^2 K \frac{\delta n}{n} \left(\frac{1}{n-1} + 1 \right) \\ &= h^2 K \frac{\delta n}{n} \frac{n}{n-1} \\ &= h^2 \frac{K}{V}, \end{aligned} \tag{7.18}$$

since

$$V - \frac{n-1}{\delta n}. \tag{5.4}$$

This result could also have been deduced from the results discussed in Chapter 5, when it was shown in Eqs. (5.3) and (5.4) that the chromatic variation in power is *K*/*V*. From this result, and knowing that a longitudinal focal shift will produce a wavefront aberration proportional to h^2, Eq. (7.18) can be derived.

7.2.7 Lateral color

For a single surface, Seidel coefficient of lateral color is

$$C_2 = \overline{A}h\delta\left(\frac{\delta n}{n}\right). \tag{6.50}$$

At the stop, $\overline{A} = -\frac{H}{h}$, and so for a single surface at the stop,

$$C_2 = -H\delta\left(\frac{\delta n}{n}\right). \tag{7.19}$$

So, for a thin lens in air at the stop,

$$C_2 = -H\left[\delta\left(\frac{\delta n}{n}\right)_1 + \delta\left(\frac{\delta n}{n}\right)_2\right] = 0. \tag{7.20}$$

This result, like the result for distortion, could have been predicted from the fact that a chief ray passing through the center of a thin lens is undeviated, and it is therefore not dispersed either.

7.3 Discussion of the thin-lens Seidel aberrations

These seven equations summarize the properties of the Seidel aberrations of a thin lens at the stop:

$$S_1 = \frac{h^4K^3}{4}\left\{\frac{n^2}{(n-1)^2} + \frac{n+2}{n(n-1)^2}\left(B + \frac{2(n^2-1)C}{n+2}\right)^2 - \frac{nC^2}{n+2}\right\} \tag{7.5}$$

$$S_2 = -\frac{h^2K^2H}{2}\left\{\frac{(n+1)B}{n(n-1)} + \frac{(2n+1)C}{n}\right\} \tag{7.6}$$

$$S_3 = H^2K \tag{7.10}$$

$$S_4 = H^2 \frac{K}{n} \tag{7.12}$$

$$S_5 = 0 \tag{7.14}$$

$$C_1 = h^2 \frac{K}{V} \tag{7.18}$$

$$C_2 = 0 \tag{7.20}$$

These are, after the seven equations that give the Seidel aberrations of a single surface [Eqs. (6.27) to (6.31), (6.49) and (6.50)], perhaps the most important group of equations in lens design.

Although real lenses must be made with reasonable non-zero thicknesses, the thin-lens equations are extremely valuable in predicting which aberrations are correctable by a given lens or group of lenses, and we will discuss these results in the next few sections. For example, it would be useful to know the type of dependence that the shape and conjugate factors have on the various aberrations. Figure 7.4 illustrates the relationship between the monochromatic aberrations of a thin lens at the stop, and its shape, represented by the shape factor B. We will now discuss these relationships, and their implications, in much more detail.

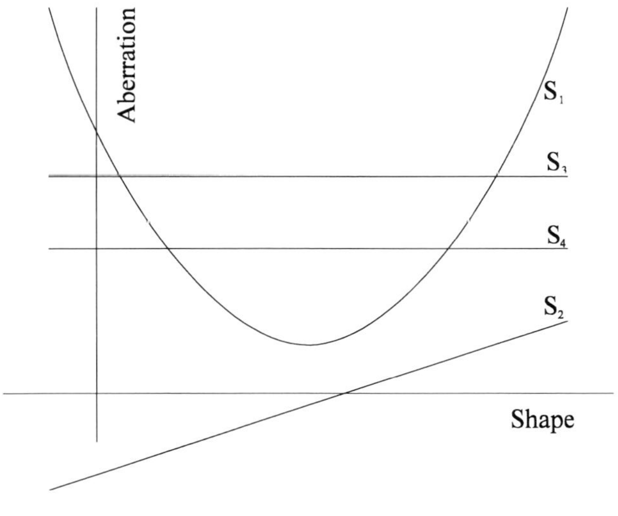

Figure 7.4. The Seidel aberrations of a thin lens at the stop.

7.3.1 Spherical aberration

Because the correction of S_1 is so important, its variation with shape factor should be examined first. By inspection of Eq. (7.5) we can see that S_1 must be a quadratic function of shape, B, since this equation contains the term

$$\left(B+\frac{2\left(n^2-1\right)C}{n+2}\right)^2. \tag{7.21}$$

We will consider the case of a lens with positive power, because that is the most common situation. If we consider the coefficient of B^2, we can see by inspection of Eq. (7.5) that B^2 has a positive coefficient, $(n+2)/n(n-1)^2$. Therefore S_1 is positive for large values of the shape factor, B, as shown in Fig. 7.4. The important question that we now consider is whether S_1 can ever be zero or negative, and, if so, under what conditions?

7.3.1.1 Bending for minimum spherical aberration

If we differentiate Eq. (7.5) with respect to B, we get

$$\frac{dS_1}{dB}=\frac{h^4K^3}{2}\frac{n+2}{n(n-1)^2}\left\{B+\frac{2\left(n^2-1\right)C}{n+2}\right\}, \tag{7.22}$$

which is zero when

$$B=\frac{2\left(1-n^2\right)C}{n+2}. \tag{7.23}$$

As an example, when the object is at infinity (zero magnification), then C is negative, and, since (1 ! n^2) is also negative, then the optimum value of B is positive. In other words, at the turning point of the curve the lens is convex to the incident light. Substituting this value of B into Eq. (7.5), we find that the minimum spherical aberration,

$$S_{1\min}=\frac{h^4K^3}{4}\left\{\frac{n^2}{(n-1)^2}-\frac{nC^2}{n+2}\right\}. \tag{7.24}$$

This expression can only be zero or negative for large values of C. For example, if $S_{1\min}$ is to be zero, C is given by

$$C^2 = \frac{n(n+2)}{(n-1)^2}. \tag{7.25}$$

For typical values of n such as 1.6, this means that $C = \pm 4$. From Eq. (7.5) the magnification is therefore either 5/3 (1.66666) or 3/5. In either case the magnification is positive, which means that either the object or the image is virtual.

We have already seen an example of a lens which has zero spherical aberration. This was in the second element of the microscope objective in Fig. 6.9. In that case the first surface was concentric with the on-axis object point and the second surface was aplanatic. It was shown that spherical aberration as well as the coma is zero for both surfaces. Because the first surface has no effect on the ray angle, the only refraction is at the second surface. It can be shown that in such a case the magnification of the thin lens will be equal to the refractive index of the lens. In both of these instances the magnification is positive, indicating a virtual image.

For the special case of an equiconvex lens ($B = 0$) at unit magnification (i.e., a magnification of -1, giving $C = 0$), the spherical aberration will be given by

$$S_1 = \frac{h^4 K^3}{4}\left(\frac{n}{n-1}\right)^2, \tag{7.26}$$

which is positive when K is positive. Note that the spherical aberration depends on h^4, as it should, and that it also has to depend on K^3, for the equation to be dimensionally correct. Note that if B is made non-zero, i.e. the lens is bent away from the equi-convex case without a change of C, the spherical aberration, which was already positive, will increase.

7.3.1.2 Effect of refractive index

If we consider the case of an object at infinity ($C = -1$), the minimum spherical aberration will be given by

$$S_{1\min} = \frac{h^4 K^3}{4}\left\{\frac{n^2}{(n-1)^2} - \frac{nC^2}{n+2}\right\}. \tag{7.27}$$

With $C^2 = 1$, the term inside the brackets,

$$\frac{n^2}{(n-1)^2} - \frac{n}{n+2}, \tag{7.28}$$

depends on the refractive index in a simple manner. We can calculate the shape factors and the (normalized) minimum spherical aberration for a few values of refractive index:

Refractive index	B for minimum S_1	Minimum S_1
1.5	0.71	1.00
1.7	1.02	0.64
2.0	2.25	0.41
3.0	3.2	0.19
4.0	5.0	0.13

This shows two things. For low refractive indices, the optimum shape factor is given by a biconvex lens ($B < 1$), but as the refractive index increases, the optimum shape becomes plano-convex ($B = 1$ at refractive index $n = 1.69$) and then meniscus ($B > 1$) for larger refractive indices. At a refractive index of $n = 4$ (e.g., germanium), the optimum shape is strongly meniscus; this is commonly observed in lenses for thermal imagers. Furthermore, the minimum spherical aberration is greatly reduced if very high index materials are available, although the improvement for indices between $n = 1.5$ to 1.8, the normal range for visible-band optical glasses, cannot be dramatic.

7.3.1.3 Effect of change of conjugates

We can also look at Eq. (7.24) to see that the minimum spherical aberration is affected by a change of conjugates. Taking a typical value of n to be 1.6, we can tabulate the minimum spherical aberration as a function of C:

Conjugate factor, C	Magnification	Minimum S_1
0	– 1.0	1.00
– 1	0.0	0.94
– 2	+0.33	0.75
– 3	+0.5	0.44
– 4	+0.6	0.00
– 5	+0.66	– 0.56

This shows a rapid reduction in spherical aberration as the conjugate factor is increased. (Remember that the minimum spherical aberration is independent of the sign of C).

7.3.1.4 Correction of spherical aberration with two positive lenses

It can be seen, from the above, that if two positive lenses are arranged together, each bent for minimum spherical aberration, the possibility of correcting the total spherical aberration will depend on the refractive index. If the refractive index is low, the minimum spherical aberration is not greatly affected by the fact that the second lens operates in a converging beam, and the spherical aberration is approximately proportional to h^4K^3. If we replace one single lens by two lenses of power $K/2$, the total spherical aberration is proportional to

$$h^4\left[\left(\frac{K}{2}\right)^3+\left(\frac{K}{2}\right)^3\right]=\frac{1}{4}h^4K^3. \tag{7.29}$$

However if the refractive index is high enough (about $n = 3$), correction of spherical aberration is possible with two lenses, as shown by Kingslake.[1] It has also been shown that correction is possible with three lenses of refractive index $n = 2$, and with four lenses with refractive index $n = 1.518$ (e.g., Schott BK7). The forms of these lenses are shown in Fig. 7.5.

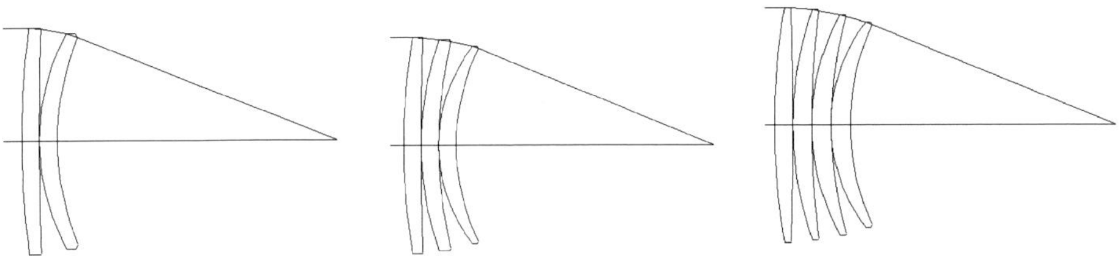

2 lenses, $n = 3$ 3 lenses, $n = 2$ 4 lenses, $n = 1.518$

Figure 7.5. Lenses corrected for spherical aberration with different refractive indices.

7.3.1.5 Correction of spherical aberration with positive and negative lenses

Normally, spherical aberration is corrected by combining positive and negative lenses. Since the combination of a positive lens and a negative lens allows us to correct axial color as well as spherical aberration, for most purposes it is the preferred method. In addition, the combination of a positive and negative lens is not significantly dependent on the values of refractive index, so spherical aberration correction is possible with almost any combination of two low-index glasses.

Of course, if the doublet is cemented, it is necessary to have a suitable refractive index difference between the lenses (the crown glass used for the positive component must have a lower refractive index than the flint glass used for the negative component), but if there is an air space, the refractive index difference is not critical, as stated above.

7.3.1.6 Seidel aberrations of thin lenses not at the stop

The equations listed above apply both to single surfaces and to complete systems, so they can be applied to single thin lenses. The most significant monochromatic aberration here is probably astigmatism. We have already seen that the astigmatism of a thin lens at the stop is given by $S_3=H^2K$, and is therefore uncorrectable, even if a single thin lens is replaced by several thin lenses at the stop. However, astigmatism can be corrected by using the stop-shift effects described in Chapter 6. All lenses can be corrected for astigmatism using this principle.

Another general principle that follows from the stop-shift equations is that, in a system in which the third-order aberrations are corrected, moving the stop will not introduce any change in the third-order aberrations. Similarly, in a system in which axial color is corrected, lateral color is also not affected by a stop-shift.

7.3.2 Correction of coma

For the case of a thin lens at the stop the coma is given by

$$S_2=-\frac{h^2K^2H}{2}\left[\frac{(n+1)B}{n(n-1)}+\frac{(2n+1)C}{n}\right]. \tag{7.6}$$

This shows that there is always a shape factor, B, that gives zero coma, and that this shape factor is dependent on the conjugate parameter, C. For zero coma, the relation between them is

$$B=-\frac{(n-1)(2n+1)C}{(n+1)}. \tag{7.30}$$

In the table below we compare the shape factors for minimum spherical aberration and zero coma, again for the case when $C=-1$ (the object at infinity):

Refractive index n	Shape B for zero coma	Shape B for minimum spherical aberration
	Eq. (7.23)	Eq. (7.20)
1.5	0.80	0.71
2.0	1.67	1.50
2.5	2.57	2.33
3.0	3.50	3.20
3.5	4.44	4.09
4.0	5.40	5.00

The shape factor for minimum spherical aberration is always a little smaller than the shape factor for zero coma. By choosing a refractive index of between 1.6 and 1.7, we can have either zero coma **or** minimum spherical with a plano-convex lens ($B = 1$). By inspection of Eqs. (7.24) and (7.30), it is clear that in the unit magnification case ($m = -1$, $C = 0$) an equiconvex lens has **both** zero coma and minimum spherical aberration. Further, for the extreme case of a germanium lens for infrared wavelengths, where n is 4, the shape for both minimum spherical aberration and zero coma is strongly meniscus. We do, in fact, see this in almost all thermal imaging lenses.

7.3.3 Correction of astigmatism

We have seen that the astigmatism of a thin lens in contact with the aperture stop is given by:

$$S_3 = H^2 K. \tag{7.10}$$

If we combine thin lenses together at the aperture stop, the total power of the lenses will be equal to the sum of the individual powers ($K = K_1 + K_2$), and the astigmatism of the complete system will be unchanged. Therefore, if astigmatism is to be corrected, it is essential to use the stop-shift formula, Eq. (6.93), in some way:

$$\Delta S_3 = 2(H\Delta E)S_2 + (H\Delta E)^2 S_1. \tag{6.93}$$

This equation gives the change of astigmatism introduced by non-zero amounts of coma or spherical aberration. It is, for example, possible to design a single lens, not in contact with the stop, in which the astigmatism is corrected by the deliberate introduction of coma.

Note that the astigmatism of a single lens will not be improved by the introduction of spherical aberration, because S_1 will be positive, and that would make matters worse. Of course, the resulting system then has non-zero coma, so the performance will not be very good. In general, if we are to correct S_1, S_2, and S_3, we require more than one lens, and there must be a significant separation between the lenses.

7.3.4 Correction of field curvature

It is clear from Eqs. (6.30) and (7.12) that the field curvature term, S_4, is proportional to H^2:

$$S_4 = -H^2 c\delta(1/n), \tag{6.30}$$

$$S_4 = H^2 K / n. \tag{7.12}$$

In Chapter 6 the Petzval sum for a system of several surfaces was defined as

$$P = \Sigma S_4/H^2, \tag{6.94}$$

so that for a single surface

$$P = -c\,\delta\,(1/n), \tag{7.31}$$

and for a thin lens

$$P = K/n. \tag{7.32}$$

In the absence of astigmatism, the sagittal and tangential surfaces collapse onto the Petzval surface.

In any lens that is required to cover a reasonable field of view it is usually necessary for the image to be flat, which requires that the Petzval sum be very small. Since a single thin lens with positive power cannot give zero Petzval sum, a system must be more complicated. There are basically three techniques for correction of Petzval sum with lenses.

7.3.4.1 Different refractive indices

If we combine positive lenses with a high refractive index with negative lenses with a low refractive index, it is possible, in principle, to reduce the Petzval sum. For example, if we have a positive lens with power $K_1 = 4$, index $n_1 = 1.8$, and a negative lens with power $K_2 = -3$, index $n_2 = 1.5$, then the Petzval sum [Eq. (7.10)] for two thin lenses together will be

$$K_1/n_1 + K_2/n_2 = 4/1.8 - 3/1.5 = 0.222.$$

This is better than the equivalent single lens with a refractive index of $n = 1.8$ which would have a Petzval sum of $1/1.8 = 0.555$. However, the powers K_1, K_2 are quite large, which implies that the surfaces are steeply curved. In practice, therefore, this technique for reducing the Petzval sum is rarely used.

7.3.4.2 Separated lenses

Suppose that we have two lenses with the same refractive index, of power $+K_1$ and $-K_1$, separated by a distance of d, as shown in Fig. 7.6.

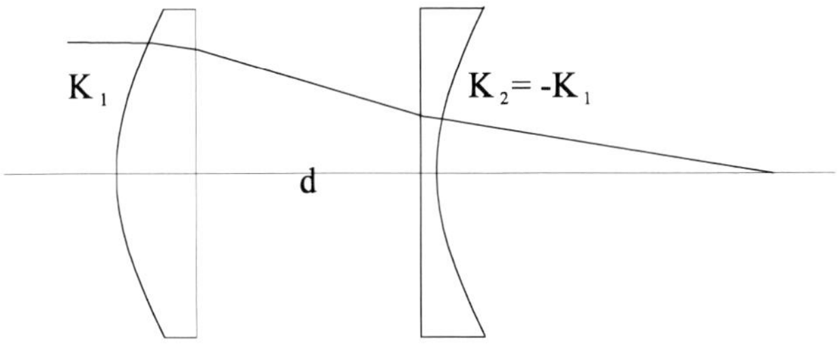

Figure 7.6. Two separated lenses with zero Petzval sum.

From Eq. (2.53), with $n' = 1$, the combined power of these lenses will be given by

$$K = K_1 + K_2 - dK_1K_2. \tag{7.33}$$

In the case where the powers are equal and opposite, then $K_2 = -K_1$,

$$\begin{aligned} K &= K_1 - K_1 + d\,K_1\,K_1 \\ &= d\,K_1^2, \end{aligned} \tag{7.34}$$

which is positive, regardless of the sign of K_1. Therefore, if we combine separated positive and negative lenses, with either lens first, we have a system with zero Petzval sum and positive power.

In general, if we have a design with several lenses, some with positive power and some with negative power, with air spaces between them, it is possible to arrange for the total power to be positive while the Petzval sum is zero, or, at least, small.

7.3.4.3 Thick meniscus lens

First, recall that the power of a thick lens with curves c_1 and c_2 (Fig. 7.7), refractive index n, and thickness d is given by

$$K = (n-1)\left[c_1 - c_2 + \frac{(n-1)dc_1c_2}{n}\right]. \tag{2.58}$$

If the lens has the same curvatures on both surfaces, the power of each surface is the negative of the other, so the Petzval sum is zero, but the lens will still have power in the amount

$$K = \frac{(n-1)^2 dc_1^{\,2}}{n}, \tag{7.35}$$

which is positive if c_1 is non-zero. Clearly the Petzval sum is zero, so we have a method of designing a lens with positive power and zero Petzval sum. Alternatively we can design a lens with zero power and negative Petzval sum, which can be used to reduce the inherent Petzval sum of a normal system with positive power. The disadvantages are fairly obvious. The power will be small unless the lens is very thick and expensive, and thick meniscus components can be difficult to manufacture economically. However, several important designs are constructed on this principle.

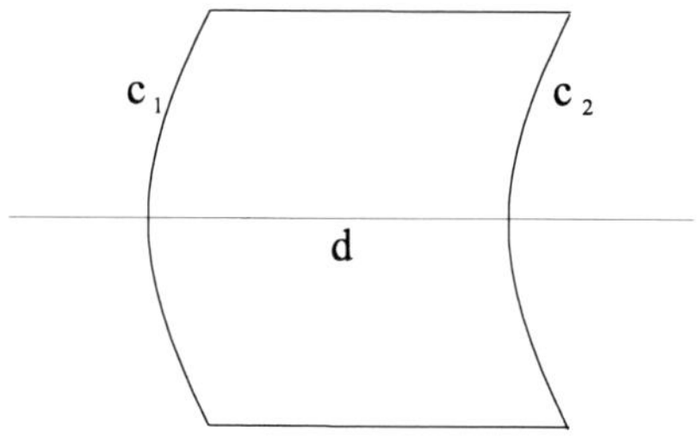

Figure 7.7. Meniscus lens with positive power and zero Petzval sum.

7.3.5 Reduction of aberrations by splitting lenses into two

Suppose we have a single lens of power *K*, with a semi-aperture *h*. The third-order spherical aberration S_1 is proportional to h^4K^3. So for the special case that was discussed in Sec. 7.3.1.1 on bending for minimum spherical aberration (unit magnification and equi-convex),

$$S_1 = \frac{h^4K^3}{4}\left(\frac{n}{n-1}\right)^2. \tag{7.26}$$

Taking this equation, we can consider the effect of replacing the single lens of power *K* with two lenses of power *K*/2. The spherical aberration is then proportional to

$$h^4\left[\left(\frac{K}{2}\right)^3+\left(\frac{K}{2}\right)^3\right]=\frac{1}{4}h^4K^3. \tag{7.36}$$

Clearly, the spherical aberration has been reduced by a factor of 4.

Since Eq. (7.36) is approximate, we actually find that the improvement is larger than this, particularly with high-index materials. With low-index materials, the above result gives a reasonably good indication of the effect of splitting a lens into two.

7.3.6 Seidel aberrations of a thin lens that is not at the stop

The changes in the thin lens aberrations produced by a change in E are, from Chapter 6:

$$\Delta S_2 = H\,\Delta E\, S_1 \tag{6.61}$$

$$\Delta S_3 = 2(H\Delta E)S_2 + (H\Delta E)^2 S_1 \tag{6.65}$$

$$\Delta S_5 = H\Delta E(3S_3 + S_4) + 3(H\Delta E)^2 S_2 + (H\Delta E)3S_1 \tag{6.66}$$

$$\Delta C_2 = H\,\Delta E\, C_1 \tag{6.67}$$

Note that these equations are not limited to small changes in E.

The equations apply both to single surfaces and to complete systems, and are independent of the nature of the surface, as long as the system has rotational symmetry. Note, in particular, that these stop-shift results also apply to aspheric surfaces. Eq. (6.67) gives us a very simple result for the lateral color (C_2) of a thin lens not located at the stop. Since C_2 is zero for a thin lens at the stop, Eq. (6.67) gives us the actual value of C_2 for a lens away from the stop:

$$C_2 = HEC_1 = \frac{\bar{h}}{h}h^2\frac{K}{V} = \bar{h}h\frac{K}{V}. \tag{7.37}$$

Had the lens been the stop, then $\bar{h}$ would be zero. This equation shows that the lateral color is a product of the chief and axial ray heights at the lens. So the farther the lens is removed from the stop and the steeper the chief ray angle, the greater is the lateral color, as one might suspect. Also the greater the dispersion, the smaller the V number, again adding to lateral color.

7.3.7 Correction of axial and lateral color

A special result that follows from the stop-shift equations is useful when considering systems in which axial color and lateral color are both corrected. Consider a system of two components, A and B (Fig. 7.8), which may or may not be single lenses.

The axial color is given by

$$\sum C_1 = C_{1_A} + C_{1_B} = 0, \tag{7.38}$$

and the lateral color is given by

$$\sum C_2 = C_{2_A} + C_{2_B} = HE_A C_{1_A} + HE_B C_{1_B} = 0. \quad (7.39)$$

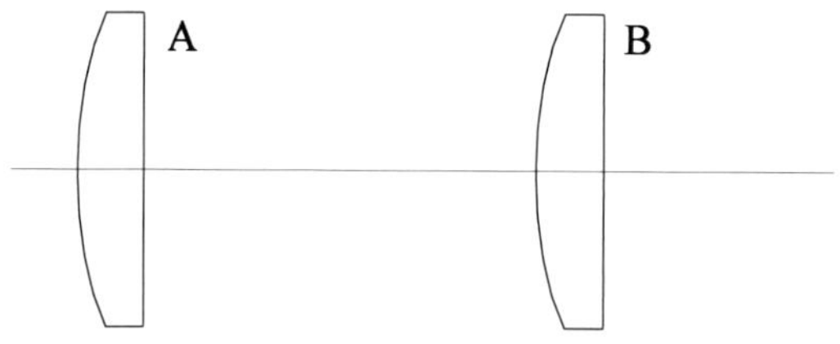

Figure 7.8. System for correcting axial and lateral color.

For both of these equations to be satisfied together, it is necessary that either:

$C_{1_A} = C_{1_B} = 0$, which means that components A and B are both achromatic

or

$E_A = E_B$, which implies that the two lenses are in contact.

In the first case ($C_{1_A} = C_{1_B} = 0$), we have a simple Petzval lens or a simple telephoto lens (Fig. 7.9), while the second case ($E_A = E_B$) is represented by a doublet, where the two lenses are effectively in contact.

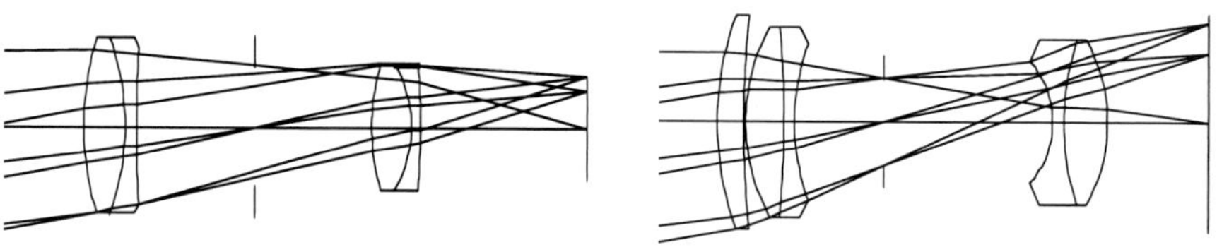

Petzval Lens Telephoto Lens

Figure 7.9. Two designs corrected for axial and lateral color.

7.4 Shape-dependent and shape-independent aberrations

We have seen that, for thin lenses **at the stop**, only spherical aberration and coma are shape-dependent. Now that we can see that astigmatism and distortion are affected by stop-shift effects, it should be clear that they are also shape-dependent in the general case where the lens is not at the stop. We can therefore draw the useful distinction between shape-dependent and shape-independent aberrations.

So, for thin lenses away from the stop, the following aberrations are **shape-dependent**:

Spherical aberration
Coma
Astigmatism
Distortion

while the following aberrations are **shape-independent**:

Field curvature (Petzval sum)
Axial color
Lateral color

In many simple systems, the basic layout is determined by the need to correct the three shape-independent aberrations.

The shape-independent aberrations are also unaffected by aspherizing, so it follows that if a lens layout is determined by the need to correct Petzval sum, axial color, or lateral color, the layout will not be affected by the addition of aspheric surfaces.

7.5 Aspheric surfaces

In order to calculate the Seidel aberrations of aspheric surfaces, we first imagine the effect of an aspheric plate, as shown in Fig. 7.10.

Clearly, an aspheric plate, as shown in Fig. 7.10, will introduce an aberration into the output wavefront, and if the asphericity of the aspheric is given by

$$z = a_4 r^4 \tag{7.40}$$

the wavefront aberration is

$$W = a_4 r^4 (n' - n). \tag{7.41}$$

This is, of course, a third-order (Seidel) aberration and can be related to S_1 by the equation

$$S_1 = 8\, a_4 r^4 (n' - n). \tag{7.42}$$

If the aspheric plate is in contact with the aperture stop, there are no other third-order aberrations, since the effect of the obliquity of the beam is to change

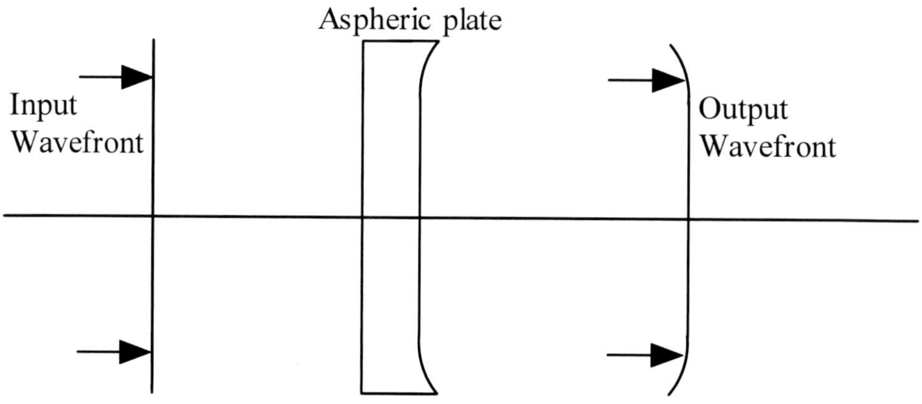

Figure 7.10. Deformation of a wavefront by an aspheric plate.

only higher-order terms. Similarly, the effect of using an aspheric surface on a zero-power meniscus lens is the same as on a plane parallel plate, to the third-order approximation.

When we have a lens with an aspheric surface, we visualize a lens with spherical surfaces, with an aspheric plate in contact with one of the spherical surfaces. The Seidel aberrations of the aspheric lens are then found by computing the aberrations of the spherical lens, and then adding the contributions of the aspheric plate, as shown in Fig. 7.11.

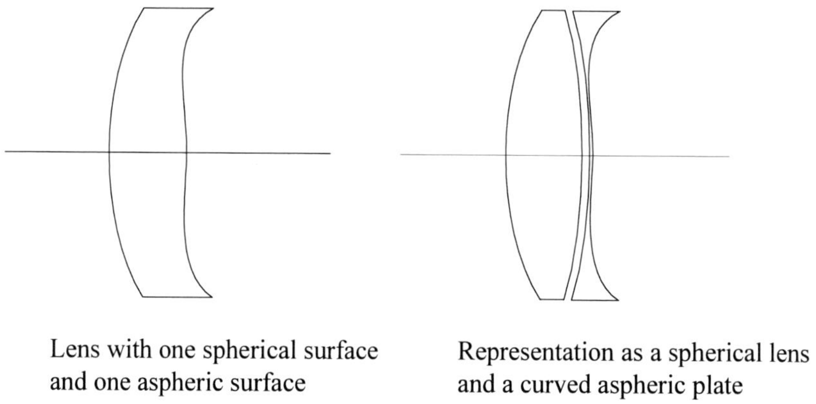

Figure 7.11. Representation of an aspheric lens for Seidel calculations.

Note, however, that there is no need to use this visualization process when actually ray tracing; it is simpler to calculate the geometry of an aspheric surface.

7.5.1 Third-order off-axis aberrations of an aspheric plate

If the aspheric plate is **not at the aperture stop**, the off-axis beams pass obliquely through the aspheric, as shown in Fig. 7.12, and the normal stop-shift effects apply.

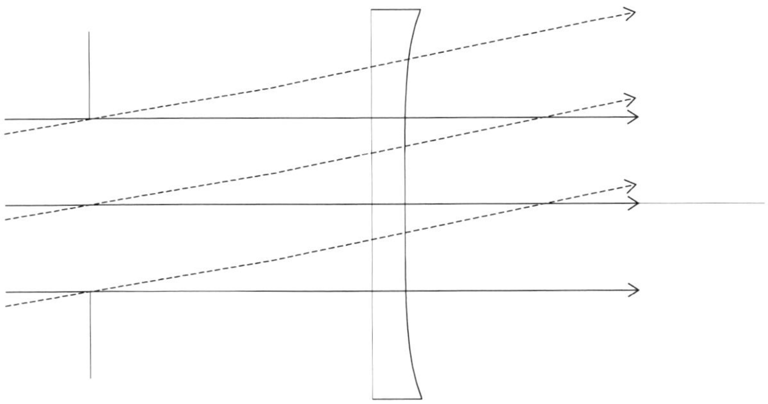

Figure 7.12. Off-axis beam passing through an aspheric plate.

First, recall the quantity, *E*, described in Sec. 2.5.7 on the Seidel difference formulae. If we have a paraxial chief ray with height $\bar{h}$ at a given surface, a paraxial marginal ray with height *h*, and a Lagrange invariant *H*, the stop-shift term *E* is defined by

$$HE = \frac{\bar{h}}{h}. \tag{2.78}$$

Since *E* is a measure of the distance of a surface from the aperture stop it will, in general, be different at every surface in the system. Then, if we have an aspheric surface at a non-zero distance from the aperture stop, so that the quantity *E* is non-zero, the aspheric will introduce off-axis aberrations as follows:

$$\begin{aligned} S_2 &= (HE)S_1 = (HE)8a_4r^4(n'-n) \\ S_3 &= (HE)^2 S_1 = (HE)^2 8a_4r^4(n'-n) \\ S_5 &= (HE)^3 S_1 = (HE)^3 8a_4r^4(n'-n). \end{aligned} \tag{7.43}$$

We can, of course, write these equations in terms of $\bar{h}/h$, to get

$$S_2 = \left(\frac{\bar{h}}{h}\right) S_1 = \left(\frac{\bar{h}}{h}\right) 8a_4 r^4 (n' - n)$$

$$S_3 = \left(\frac{\bar{h}}{h}\right)^2 S_1 = \left(\frac{\bar{h}}{h}\right)^2 8a_4 r^4 (n' - n) \qquad (7.44)$$

$$S_5 = \left(\frac{\bar{h}}{h}\right)^3 S_1 = \left(\frac{\bar{h}}{h}\right)^3 8a_4 r^4 (n' - n).$$

Note that the introduction of an aspheric surface has no effect on either the field curvature term, S_4, or on the first-order chromatic aberration terms.

7.5.2 Chromatic effects

Since the effect of the aspheric is essentially to change the spherical aberration, and possibly to generate stop-shift effects, there are no changes to the Seidel chromatic aberration coefficients, C_1 and C_2. There are, however, other chromatic effects when the aspheric surface is refracting because the refractive index difference $(n' - n)$ will, in general, be a function of wavelength. Therefore, in the case of a refracting aspheric surface there will be chromatic variations of S_1, S_2, S_3 and S_5. The chromatic variation of spherical aberration is important in the design of Schmidt cameras, as we shall see in Chapter 13.

7.6 The sine condition

A theorem that was very important in the early development of lens design, but has been somewhat neglected, is known as the Abbe sine condition.

7.6.1 Sine condition in the finite conjugate case

For a lens corrected for spherical aberration, coma is also corrected if

$$\frac{\sin U'}{u'} = \frac{\sin U}{u}, \qquad (7.45)$$

where u' and u are paraxial ray angles, and U' and U are real ray angles. If $n' = n$, the (paraxial) magnification is given by

$$\text{magnification} = \frac{\eta'}{\eta} = \frac{u}{u'}, \qquad (2.33)$$

so it follows that $\sin U / \sin U'$ must equal the magnification, or

$$\sin U = \frac{\sin U'}{\text{magnification}} \tag{7.46}$$

for all values of U' and U. Suppose that we have a lens with a magnification of –0.1. If we trace a ray from the axial object point (Fig. 7.13), with $\sin U = 0.025$, and if the spherical aberration is corrected, then, if $\sin U' = 0.025/{-0.1} = -0.25$, it follows from the sine condition that the coma is corrected.

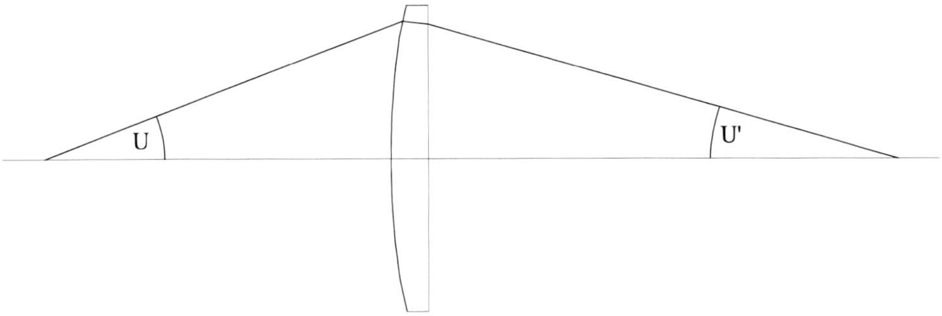

Figure 7.13. The sine condition—finite-conjugate case.

7.6.2 The sine condition with the object at infinity

If the object is at infinity, the sine condition takes the form

$$\sin U' = \frac{-y}{F}, \tag{7.47}$$

where F is the focal length and y is the height of the ray in the entrance pupil.

One case where this is important is in astronomical telescopes. As we shall see in Chapter 13, a classical Cassegrain telescope has a paraboloidal primary mirror and a hyperboloidal secondary mirror. Both mirrors are corrected for spherical aberration. If we trace a marginal ray through such a telescope, we will see immediately that although the spherical aberration is corrected, the sine condition is not satisfied, and that the coma is therefore not corrected.

For this reason, in Ritchey-Crétien telescopes (such as the Hubble Space Telescope) the primary mirror is given non-zero spherical aberration, so that the angle of the marginal ray is adjusted to satisfy the sine condition. In order to correct the total spherical aberration of the telescope, the secondary mirror also has to have non-zero spherical aberration.

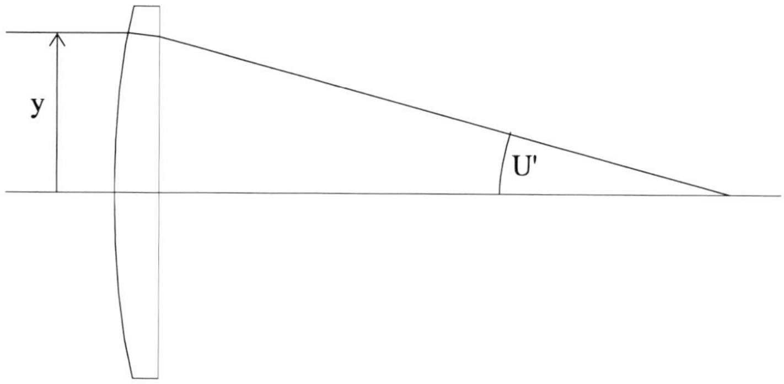

Figure 7.14. Sine condition—infinite object.

7.6.3 The sine condition for the afocal case

In the case of an afocal system (Fig. 7.15), the sine condition states that coma is corrected if, and only if

$$y \text{ (exit pupil)} = y \text{ (entrance pupil)} \times \text{ expansion factor}, \tag{7.48}$$

where the expansion factor is the ratio of the paraxial ray heights in the exit and entrance pupils.

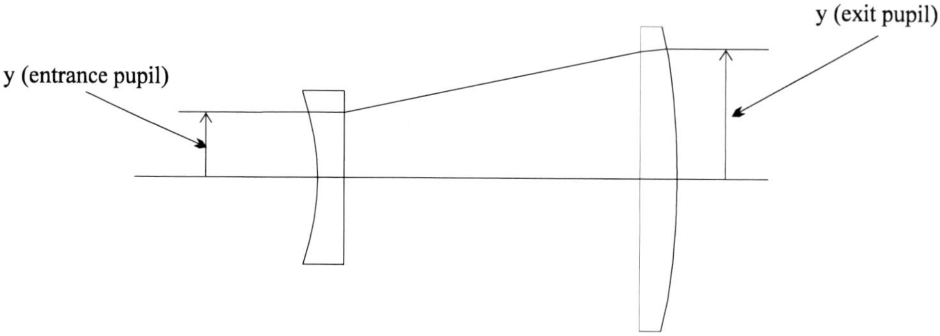

Figure 7.15. Sine condition for an afocal system.

7.7 Other design strategies

In addition to aberration reduction by changing the prescription parameters of the lens, a designer can approach the problem by using symmetry. Just as in the case of aplanatic surfaces, some rays will not contribute to certain aberrations when the elements are designed to take advantage of special geometries.

7.7.1 Monocentric systems

A method of avoiding some off-axis aberrations is to design a lens so that all surfaces are monocentric, with the same center of curvature as the aperture stop, as shown in Fig. 7.16.

It then follows that the aberrations off-axis are essentially the same as on-axis, apart from a reduction of the beam width. Therefore, coma, astigmatism, and lateral color are automatically zero. Distortion is zero in the sense that the image size (measured along the spherical image surface) is proportional to the field angle.

Spherical aberration and axial color are not necessarily corrected (although they can be) but the image surface is necessarily concentric with the other surfaces, so field curvature is uncorrectable, unless the image is at infinity.

It is interesting to compare this arrangement to that of the human eye which, although not exactly monocentric, makes use of a spherical image surface in a similar way.

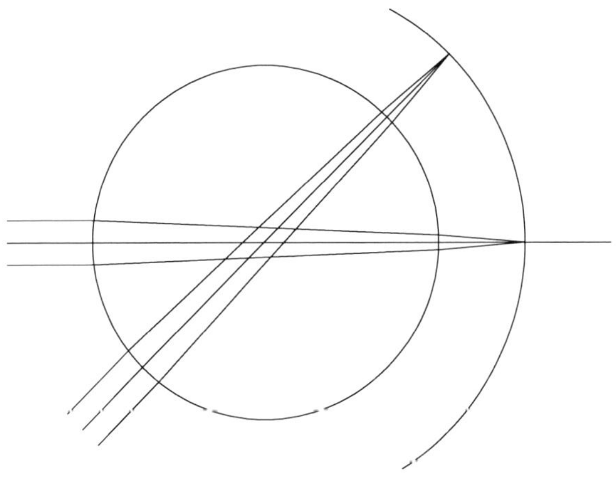

Figure 7.16. A monocentric lens.

7.7.2 Use of front-to-back symmetry

There are some ways in which aberrations can be automatically corrected. The first of these is the use of front-to-back symmetry, as shown in Fig. 7.17.

If a lens is symmetrical, such that the rear half of the lens is a mirror image of the front half, and if it works at unit magnification, we can see from Fig. 7.17 that the path of ray A is exactly equivalent to the path of ray B. Therefore, it follows that the optical paths of the two rays are identical, and therefore that the wavefront aberrations are the same.

From this we can see that the only aberrations that can be present are terms that are even powers in *y*, and the terms that are odd powers in *y* (e.g., coma, distortion, lateral color) must be zero. If the magnification were not unity, the odd terms would no longer be zero, but in most cases they would still be small, and they could usually be corrected by a slight change from the exactly symmetrical condition.

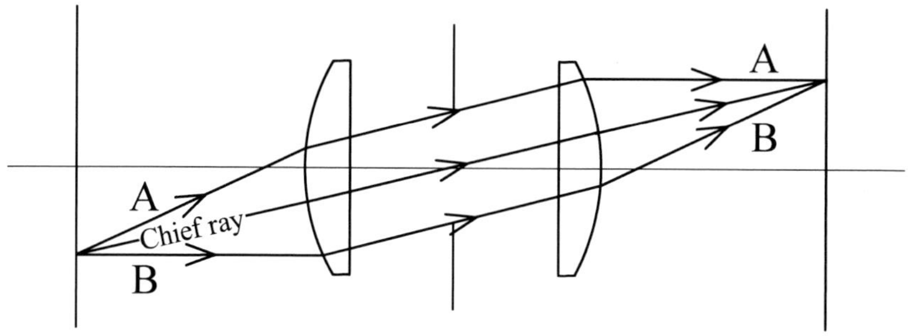

Figure 7.17. Use of front-to-back symmetry, to correct odd aberrations.

References

1. R. Kingslake, *Lens Design Fundamentals*, Academic Press (1978).

CHAPTER 8
ACHROMATIC DOUBLET OBJECTIVES

Doublets are suitable for many applications when the aperture is fairly small (up to about *f*/4) and when the field angle is limited to a few degrees. The two components of a doublet may either be cemented together or separated by an air space. In most cases the cemented solution is preferred because its tolerances are easier to meet and it is obviously more robust.

In addition to being used alone, doublets are used in collimators and as objectives of telescopes. Many lens systems contain one or more doublets, and most of the comments made below concerning "stand-alone" doublets will also apply to doublets that are used in these complex systems.

8.1 Seidel analysis

8.1.1 Correction of chromatic aberration

We have seen in Chapters 5 and 7 that setting the terms in Eq. (7.18) to zero in the following conditions describes axial color correction for two lenses:

$$\frac{C_1}{h^2} = \frac{K_1}{V_1} + \frac{K_2}{V_2} = 0, \tag{5.6}$$

while requiring the sum of the individual lens powers be equal to the overall power of the lens K:

$$K = K_1 + K_2, \tag{5.7}$$

and recalling that the values of K_1 and K_2 that satisfy these conditions are:

$$K_1 = \frac{KV_1}{V_1 - V_2} \quad \text{and} \quad K_2 = \frac{-KV_2}{V_1 - V_2}. \tag{5.8}$$

A look at the glass map for commercial glasses (e.g., Fig. 5.3) will show that the range of available refractive indices is from 1.44 to 2.02, and the range of values of V is from 91 to 20. However, in normal instruments the refractive indices will be between 1.5 and about 1.8, and values of V will vary from 64 to about 25. Typically, in a doublet objective, V_1 might be 60 and V_2 might be 36. Inserting these values into Eq. (8.3), we obtain $K_1 = 2.5$ K and $K_2 = -1.5$ K.

8.1.2 Astigmatism and field curvature

As we have seen in Chapter 6, in the presence of third-order field curvature and astigmatism the transverse ray aberrations are given by

$$\delta\eta' = \frac{3S_3 + S_4}{2n'u'}, \tag{6.90}$$

$$\delta\xi' = \frac{S_3 + S_4}{2n'u'}. \tag{6.91}$$

Since, for a thin lens at the stop, $S_3 = H^2K$ and $S_4 = H^2K/n \approx 0.6H^2K$ for $n = 1.5$, these equations simplify to

$$\delta\eta' = \frac{3.6H^2K}{2n'u'} = 1.8\eta'^2u'K \tag{8.1}$$

$$\delta\xi' = \frac{1.6H^2K}{2n'u'} = 0.8\eta'^2u'K. \tag{8.2}$$

These results apply for any thin lens at the stop, as long as the refractive indices are not very different from 1.5. The largest aberration is determined by the factor of 3.6 in Eq. (8.1). This is due more to the astigmatism coefficient, S_3, and less to the field curvature coefficient, S_4.

8.1.3 Comparison with the actual aberrations of a doublet

The example that we use has the following specifications:

$f' = 100$ mm, therefore $K = 0.01$ mm^{-1}
Aperture $= f/5$, therefore numerical aperture in air, $u' = 0.1$

If the semi-field angle is set to 1 deg, then the image size is $\eta' = f' \tan(1 \text{ deg}) = 1.74$ mm.

Therefore, $\delta\eta' = 1.8\,\eta'^2 u'K = 1.8\,(1.74)^2\,(-0.1)(0.01) = -0.0054$ mm.

Similarly, $\delta\xi' = 0.8\,\eta'^2 u'K = 0.8\,(1.74)^2\,(-0.1)(0.01) = -0.0024$ mm.

The aberration curves in this chapter show that this result is in close agreement with the actual results obtained from ray tracing a reasonably well-designed doublet. Also, at this numerical aperture, because third-order aberration theory gives good agreement, we can assume that higher-order aberrations are not significant.

8.1.4 Correcting both Petzval sum and axial color in doublets

It is not always possible to correct all the aberrations in a lens. This is particularly true for a doublet designed for visible wavelengths. A doublet has too few design parameters to vary and the glasses are not always available with the necessary parameter ranges to build a reasonable lens. First, remember that the field curvature of a combination of two thin lenses in contact is given by

$$S_4 = H^2\left(\frac{K_1}{n_1}+\frac{K_2}{n_2}\right), \tag{8.3}$$

and the axial color is

$$C_1 = h^2\left(\frac{K_1}{V_1}+\frac{K_2}{V_2}\right). \tag{8.4}$$

These two equations are very similar, and it should be clear that if we can find a pair of glasses such that

$$\frac{n_1}{n_2}=\frac{V_1}{V_2}, \tag{8.5}$$

it would be possible to correct S_4 and C_1 simultaneously. However, inspection of the glass map (Fig. 5.3) shows that there are no suitable materials, at least in the visible waveband. For example, we could choose glasses with

$$n_1 = 1.75,\ n_2 = 1.55,\ V_1 = 52,\ V_2 = 46,$$

but with these V-values we would obtain powers for the two components of 8.66 K and −7.66 K, to be compared with the values of 2.5 K and −1.5 K that we found in Sec. 8.1.1. These individual powers are much too large to be satisfactory. In any case, we can see from Eqs. (6.90) and (6.91) that correcting the Petzval sum, S_4, alone, will not greatly affect the aberrations.

The Petzval sum must therefore be regarded as not correctable by a doublet in the visible band, although in other wavebands there are suitable materials that permit a solution with modest powers for the separate elements.

8.1.5 Possibilities of aberration correction in doublets

The possibilities for the correction of third-order aberrations of a cemented doublet at the stop are summarized in the following table:

Table 8.1. Aberration correction for doublets.

CORRECTABLE:	Spherical aberration	S_1
	Coma	S_2
	Axial color	C_1
UNCORRECTABLE:	Astigmatism	S_3
	Field curvature	S_4
ZERO:	Distortion	S_5
	Lateral color	C_2

8.2 The cemented doublet

Assume that the two glass types have been chosen for a cemented doublet (Fig. 8.1). Once the powers of the components have been determined, there is only one free variable remaining and this is the shape of the doublet. (In general, the lens thicknesses are not freely variable, or, rather, they can be varied, but they are not very useful parameters for changing aberrations to any large extent).

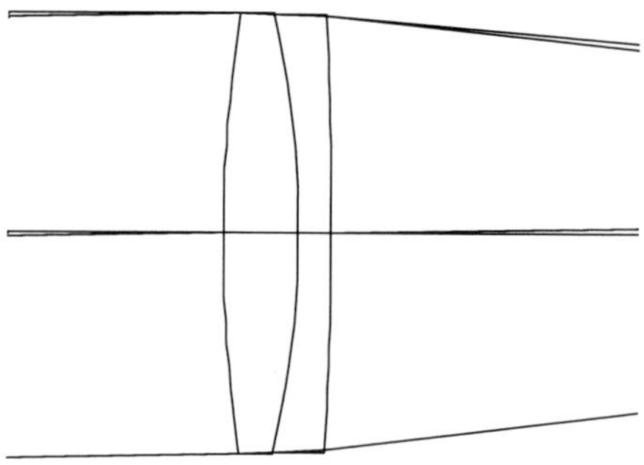

Figure 8.1. A cemented doublet.

If the stop is in contact with the lens, as the doublet is bent the coma will be a linear function of lens shape, and the spherical aberration will be a quadratic function of lens shape (see Sec. 7.3, Fig. 7.4). It is therefore always possible to find a solution for the lens shape that corrects the coma, but spherical aberration will only be corrected if the glasses have been correctly chosen.

Figure 8.2 shows a plot of suitable crown glasses for an achromatic aplanatic (spherical aberration and coma corrected) doublet, for a particular flint glass. Crown glasses above this curve will have positive spherical aberration when the coma is corrected; below the curve the spherical aberration will be negative. At the left-hand end of the curve there are doublets with low individual powers and therefore shallow surface curvatures. As the choice of crown glass approaches the flint glass (upper right-hand corner of Fig. 8.2), the powers will increase, and the surface curvatures will become steeper, until a point is reached where they are too steep to be manufacturable.

The negative component of a doublet has two functions: to introduce negative spherical aberration and to introduce negative axial color. It turns out that a lens pair with powers of about +2.5 and −1.5 is usually the optimum for correcting spherical aberration with reasonably shallow surfaces.

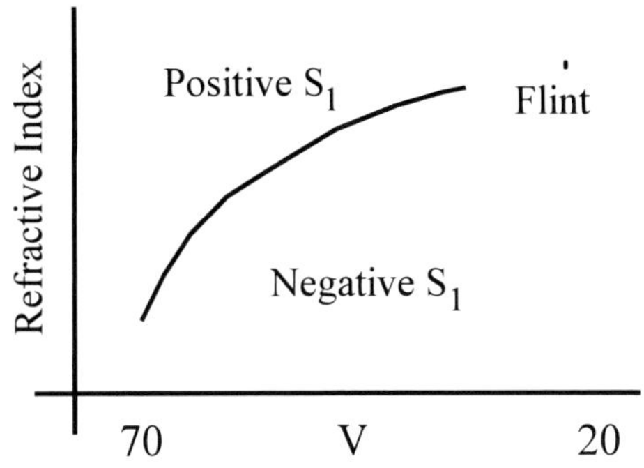

Figure 8.2. Locus of crown glass parameters for zero spherical aberration.

8.2.1 Optimization of cemented doublets

If the glass choice appears reasonable, as described above, we can optimize using the three surface curvatures of the lens as variables. For the case of a lens with the first element made of crown glass, called a "crown-first doublet," one can start a design with an equi-convex positive component with approximately the correct power and then determine power of the flint component needed to produce the correct total power.

If the glass choice is not reasonable, one of the glasses (either will do) must be changed to give the extra variable needed to satisfy the four conditions ($S_1 = 0$, $S_2=0$, $C_1 = 0$, $f' = 100$). The designer can either change one glass type manually, or let an optimization program vary one glass type. If the glass is changed by the optimization, most programs will treat the refractive index and dispersion as continuous variables, leaving the designer to choose the nearest real glass from the catalog at the end. In principle it will not matter whether the crown or flint glass is to be variable, unless factors such as availability, delivery times, etc., impose external constraints.

Although one might be tempted to change the thicknesses, this is rarely a good idea in the case of doublets. Thicknesses do not have a great effect on the aberrations of a cemented doublet objective since thin-lens theory is still more or less valid, so that large changes are required in order to achieve anything significant.

8.2.2 Crown-first doublet

In the prescription below, and Fig. 8.3, the first lens is an SK11 crown glass and the second is an SF19 flint. This is a typical example of a crown-first doublet, usually known as a Fraunhofer doublet. The transverse ray aberration curves are given in Fig. 8.4.

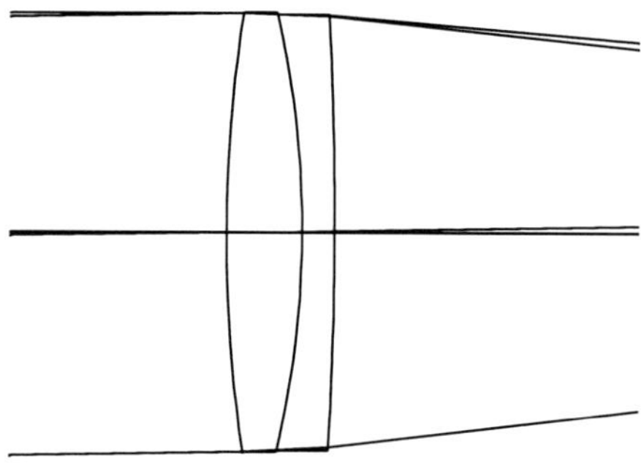

Figure 8.3. Fraunhofer doublet.

```
EFL =    100.000

WAVELENGTHS   [nm]
      587.60        656.30        486.10

SURFACES
  # SURF SPACE      RADIUS         SEPN      INDEX1       V       CLR RAD   GLASS
  0                              0.00000   1.000000
  1#S              64.10000      3.50000   1.563838    60.77     10.000   S-SK11
  2 S             -43.24900      1.50000   1.666796    32.99      9.926   S-SF19
  3 S            -183.68500     97.44950   1.000000               9.860
  4 S               Plane                                         1.763

LAGRANGE INVARIANT =     -0.1746

 #      H          U          HBAR       UBAR      D(U/N)       A          ABAR
 1  10.00000   0.00000    0.00000    0.01746   -0.03597   0.15601    0.01746
 2   9.80313  -0.05625    0.03907    0.01116    0.01271  -0.44243    0.01604
 3   9.74498  -0.03877    0.05486    0.01053   -0.07674  -0.15305    0.01705
             -0.100000              0.017349

 #     S1         S2         S3         S4         S5         C1         C2
 1  0.008754   0.000979   0.000110   0.000171   0.000031   0.009255   0.001036
 2 -0.024383   0.000884  -0.000032  -0.000028   0.000002  -0.026861   0.000974
 3  0.017518  -0.001952   0.000217   0.000066  -0.000032   0.018085  -0.002015

Sum 0.001889  -0.000088   0.000295   0.000210   0.000002   0.000480  -0.000005
```

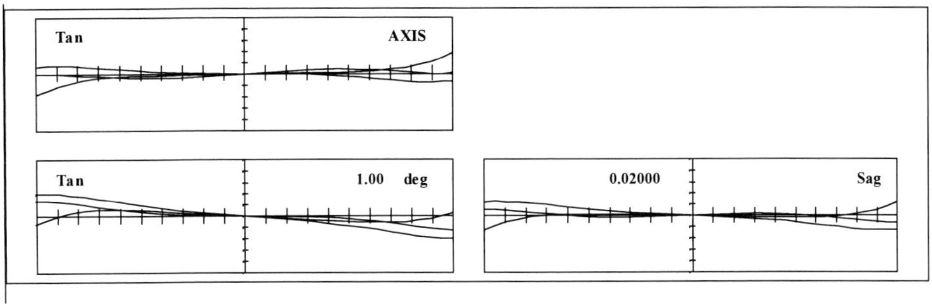

Figure 8.4. Transverse aberration curves for a Fraunhofer doublet.

The fact that the on-axis aberration plot is somewhat flatter than the standard S-shaped spherical aberration curve indicates that there is a small amount of compensating high-order spherical aberration. The chromatic variation of spherical aberration, often called "sphero-chromatism" is also present because the different wavelength curves are not all the same shape. Although it is not possible to show easily in a black and white plot, the curves for the long and short wavelengths are closer to each other than to that of the center wavelength, telling us that there is also some secondary spectrum. This indicates that the correction at the extreme wavelengths does not coincide with the central wavelength.

A comparison between the slopes at the origin of the tangential and sagittal 1-deg curves indicates that there is only a little astigmatism. The different slope of the axial and tangential 1-deg curves says that there is some field curvature, which would be expected for a single positive lens.

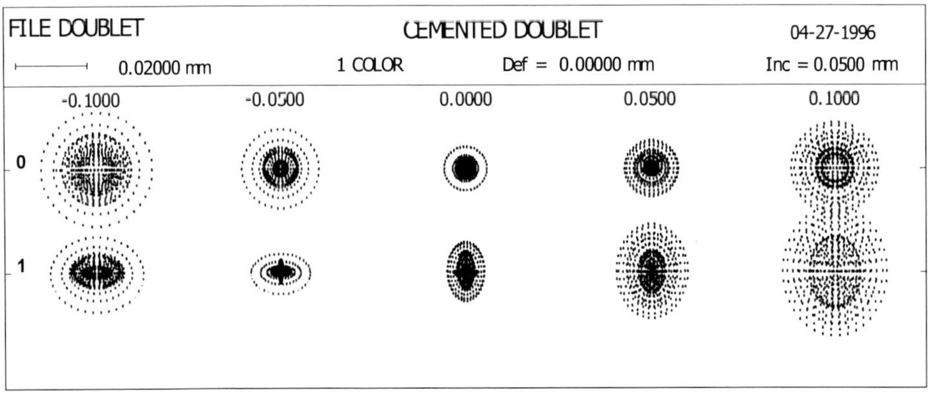

Figure 8.5. Spot diagrams for a Fraunhofer doublet.

The spot diagrams (Fig. 8.5) reveal what is hard to see in the sectional transverse aberration plots of Fig. 8.4. These give the ray intersection patterns, on axis, and at 1-deg semi-field, both at the image plane and at four other focal planes, defocused at intervals of 0.05 mm. The scale bar on the left indicates a distance of 0.02 mm. Even at 1-deg off-axis, one can see that aberrations are affected by astigmatism in that the ray patterns are elongated when one compares the on-axis (0 deg) to the off-axis (1 deg) patterns.

In the spot diagrams, the three wavelengths (*d*, *C*, and *F*) are superimposed, so it is not easy to tell what the individual monochromatic spot diagrams would be like. Note that there is very little evidence of coma, which would show as a quadratic component in the transverse ray aberration curve, and as a "comet" shape in the off-axis spot diagram.

Given that the maximum transverse ray aberration appears to be about 0.005 mm, we can expect a total blur circle at the edge of the field of diameter twice this, i.e. about 0.01 mm. This indicates that the off-axis MTF will fall to a small value at a spatial frequency of about the reciprocal of this figure—namely at 100 cycles/mm. The tangential MTF curve at 1 deg, shown in Fig. 8.6, agrees with this prediction.

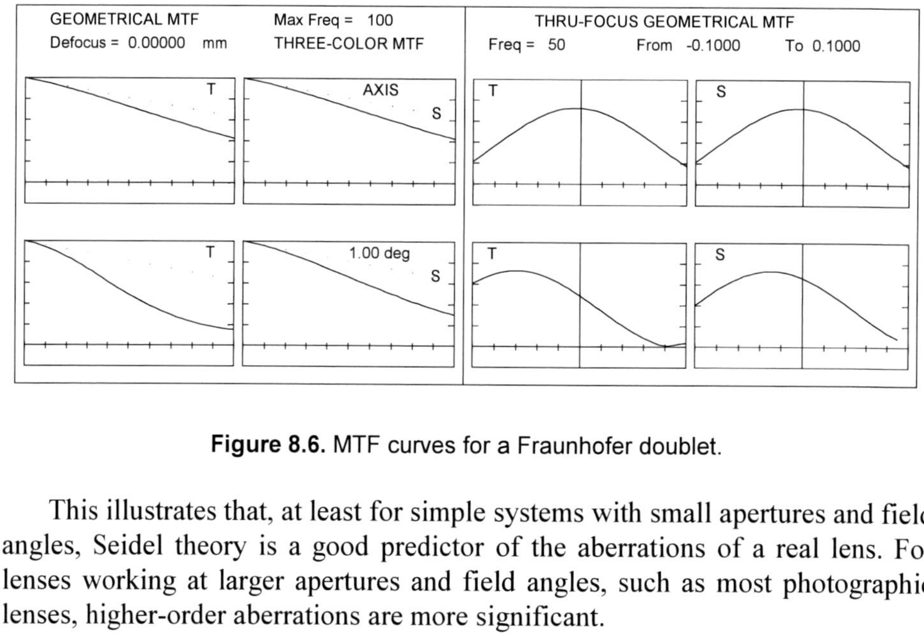

Figure 8.6. MTF curves for a Fraunhofer doublet.

This illustrates that, at least for simple systems with small apertures and field angles, Seidel theory is a good predictor of the aberrations of a real lens. For lenses working at larger apertures and field angles, such as most photographic lenses, higher-order aberrations are more significant.

8.2.3 Flint-first doublet

There is no reason why the crown component has to always be at the front. In Fig. 8.7 we show a flint-first doublet, often known as a Steinheil doublet. The aberrations of these two types of cemented doublets are not very different, but the Fraunhofer type is almost always preferred because of the shallower curvature of the cemented surfaces. In manufacturing lenses with smaller curvatures (larger radii), it is possible to polish more glass components at the same time, giving an economic benefit. As before, the table below gives a prescription of a typical doublet, a paraxial ray analysis, and a table of the third-order aberrations. If you compare transverse ray aberration curves (Fig. 8.8), spot diagrams (Fig. 8.9), and MTF curves (Fig. 8.10), you will find that the performance of this crown-first doublet is similar to that of the Fraunhofer doublet.

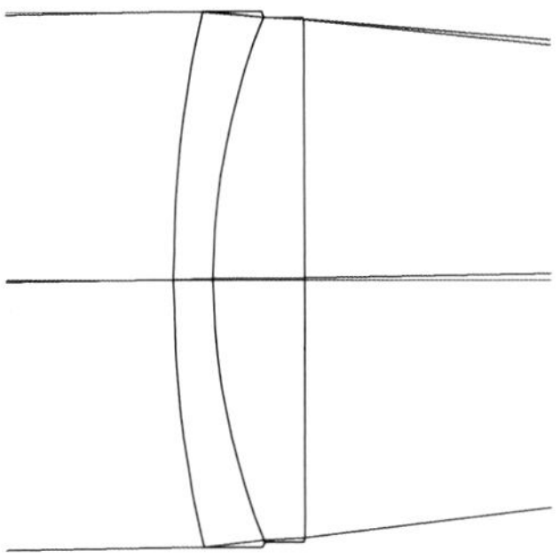

Figure 8.7. Steinheil doublet.

EFL = 100.000

WAVELENGTHS [nm]

587.60 656.30 486.10

SURFACES

#	SURF SPACE	RADIUS	SEPN	INDEX1	V	CLR RAD	GLASS
0			0.00000	1.000000			
1	S	44.65500	1.50000	1.672693	32.20	10.003	S-SF5
2	S	25.19100	3.50000	1.539958	59.68	9.799	S-BAK2
3	#S	-3892.33500	96.40666	1.000000		9.688	
4	S	Plane				1.763	

LAGRANGE INVARIANT = -0.1746

#	H	U	HBAR	UBAR	D(U/N)	A	ABAR
1	10.00000	0.00000	0.00000	0.01746	-0.05384	0.22394	0.01746
2	9.86491	-0.09006	0.01565	0.01044	0.01224	0.50439	0.01849
3	9.64067	-0.06407	0.05551	0.01139	-0.05840	-0.10248	0.01752
		-0.100000		0.017530			

#	S1	S2	S3	S4	S5	C1	C2
1	0.027001	0.002105	0.000164	0.000274	0.000034	0.027972	0.002180
2	-0.030712	-0.001126	-0.000041	-0.000062	-0.000004	-0.032920	-0.001207
3	0.005912	-0.001011	0.000173	0.000003	-0.000030	0.005804	-0.000992
Sum	0.002201	-0.000032	0.000295	0.000215	0.000000	0.000855	-0.000019

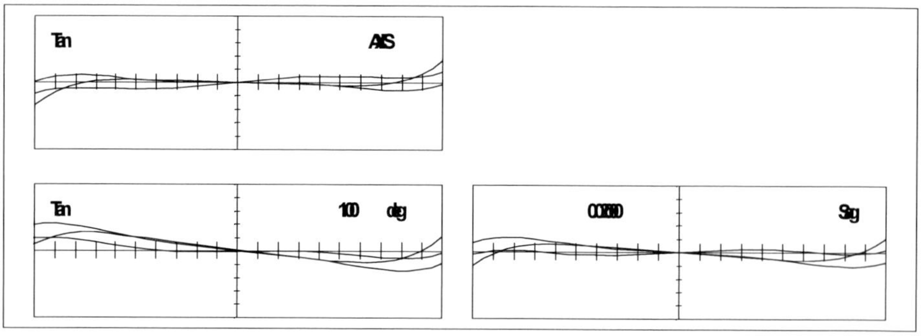

Figure 8.8. Transverse aberration curves for a Steinheil doublet.

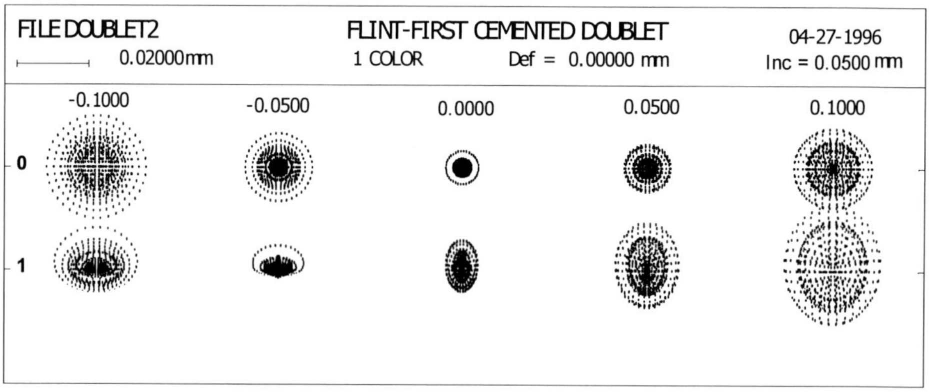

Figure 8.9. Spot diagrams for a Steinheil doublet.

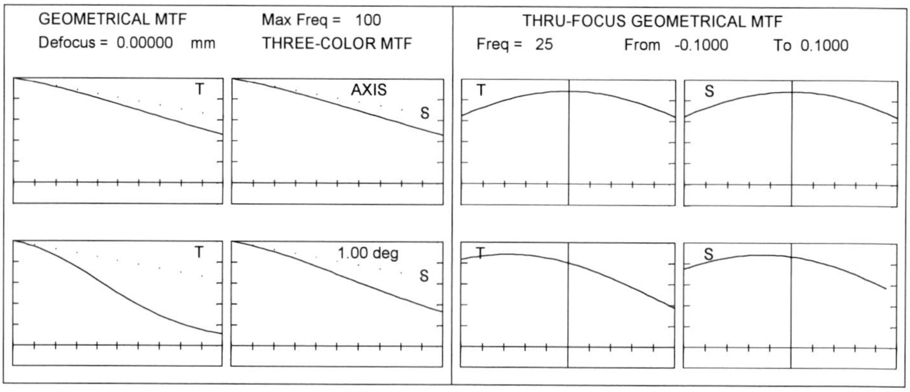

Figure 8.10. MTF curves of a Steinheil doublet.

8.3 The split doublet

If the two inner surfaces are not cemented they can, of course, have different curvatures, so that an extra design parameter becomes available. The effect of this is that the two shapes can be used independently to control spherical aberration and coma. Also, the glass choice becomes less critical.

It is not difficult to find analytic solutions in this case, using the equations for the spherical aberration and coma of a thin lens. One advantage of the analytic approach is that all possible solutions are found. In this case we find that there are two solutions for the crown-before-flint choice of glass types (Fig. 8.11). One of these is known as the Fraunhofer doublet, by analogy to its cemented equivalent. The other, in which the flint has both surfaces convex to the front, is known as the Gauss doublet. It can be seen that this latter solution has much stronger curves, and in general, steep curves imply large higher-order aberrations. For the Gauss doublet this is indeed the case.

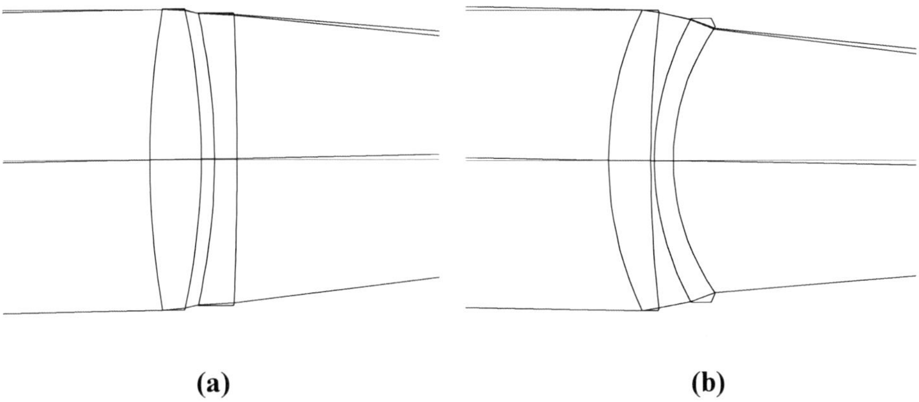

Figure 8.11. Split doublets: (a) Fraunhofer, (b) Gauss.

The optimization of split doublets is usually very simple, as all four curves can be variable, and this is generally sufficient to ensure that the four conditions ($S_1 = 0$, $S_2 = 0$, $C_1 = 0$, $f' = 100$) are satisfied. In some cases the air space may also usefully be treated as a variable; this sometimes helps to reduce higher-order spherical aberration and spherochromatism.

8.3.1 The split Fraunhofer doublet

The design for a split Fraunhofer doublet is given below. Note that the transverse ray aberration curves, plotted at the same scale as the earlier two doublets, are flatter and the spot diagram at the image plane is tighter and more symmetrical than in the cemented doublet cases.

The aberrations shown in Figs. 8.12 and 8.13 are very similar to the aberrations of the cemented doublet, but the high-order spherical aberration is smaller and the spherochromatism is perhaps a little less. The high-order spherical aberration is reduced because the negative lens operates in the aberrated beam of the positive lens. The effect of the air space is to change the ray heights at the negative lens so that the aberration in the wavefront on the negative lens can be canceled. This is an example of what is known as an "induced aberration," as opposed to an "intrinsic aberration."

Figure 8.14 shows the effect of these aberrations on MTF.

EFL = 100.001

WAVELENGTHS [nm]
587.60 656.30 486.10

SURFACES

# SURF SPACE	RADIUS	SEPN	INDEX1	V	CLR RAD	GLASS
0		0.00000	1.000000			
1#S	57.36300	3.50000	1.563838	60.77	10.000	S-SK11
2 S	-45.66600	0.90000	1.000000		9.919	
3 S	-43.88000	1.50000	1.666796	32.99	9.719	S-SF19
4 S	-238.80500	95.16853	1.000000		9.639	
5 S	Plane				1.763	

LAGRANGE INVARIANT = -0.1746

#	H	U	HBAR	UBAR	D(U/N)	A	ABAR
1	10.00006	0.00000	0.00000	0.01746	-0.04019	0.17433	0.01746
2	9.78008	-0.06285	0.03907	0.01116	-0.17886	-0.43321	0.01612
3	9.58293	-0.21905	0.05434	0.01697	0.19262	-0.43744	0.01573
4	9.51685	-0.04405	0.07036	0.01068	-0.07357	-0.13985	0.01731
		-0.100000		0.017602			

#	S1	S2	S3	S4	S5	C1	C2
1	0.012215	0.001223	0.000122	0.000192	0.000031	0.010342	0.001036
2	0.328284	-0.012213	0.000454	0.000241	-0.000026	0.025135	-0.000935
3	-0.353207	0.012705	-0.000457	-0.000278	0.000026	-0.050830	0.001828
4	0.013694	-0.001695	0.000210	0.000051	-0.000032	0.016139	-0.001997
Sum	0.000986	0.000019	0.000330	0.000205	-0.000000	0.000786	-0.000069

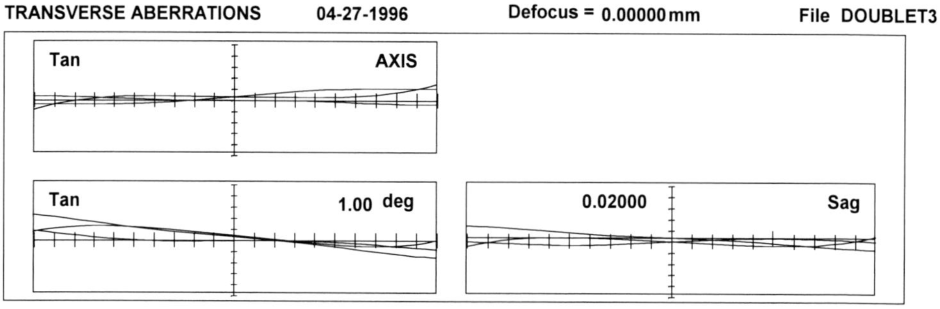

Figure 8.12. Transverse ray aberrations of a split Fraunhofer doublet.

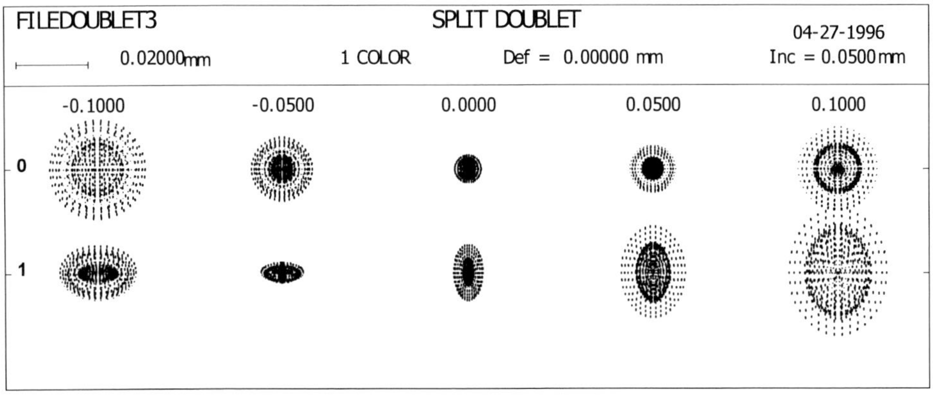

Figure 8.13. Spot diagrams of a split Fraunhofer doublet.

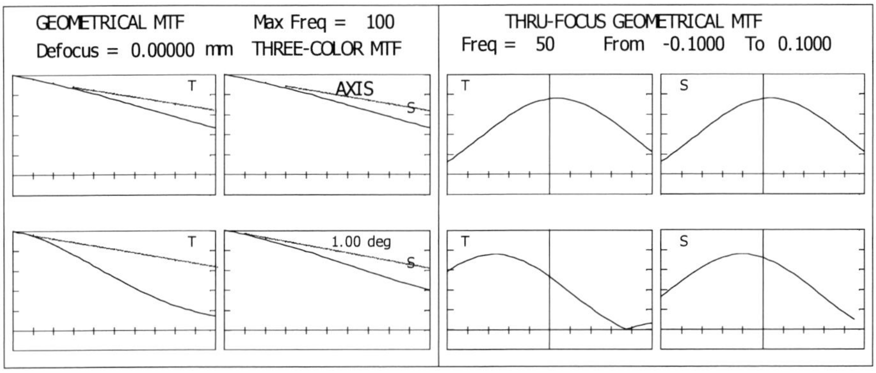

Figure 8.14. MTF curves of a split Fraunhofer doublet.

8.3.2 The split Gauss doublet

An example of the other air-spaced doublet solution, the Gauss doublet, is shown below in Fig. 8.15.

From the value for S_2, the very pronounced U-shaped tangential curve (Fig. 8.16) at 1 deg, and the comet-shaped spot diagrams (Fig. 8.17) at 1 deg, it is clear that there is a large amount of coma at a very small field angle. This is difficult to correct and degrades the off-axis MTF, as shown in Fig. 8.18. Therefore, the Gauss doublet is much worse than the Fraunhofer doublet. With this choice of glasses, the coma can, in fact, be corrected, but only if the axial color is left uncorrected. A different choice of glasses may result in a better design.[1]

EFL = 99.987

WAVELENGTHS [nm]

587.60 656.30 486.10

SURFACES

# SURF SPACE	RADIUS	SEPN	INDEX1	V	CLR RAD	GLASS
0		0.00000	1.000000			
1#S	23.01300	2.85700	1.589128	61.24	9.999	S-SK5
2 S	86.67900	0.25000	1.000000		9.825	
3 S	19.37800	1.25000	1.805176	25.42	9.055	S-SF6
4 S	15.06800	90.85289	1.000000		8.217	
5 S	Plane				1.823	

LAGRANGE INVARIANT = -0.1745

#	H	U	HBAR	UBAR	D(U/N)	A	ABAR
1	9.99868	0.00000	0.00000	0.01746	-0.10136	0.43448	0.01746
2	9.53849	-0.16107	0.03138	0.01098	-0.08978	-0.08109	0.01803
3	9.49071	-0.19113	0.03580	0.01767	0.01146	0.29863	0.01952
4	9.08529	-0.32434	0.04700	0.00896	0.07967	0.50295	0.02181
		-0.100000		0.018693			

#	S1	S2	S3	S4	S5	C1	C2
1	0.191312	0.007686	0.000309	0.000491	0.000032	0.026297	0.001056
2	0.005631	-0.001252	0.000278	-0.000130	-0.000033	0.004682	-0.001041
3	-0.009703	-0.000634	-0.000041	0.000701	0.000043	0.049729	0.003250
4	-0.183099	-0.007941	-0.000344	-0.000902	-0.000054	-0.080175	-0.003477
Sum	0.004140	-0.002141	0.000201	0.000160	-0.000012	0.000534	-0.000212

Fig. 8.15. The Gauss doublet.

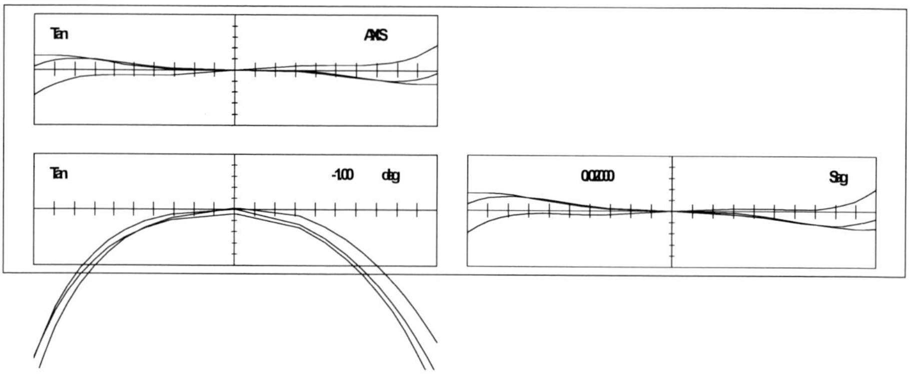

Figure 8.16. Transverse ray aberration curves for the split Gauss doublet.

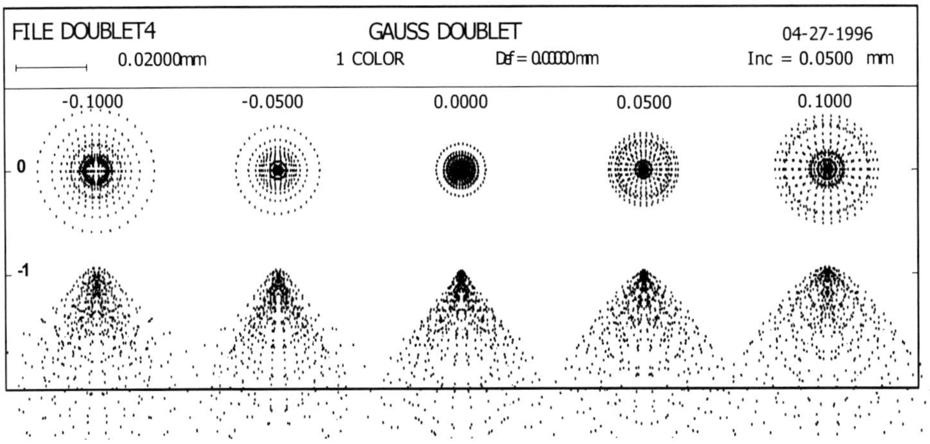

Figure 8.17. Spot diagrams for the split Gauss doublet.

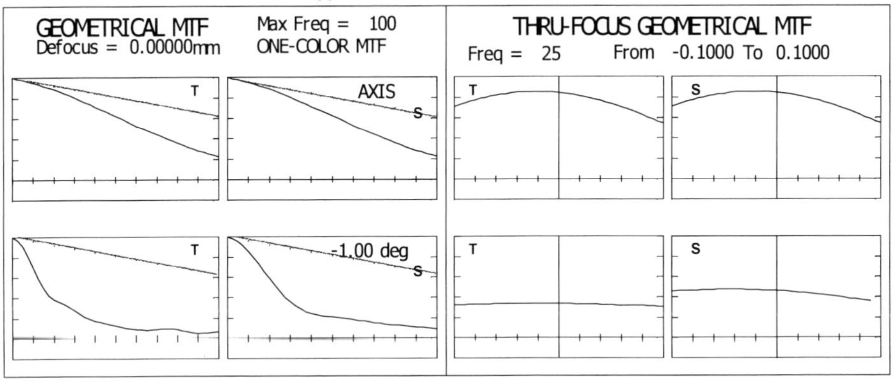

Figure 8.18. MTF curves for the split Gauss doublet.

Imagine that one were to enter some arbitrary values for the curvatures of an air-spaced doublet into a lens design program, assign weights to the aberrations, and then optimize, allowing all curvatures to vary, using a "local gradient" type of optimization method. One would end up with a solution of either the Fraunhofer or the Gauss type, depending mainly on the initial curves chosen. Although the above analysis shows the Fraunhofer doublet to be the better lens, the Gauss design is quite good. If the starting lens resembled a Gauss doublet, then that would be the result given, and no amount of further optimization would cause the second lens to bend to the shape of the Fraunhofer solution. The reason is that an optimization program of this type explores only small "steps" in parameter space before making a change. Once a local aberration minimum has been found, a better one can only be reached by passing through a region where the aberrations, and therefore the merit function, are worse. The Gauss solution to the air-spaced-doublet is a classic example of what is known as a "local minimum." Of course, in this case, the existence of the two solutions is predictable from simple thin-lens theory. In lenses with more elements, however, the old strategies based on experience or intuition are to some extent being replaced by what are known as "global" optimization methods.

8.4 General limitations of doublets

Apart from astigmatism and field curvature, which have been discussed above, doublet objectives may also suffer from secondary spectrum or fifth-order spherical aberration. Since the wavefront aberration associated with fifth-order spherical aberration depends on sixth power of the aperture, lenses with high apertures may be affected by fifth-order spherical aberration. On the other hand, lenses with low apertures operating over broad spectral wavelength bands may be limited by secondary spectrum, particularly if the focal length is large, unless unusual glass types are employed.

As we have seen, the split doublet is easier to design because this gives one more variable than in a cemented doublet. In addition, the split doublet can have a higher performance because varying the air space can be used to correct, or at least reduce, fifth-order spherical aberration. However, from a manufacturing point of view the cemented doublet is preferable, because the cemented surface has much smaller aberrations than the air-glass surfaces in the split doublet. This implies that the manufacturing tolerances will be significantly easier. In addition, there is the obvious point that a cemented doublet is more durable, and less liable to go out of adjustment; the central air space of a split doublet is usually a very sensitive design parameter and subject to loss of alignment during service under conditions of shock or vibration.

References

1. W. J. Smith, *Modern Lens Design, 2nd Ed.,* SPIE Press, Bellingham, Washington (2004).

CHAPTER 9
PETZVAL LENSES AND TELEPHOTO OBJECTIVES

The doublets described in the last chapter are limited to small apertures (up to about *f*/4) and a small field of view. If we require a larger field angle and aperture then the Petzval lens, which consists of two positive groups separated by an air space, is a simple solution. Joseph Petzval designed his famous "Portrait Lens" in 1839. The front lens was a cemented doublet, the rear component an air-spaced doublet. This was probably the first lens ever to be designed by computation, in this case by logarithmic tables.

The limitation of the doublet objective is that astigmatism cannot be corrected in a single component, and this severely limits the field angle that it can cover. We have seen from earlier chapters that for a thin lens at the stop, the astigmatism is given by

$$S_3 = H^2K, \tag{7.8}$$

which must always be non-zero. Hence, astigmatism can only be corrected if there are one or more lenses that are **not** at the stop. Thus, two separated lenses, as shown in the Fig. 9.1, may be employed to generate equal and opposite amounts of astigmatism. In practice, these will not always be single lenses, but achromatic doublets or more complex combinations.

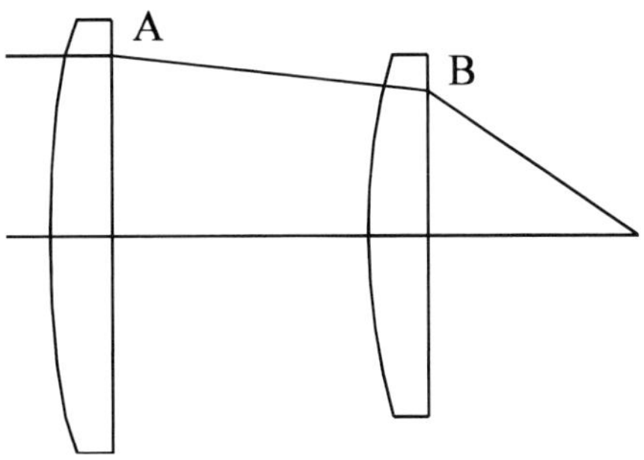

Figure 9.1. A simple Petzval lens.

9.1 Seidel analysis

If we assume that the stop is in contact with the first lens and that the second lens is separated from it, then there will be stop-shift effects, as described in Chapter 6. As long as the second lens is not corrected for both spherical aberration and coma, moving the second lens away from the stop can result in sufficient astigmatism to cancel that of the first lens.

Although we can obtain almost any value of astigmatism, S_3, the two positive lenses will each contribute to the field curvature, so that P, the Petzval sum of the Petzval lens, will inevitably be positive. This means that the image surface will be concave toward the lens. If, as is often the case, we want zero astigmatism, then by making the total S_3 to be zero, the field curvature would cause the both tangential and sagittal foci to lie on a curved surface. However, other options are available. From the equation for the sagittal image error, Eq. (6.91), we can make the sagittal image surface to be flat, provided $S_3 = -S_4$. Alternatively if we choose $S_3 = -S_4/3$, the tangential image surface will be flat, as shown in Fig. 6.15.

A good rule of thumb to use in the design of a Petzval lens is to make the powers $K/2$ for the first lens (A) and K for the second (B), where the total power is K; and make the separation between them $1/K$. It can be shown that this will more or less equalize the deviation of the axial marginal ray at the two lenses, hence minimizing the spherical aberration, coma, and astigmatism contributed by either lens.

Chromatic aberration must also be corrected. One strategy that works well is to correct each lens individually, by making each component an achromatic doublet.

Suppose that we design each component to have zero S_1. Then the total coma will be given by

$$\Sigma S_2 = S_{2\mathrm{A}} + S_{2\mathrm{B}}; \tag{9.1}$$

and, if the stop is at the first lens the total astigmatism will, using the stop-shift equations, be given by

$$\Sigma S_3 = H^2 K_\mathrm{A} + H^2 K_\mathrm{B} + 2(HE_\mathrm{B})\, S_{2\mathrm{B}}, \tag{9.2}$$

Suppose now that we want $\Sigma\, S_3$ to be zero. Then,

$$S_{2\mathrm{B}} = -H(K_\mathrm{A} + K_\mathrm{B})/2E_\mathrm{B} = -0.75\ HK/E_\mathrm{B}. \tag{9.3}$$

Because we want ΣS_2 to be zero, we must make $S_{2\mathrm{A}} = -S_{2\mathrm{B}}$. We, therefore, need two different doublets, each with zero S_1 and with equal and opposite amounts of S_2. The problems of designing these doublets are similar to the problems of designing doublet objectives described in Chapter 8. If the components are cemented, the choice of glasses will be fairly critical; but if they are split, a larger variety of glass types can be used.

Note that once we have achieved a design with zero S_1 and S_2, there is no need for the stop to be at the first lens. However, in practice it is quite a good position, since minimizing the diameter of the first lens is often a design constraint, since it is the front lens that is most often exposed to the vicissitudes of the outside environment.

9.1.1 Calculation of predicted transverse aberrations from Seidel coefficients

Let us take, as an example, an objective in air with focal length $f' = 100$ mm ($K = 0.01$ mm^{-1}), a semi-field angle of 10 deg, and an aperture of $f/2$. In this case, the Lagrange invariant is

$$H = n'u'\eta' = (1)\,(0.25)\,(100 \tan 10\ \text{deg}) = 4.4\ \text{mm}.$$

Then, the field curvature coefficient, S_4, will be given by

$$\begin{aligned} S_4 &= H^2 \Sigma\, K/n \\ &= H^2\,(K_A + K_B)/n \\ &= H^2(K/2 + K)/n \\ &= H^2(K/2 + K)/1.5 \\ &= H^2K. \end{aligned} \tag{9.4}$$

If we correct the tangential astigmatism, then

$$(3\,S_3 + S_4) = 0$$

or

$$S_3 = -\,S_4/3 = -\,H^2K/3. \tag{9.5}$$

Then, the transverse ray aberration in the sagittal section will be given by

$$\begin{aligned} \delta\xi' &= (S_3 + S_4)/(2n'u') \\ &= (S_3/\,3 + S_4)/(2u') \\ &= (2/3)\,H^2K/(2u') \\ &= u'\eta'^2K/3 \\ &= -(0.25)\,(17)2\,(0.01)/3 \\ &= -0.24\ \text{mm}. \end{aligned} \tag{9.6}$$

We should therefore expect an image blur of twice this (i.e. 0.48 mm). This is not very good, but it is much better than the blur of an equivalent doublet, considering that the doublets in the previous chapter had similar blur spot dimensions for a 1 deg field, and the field here is 10 times as large.

9.2 Optimization

Optimization of Petzval systems is straightforward, so long as there are enough design parameters to work with. Naturally, all of the curves are normally allowed to be variable. It is also useful if the air space is variable. Note that any movement of the stop away from the front lens will increase the diameter of this lens. In general, there should be no need to vary the glass thicknesses, provided that they are thick enough not to give mechanical problems. The designer must choose between a flat field with some astigmatism, or a curved field that is anastigmatic. To obtain the latter—known as an anastigmat—either the image curvature is set to an appropriate non-zero value, or it is included as a design variable during the optimization.

9.3 Examples

9.3.1 Simple Petzval lens with two doublets

This example, shown in Fig. 9.2, has been optimized with a flat field, and the transverse ray aberration curves of Fig. 9.3 show that the Seidel aberrations do describe the real-ray performance with reasonable accuracy, despite the fact that *f*/2.3 is a reasonably high-aperture and one might have expected that high-order aberrations would be dominant.

Note that all of the curves are simple S-curves indicating that there is spherical aberration present, but very little coma or higher-order aberrations. The fact that the curves get shorter as the field increases is an indication of vignetting, which is always a function of field angle. Vignetting is the effect where two lenses are separated by a stop, and at higher angles the second lens aperture cuts off rays transmitted by the first. It should be obvious from an examination of the tangential curves that the vignetting has a beneficial effect, for if there were no vignetting the rays in the tangential section at the top and bottom of the pupil will suffer from large amounts of aberration, and the edges of the image would be poorly resolved. The trade-off is that the image at the edge of the field is less bright than the image close to the axis.

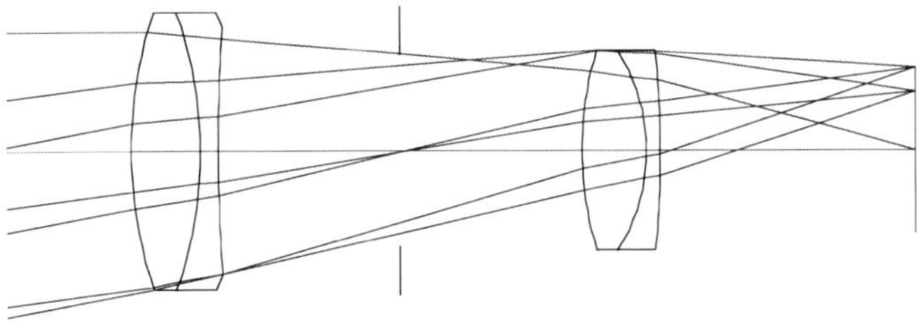

Figure 9.2. Petzval lens made of two cemented doublets.

EFL = 100.005

WAVELENGTHS [nm]

546.00 650.00 475.00

SURFACES

#	SURF SPACE	RADIUS	SEPN	INDEX1	V	CLR RAD	GLASS
0			0.00000	1.000000			
1	S	89.98583	15.00000	1.567505	40.53	27.817	S-LF8
2	S	-84.19189	4.00000	1.677649	29.80	25.976	S-SF5
3	S	376.69338	40.00000	1.000000		20.221	
4#	S	Plane	40.00000	1.000000		17.305	
5	S	71.13013	14.00000	1.626901	43.56	20.653	S-BAF8
6	S	-37.24710	3.00000	1.694169	28.84	20.697	S-SF8
7	S	-209.61078	56.14635	1.000000		17.317	
8	S	Plane				17.435	

LAGRANGE INVARIANT = -3.8003

#	H	U	HBAR	UBAR	D(U/N)	A	ABAR
1	21.55260	0.00000	-11.93256	0.17633	-0.05532	0.23951	0.04372
2	20.25190	-0.08671	-9.52510	0.16050	0.01644	-0.51298	0.42892
3	19.99099	-0.06523	-8.95497	0.14253	-0.03459	-0.02040	0.19924
4	17.05233	-0.07347	-0.03456	0.22301	0.00000	-0.07347	0.22301
5	14.11367	-0.07347	8.88585	0.22301	-0.00129	0.12495	0.34793
6	12.41106	-0.12162	10.13100	0.08894	0.01363	-0.73995	-0.29781
7	12.10039	-0.10356	10.41962	0.09621	0.15439	-0.27324	0.07878
		-0.215515		0.128485			

#	S1	S2	S3	S4	S5	C1	C2
1	0.068396	0.012485	0.002279	0.058106	0.011023	0.046107	0.008417
2	-0.087607	0.073252	-0.061249	-0.007185	0.057220	-0.048042	0.040170
3	0.000288	-0.002810	0.027446	-0.015486	-0.116821	0.005528	-0.053994
4	0.000000	0.000000	0.000000	0.000000	0.000000	0.000000	0.000000
5	0.000283	0.000789	0.002198	0.078239	0.223975	0.015601	0.043442
6	-0.092606	-0.037271	-0.015001	-0.009463	-0.009846	-0.049247	-0.019821
7	0.139481	-0.040212	0.011593	0.028231	-0.011481	0.046980	-0.013544
Sum	0.028235	0.006233	-0.032733	0.132442	0.154070	0.016927	0.004669

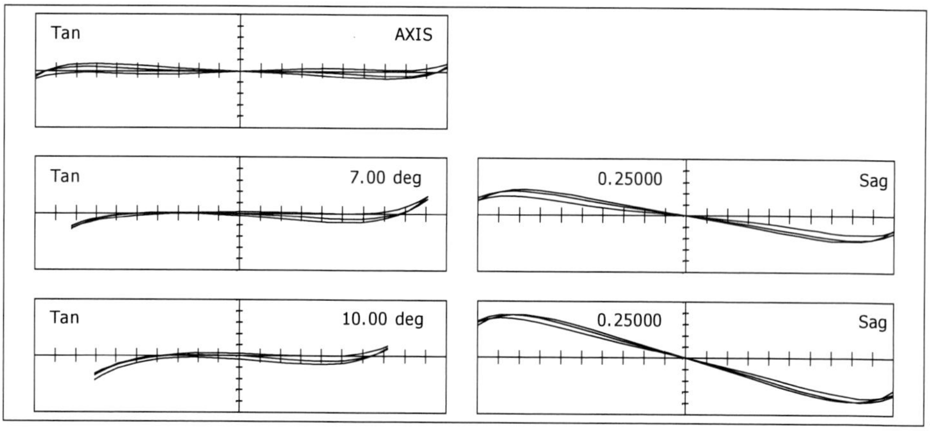

Figure 9.3. Transverse ray aberrations of a Petzval lens made of cemented doublets.

The difference between the central slopes of the tangential and sagittal section curves indicates that there is a fair amount of astigmatism. Because the slope of the tangential curves change hardly at all between the axis and 10-deg field, the field curvature here is rather modest. The tight overlap of all three-color curves tells us that there is good color correction.

Looking at the aberrations of the full-field skew rays, we can see that although the tangential image surface is quite flat, the sagittal image is curved. The aberration in this section at the edge of the pupil is about 0.20 mm, compared with the paraxial value of 0.24 mm. (This paraxial value can be obtained from bottom right-hand curve of Fig. 9.3 by projecting a straight line that passes through the origin, and with the same slope as the mean-wavelength aberration curve at the origin, and noting the intersection height of this line at the edge of the pupil). This difference between the third-order approximation and the finite ray result is due almost entirely to some high-order sagittal oblique spherical aberration (SOBSA), which reduces the aberration at full aperture. At smaller apertures the curves are almost straight lines. The meridian rays show small aberrations at low apertures, but just inside the vignetted edge there is evidence of some high-order tangential oblique spherical aberration (TOBSA).

More detailed analysis of the performance of this design is given by the modulation transfer function curves in Fig. 9.4.

Note that there are several approaches by which this design may be improved further. One is to split one or both of the doublets to provide air gaps. This will have no effect on the field curvature, S_4, so its only benefit would be to reduce the astigmatism S_3. Another approach is to try different glass types in an attempt to reduce S_4. Two other options are discussed in the sections that follow.

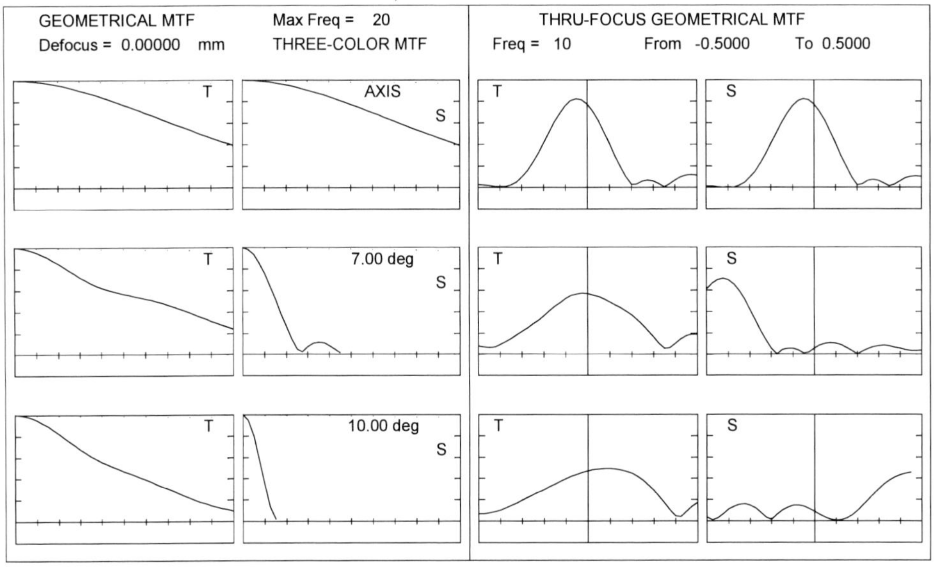

Figure 9.4. MTF curves of a Petzval lens made of cemented doublets.

9.3.2 Petzval lens with curved image surface

One solution to the problem of field curvature in Petzval lenses is to intentionally permit the image to lie on a curved surface (Fig. 9.5). For example, suppose that we have two doublets with focal lengths of 200 mm and 100 mm respectively. Assuming a mean refractive index of 1.6, the Petzval sum using a thin lens approximation will be

$$P = K_1/1.6 + K_2/1.6$$
$$= 0.005/1.6 + 0.01/1.6 \approx 0.01 \text{ mm}^{-1},$$

giving a Petzval surface with a radius of –100 mm.

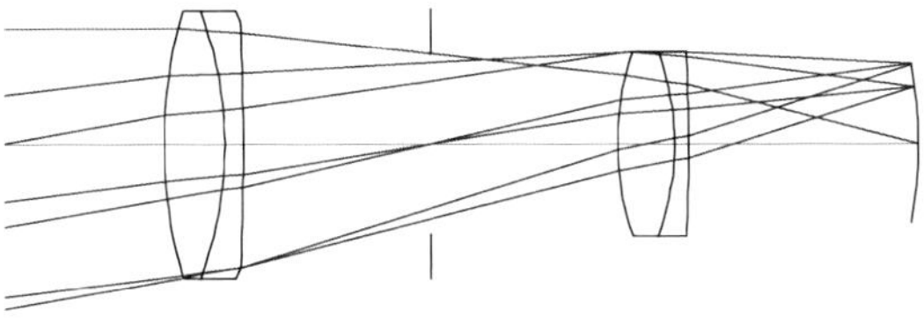

Figure 9.5. Petzval lens with curved image surface.

The example given in the table below shows a Petzval lens optimized for a curved image surface, and we see that the optimum image surface radius (listed as the radius for surface 8) is actually 93.6 mm, the minus sign indicating that it is concave toward the lens.

EFL = 100.000

WAVELENGTHS [nm]
546.00 650.00 475.00

SURFACES

#	SURF SPACE	RADIUS	SEPN	INDEX1	V	CLR RAD	GLASS
0			0.00000	1.000000			
1	S	105.34415	13.00000	1.533172	57.78	28.760	S-BK6
2	S	-82.28685	4.00000	1.668862	33.15	26.558	S-BASF2
3	S	-568.00270	40.00000	1.000000		19.456	
4#	S	Plane	40.00000	1.000000		15.685	
5	S	63.45143	12.00000	1.591429	56.94	19.893	S-SK5
6	S	-59.82517	3.00000	1.677649	29.80	19.631	S-SF5
7	S	-463.06351	48.86124	1.000000		17.173	
8	S	-93.55990				17.312	

LAGRANGE INVARIANT = -3.5784

#	H	U	HBAR	UBAR	D(U/N)	A	ABAR
1	20.29419	0.00000	-11.87648	0.17633	-0.04370	0.19265	0.06359
2	19.42326	-0.06699	-9.87170	0.15421	0.01832	-0.46461	0.42037
3	19.25384	-0.04236	-9.34401	0.13192	-0.06798	-0.12726	0.24761
4	15.51953	-0.09336	-0.09755	0.23116	0.00000	-0.09336	0.23116
5	11.78521	-0.09336	9.14892	0.23116	0.01312	0.09238	0.37535
6	10.25295	-0.12769	10.24895	0.09167	0.01329	-0.47595	-0.12675
7	9.91599	-0.11232	10.53624	0.09576	-0.13599	-0.22436	0.12248
		-0.202942		0.145237			

#	S1	S2	S3	S4	S5	C1	C2
1	0.032911	0.010863	0.003586	0.042271	0.015136	0.023529	0.007766
2	-0.076797	0.069484	-0.062868	-0.008253	0.064348	-0.054797	0.049579
3	0.021195	-0.041242	0.080247	0.009035	-0.173726	0.029624	-0.057642
4	0.000000	0.000000	0.000000	0.000000	0.000000	0.000000	0.000000
5	-0.001320	-0.005362	-0.021789	0.074999	0.216203	0.007105	0.028870
6	-0.030856	-0.008217	-0.002188	-0.006912	-0.002424	-0.034305	-0.009136
7	0.067877	-0.037057	0.020230	0.011170	-0.017142	0.030159	-0.016465
Sum	0.013010	-0.011531	0.017219	0.122311	0.102395	0.001315	0.002973

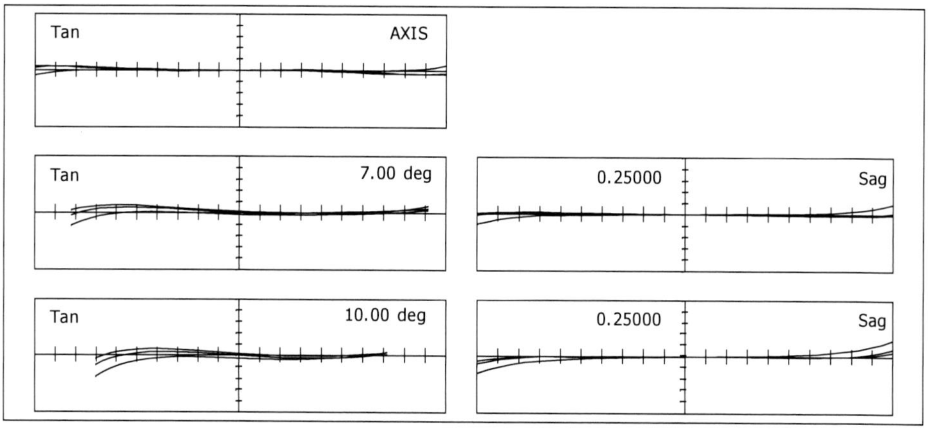

Figure 9.6. Transverse ray aberrations of Petzval lens with a curved image surface.

The aberrations of this design are shown in the curves of Fig. 9.6, and the MTF curves are given in Fig. 9.7.

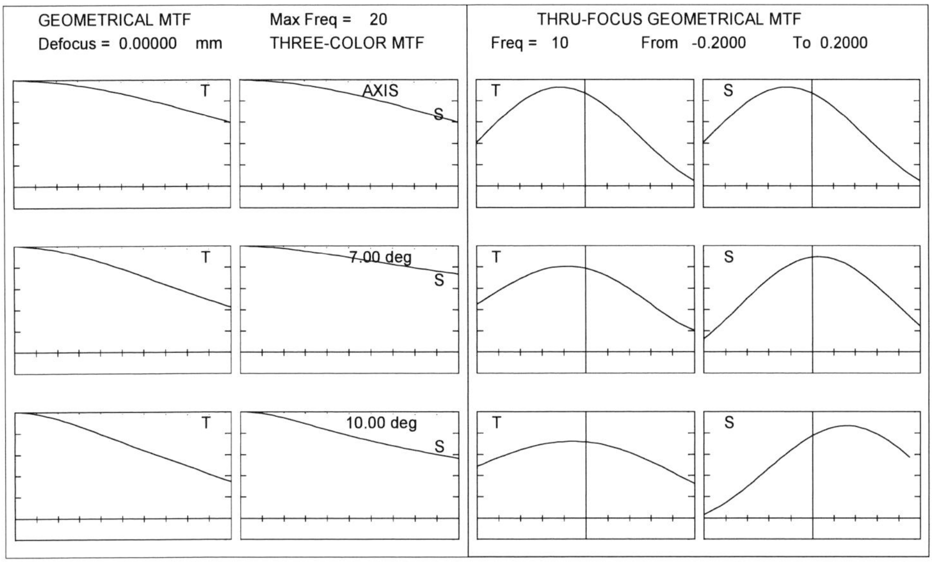

Figure 9.7. MTF curves of Petzval lens with a curved image surface.

9.3.3 Petzval lens with field flattener

The second solution to the problem of field curvature is to add a negative lens near the image plane. This will have negative Petzval sum, and because the marginal ray height is small, all the other aberrations will be small. (Note that if the field flattener were in contact with the image, any dust particles or scratches on the glass would be in focus; for this reason the field flattener is usually placed a small distance from the image surface).

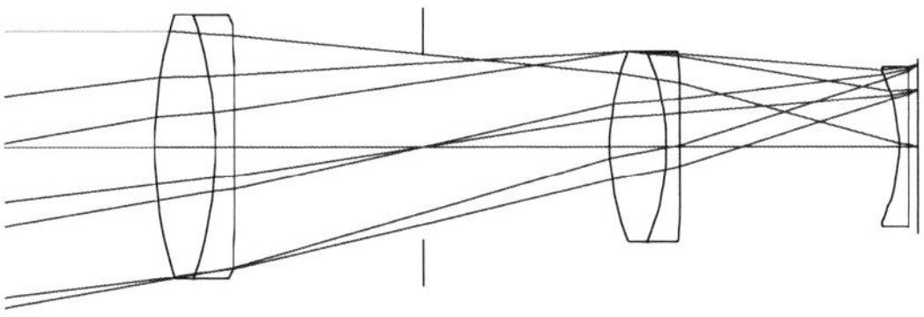

Figure 9.8. Petzval lens with field flattener.

This design is most commonly used for objectives with apertures around *f*/2, and at semi-field angles of 10 deg or less. We can see from the aberration curves in Fig. 9.9 that the aberrations are much better than those of the simple Petzval objective. It might seem at first sight that the field angle could be opened up to

15 or even 20 deg. However, the practicable field of view of the field-flattened Petzval is limited by the obliquity of edge-of-field rays on the two surfaces of the field flattener. As a general rule, large angles of incidence of any ray on a steep surface is a sure sign that the aberrations may be excessive and that manufacturing tolerances will be tight.

The aberrations of the design shown in the transverse ray aberration curves of Fig. 9.9 and the MTF curves in Fig. 9.10 are similar to those of the curved-field Petzval described in Sec. 9.3.2.

EFL = 100.000

WAVELENGTHS [nm]

546.00 650.00 475.00

SURFACES

#	SURF SPACE	RADIUS	SEPN	INDEX1	V	CLR RAD	GLASS
0			0.00000	1.000000			
1	S	99.56266	13.00000	1.518726	59.69	28.478	S-BK7
2	S	-86.84002	4.00000	1.674039	36.28	26.267	S-BASF12
3	S	-1187.63858	40.00000	1.000000		20.020	
4	#S	Plane	40.00000	1.000000		16.631	
5	S	57.47191	12.00000	1.609942	52.60	20.543	S-SK2
6	S	-54.61865	3.00000	1.677649	29.80	20.074	S-SF5
7	S	-614.68633	46.82210	1.000000		16.492	
8	S	-38.17110	2.00000	1.677649	29.80	17.297	S-SF5
9	S	Plane	1.98179	1.000000		18.940	
10	S	Plane				17.904	

LAGRANGE INVARIANT = -3.6850

#	H	U	HBAR	UBAR	D(U/N)	A	ABAR
1	20.89845	0.00000	-11.61031	0.17633	-0.04721	0.20990	0.05971
2	19.96644	-0.07169	-9.58320	0.15593	0.02110	-0.45807	0.40442
3	19.79160	-0.04371	-9.05830	0.13123	-0.05829	-0.10107	0.23245
4	16.41543	-0.08440	-0.06555	0.22482	0.00000	-0.08440	0.22482
5	13.03925	-0.08440	8.92719	0.22482	-0.00155	0.14248	0.38015
6	11.37866	-0.13838	9.89674	0.08080	0.01181	-0.55819	-0.16164
7	11.00549	-0.12439	10.15128	0.08485	-0.14667	-0.23872	0.11464
8	0.66644	-0.22082	16.29213	0.13115	0.14656	-0.23827	-0.29567
9	0.41730	-0.12457	16.79329	0.25058	-0.13473	-0.20898	0.42039
		-0.208984		0.420386			

#	S1	S2	S3	S4	S5	C1	C2
1	0.043466	0.012365	0.003518	0.046583	0.014253	0.025101	0.007141
2	-0.088381	0.078029	-0.068889	-0.009552	0.069253	-0.049165	0.043406
3	0.011785	-0.027105	0.062337	0.004604	-0.153956	0.022199	-0.051054
4	0.000000	0.000000	0.000000	0.000000	0.000000	0.000000	0.000000
5	0.000410	0.001095	0.002922	0.089513	0.246634	0.013382	0.035706
6	-0.041869	-0.012124	-0.003511	-0.006232	-0.002822	-0.040350	-0.011685
7	0.091987	-0.044174	0.021213	0.008923	-0.014472	0.035615	-0.017103
8	-0.005545	-0.006881	-0.008539	-0.143693	-0.188898	-0.002153	-0.002671
9	0.002456	-0.004939	0.009936	0.000000	-0.019987	0.001182	-0.002378
Sum	0.014309	-0.003735	0.018988	-0.009854	-0.049994	0.005811	0.001361

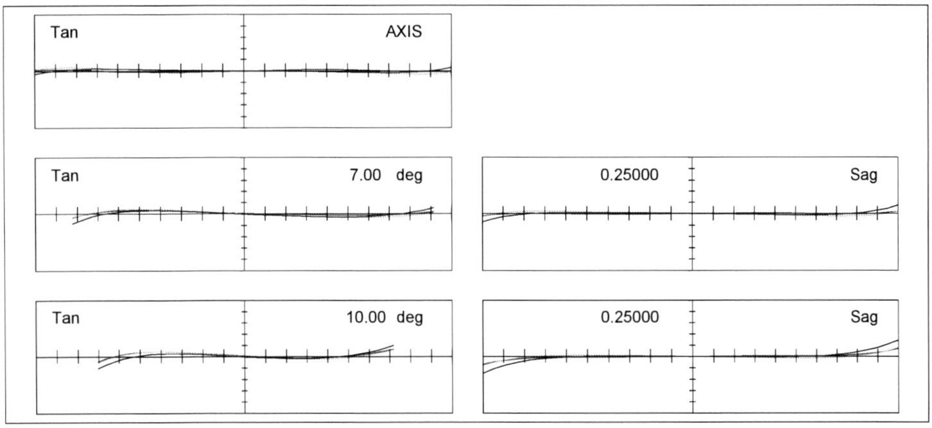

Figure 9.9. Transverse ray aberrations of Petzval lens with field flattener.

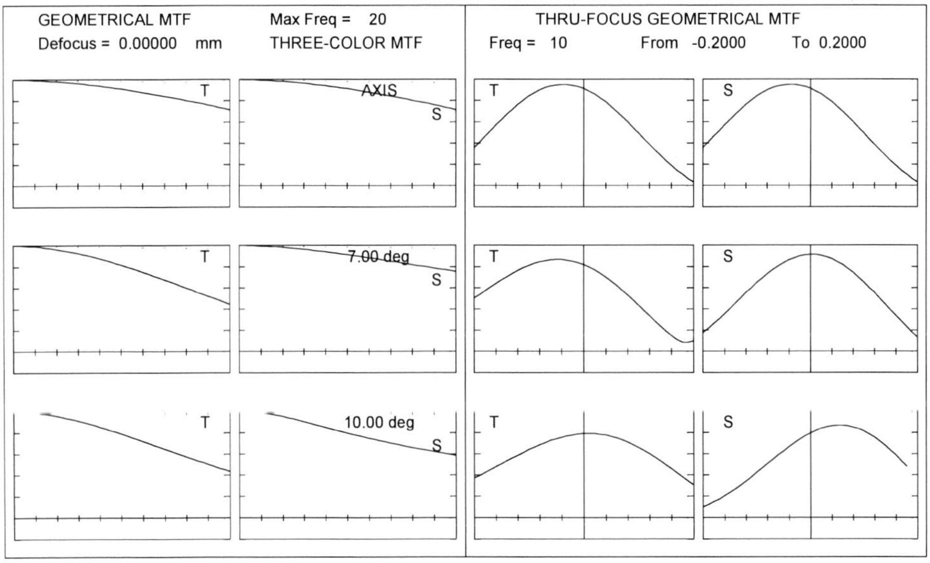

Figure 9.10. MTF curves of Petzval lens with field flattener.

9.4 The telephoto lens

We have shown from thin lens analysis that the field curvature will be zero if the sums of the powers of the lenses are nearly zero. If two thin lenses of equal and opposite power are placed in contact they will of course have no focusing power. But since field curvature is independent of the spacing of the components, it is possible to design an objective consisting of two thin lenses of the same refractive index, of equal and opposite power, as illustrated in Fig. 9.11. This will have both Petzval sum and astigmatism corrected.

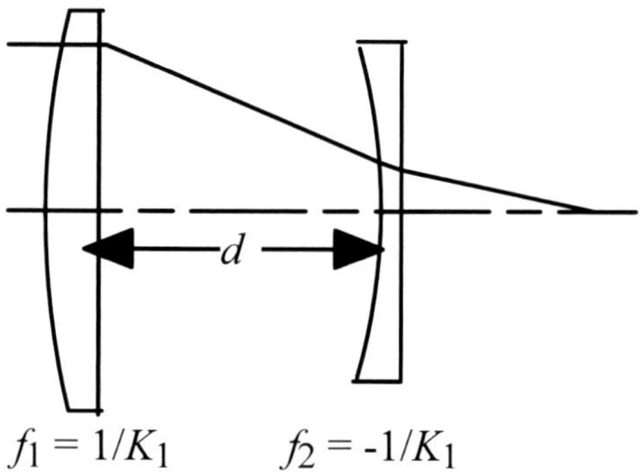

Figure 9.11. A telephoto lens.

If the thin lenses have powers $+K_1$ and $-K_1$, respectively, and a separation d, then the overall power of the objective is given by:

$$K = K_1 - K_1 + dK_1K_1 = dK_1^2. \tag{9.7}$$

In this case this will always be positive, whichever lens is placed in front.

If the front lens is positive, the combination is called a telephoto lens. If the refractive indices of the two lenses are the same, then the Petzval sum,

$$P = (K_1/ n + (-K_1)/ n)$$

will be zero. In practice, the refractive indices will not necessarily be the same, but the powers can then be adjusted to maintain a zero Petzval sum.

Furthermore, if each element in the telephoto doublet is an aplanat (that is, one that is corrected for both third-order spherical aberration and coma), then there will be no stop-shift terms from S_1 and S_2 for either element. Hence, the astigmatism given by

$$S_3 = H^2 (K_1 + K_2)$$

will also be zero. Therefore, the total astigmatism will be zero. Thus, it is clear that a telephoto lens can be designed using these rules, in which all of the monochromatic Seidel aberrations are zero. An example of this general type of lens is shown below. In this case the front lens is made as a triplet to reduce the surface curvatures, but the basic design principles have still been followed.

Note that, although this design has two components equally disposed about the stop, it does not have front-back symmetry. This means the odd aberrations (coma, distortion, and lateral color) are not automatically corrected. In addition, the first lens of a telephoto must be more powerful than the complete objective. For example, this system operates at an aperture of *f*/4, while the first component (the triplet) operates at an aperture of *f*/3. This is the factor that restricts the use of these lenses at higher apertures.

In designing a lens of this type from first principles, one approach would be to first design an aplanatic doublet or triplet to use as the primary component. The negative doublet could then be added, followed by an optimization run in which only the negative lens was varied. Finally, optimization in which both components were allowed to vary would give the final balance of aberrations.

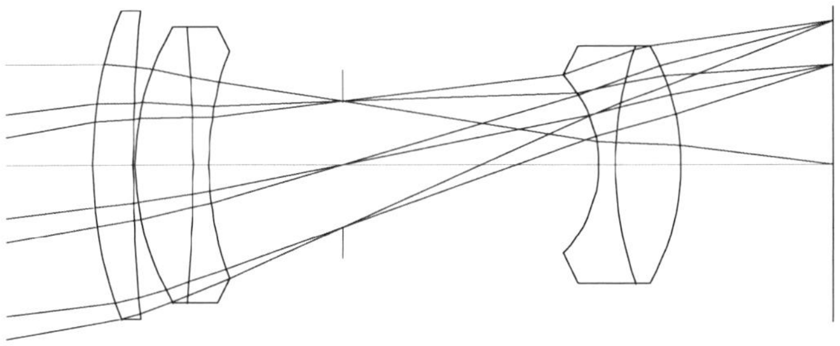

Figure 9.12. Multielement telephoto lens.

In this design, an extra single lens has been added in front of the first doublet, in order to reduce the aberrations. As an exercise, the reader might like to try the effect of removing the front lens, and re-optimizing. The result will be greatly inferior!

The data for the lens shown in Fig. 9.12 is given below.

The transverse aberration curves for the multielement telephoto lens are plotted in Fig. 9.13. Note that the scales of these aberration curves are five times more sensitive than those of the Petzval examples. Again, the MTF curves are shown in Fig. 9.14. The maximum spatial frequency plotted is 50 cycles/mm, as opposed to 20 cycles/mm limit of the curves plotted for Petzval lenses given in Figs. 9.4, 9.7, and 9.10.

EFL = 100.000

WAVELENGTHS [nm]
587.60 656.30 486.10

SURFACES

#	SURF SPACE	RADIUS	SEPN	INDEX1	V	CLR RAD	GLASS
0			0.00000	1.000000			
1	S	53.72645	5.00000	1.651597	58.49	19.206	S-LAKN7
2	S	205.54068	0.25000	1.000000		18.800	
3	S	34.32911	7.00000	1.603108	60.57	17.139	S-SK14
4	S	-227.00584	2.00000	1.672693	32.20	17.139	S-SF5
5	S	41.32751	16.42940	1.000000		14.187	
6#	S	Plane	31.38080	1.000000		7.810	
7	S	-16.44330	2.00000	1.516798	64.14	11.116	S-BK7
8	S	46.38767	8.00000	1.651128	55.87	14.816	S-LAKN22
9	S	-30.50036	18.80042	1.000000		14.816	
10	S	Plane				18.096	

LAGRANGE INVARIANT = -2.2041

#	H	U	HBAR	UBAR	D(U/N)	A	ABAR
1	12.50001	0.00000	-7.79650	0.17633	-0.05558	0.23266	0.03121
2	12.04106	-0.09179	-6.97644	0.16401	-0.05785	-0.05485	0.21483
3	12.01270	-0.11343	-6.91424	0.24877	-0.01283	0.23650	0.04736
4	10.59589	-0.20240	-5.29759	0.23095	0.01145	-0.39930	0.40765
5	10.21181	-0.19204	-4.85684	0.22037	-0.04020	0.09209	0.17204
6	7.66515	-0.15501	-0.09955	0.28956	0.00000	-0.15501	0.28956
7	2.80093	-0.15501	8.98708	0.28956	0.12590	-0.32534	-0.25699
8	2.71262	-0.04416	9.74132	0.37712	0.00166	0.02172	0.89054
9	2.35005	-0.04532	12.37616	0.32935	-0.09755	-0.20205	-0.12617
		-0.125000		0.279598			

#	S1	S2	S3	S4	S5	C1	C2
1	0.037605	0.005045	0.000677	0.035673	0.004877	0.019615	0.002631
2	0.002095	-0.008208	0.032148	-0.009325	-0.089395	0.004454	-0.017447
3	0.008619	0.001726	0.000346	0.053239	0.010730	0.017645	0.003533
4	-0.019339	0.019743	-0.020156	-0.000555	0.021145	-0.026570	0.027126
5	0.003481	0.006503	0.012149	-0.047274	-0.065619	-0.011746	-0.021944
6	0.000000	0.000000	0.000000	0.000000	0.000000	0.000000	0.000000
7	-0.037325	-0.029483	-0.023289	-0.100661	-0.097908	-0.004841	-0.003824
8	-0.000002	-0.000087	-0.003577	0.005617	0.083641	0.000103	0.004219
9	0.009359	0.005844	0.003650	0.062812	0.041503	0.003352	0.002093
Sum	0.004493	0.001084	0.001948	-0.000474	-0.091028	0.002012	-0.003612

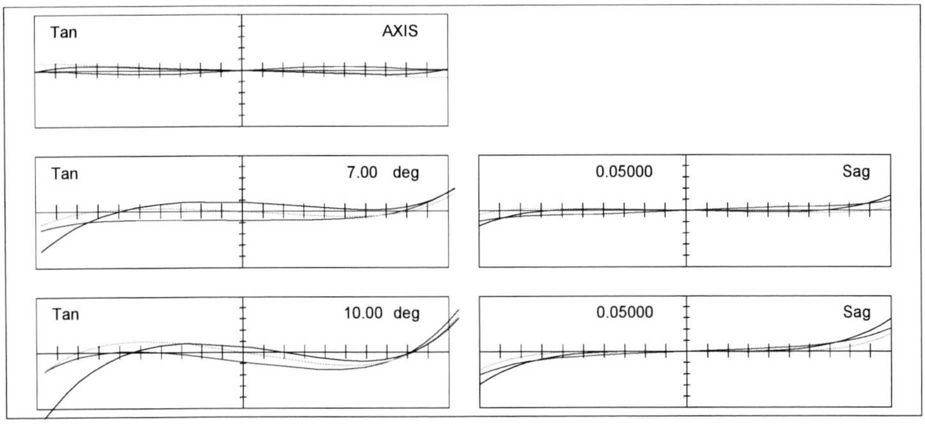

Figure 9.13. Transverse ray aberrations of a multielement telephoto lens.

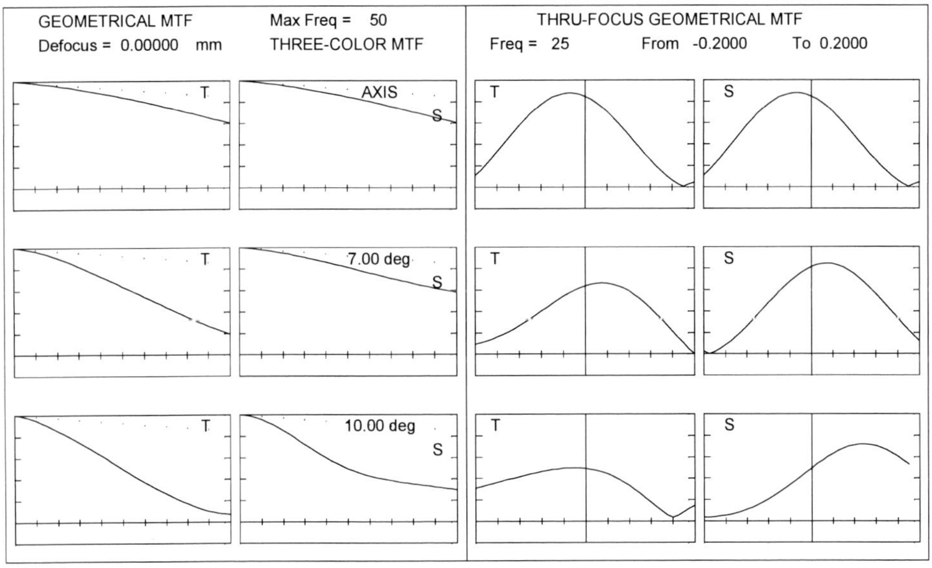

Figure 9.14. MTF curves of a multielement telephoto lens.

CHAPTER 10
TRIPLETS

The simplest design that is capable of correcting all of the seven Seidel aberrations over a wide field of view is the Cooke triplet. H. Dennis Taylor invented this in 1893, using the advances of Seidel's theory. It is named after the optical company in York, England, for which Taylor worked at the time, Cooke and Sons (later to become Cooke, Troughton and Sims). The lens is described in two very interesting United States patents, Nos. 540,132 (1895) and 568,053 (1896). Taylor's designs, despite their antiquity, are close to optimum for the aperture and field he intended, given the glass types available in his day. The triplet uses two of the principles of a good design. First, the Petzval sum is corrected by the use of spaced positive and negative lenses, as described in Chapter 9 on telephoto lenses. Secondly, it has approximate front-back symmetry about a central stop, to control the odd-order aberrations, coma, distortion, and transverse color.

10.1 Seidel theory

From the designer's point of view, the triplet (Fig. 10.1) is interesting because the initial analysis can be carried out using thin-lens theory, as will be briefly outlined below.

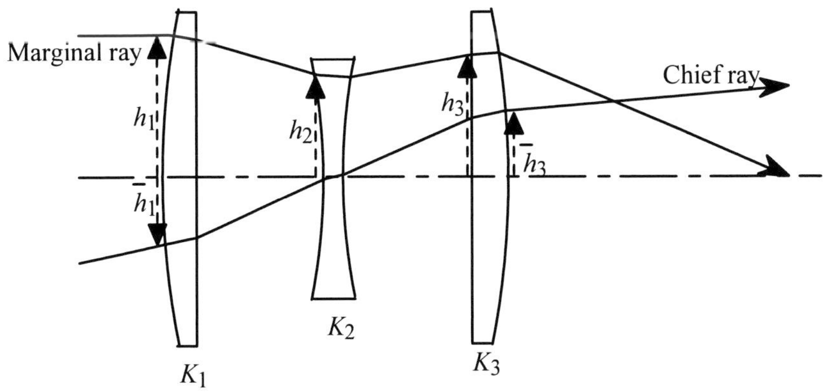

Figure 10.1. A schematic Cooke triplet.

For the sake of this discussion we will assume that the lens thicknesses are fixed. In practice, the lens thicknesses of a triplet are not useful as design parameters. Of course, the lenses must be thick enough to be mechanically rigid and easy to mount. On the other hand, they must not be too thick if glass material costs or weight are factors governing the design. Typically, lens thicknesses are chosen to be between 5% and 10% of the glass diameter, depending on the type of lens, the manufacturer, and its power.

As in all designs, choice of glass types is important, but we assume at this stage that a preliminary choice of three glasses has been made. The effect of changing glass types will be shown in Sec. 10.3.

Having made these assumptions, there are now eight variables available. Two of them are the spaces, d_1 and d_2, between the three lenses (the back focal length to the image is treated as a quantity to be calculated, rather than as a parameter). There are also the curvatures of the six surfaces, which, for this analysis, are best considered as three powers, K_1, K_2, K_3, and three shape factors. These are the eight variables that we must employ to control the seven Seidel aberrations and the focal length.

Another factor is the position of the stop. In the classical Cooke triplet photographic objective, the stop is just behind the central negative lens. For the thin-lens analysis, we can assume the stop to coincide with the second lens, although in practice, since the iris diaphragm of a camera must be located at this point, there will need to be mechanical clearance between the aperture stop and the glass component. For a well-corrected design, however, a small movement of the stop will not change the aberrations by more than a small amount.

For a given choice of three glass types, the equations for the power, the Petzval sum, and the longitudinal and transverse color (the three shape-independent aberrations) become

$$\begin{aligned}
K &= K_1 + \frac{h_2 K_2}{h_1} + \frac{h_3 K_3}{h_1} \\
P &= \frac{K_1}{n_1} + \frac{K_2}{n_2} + \frac{K_3}{n_3} \\
C_1 &= \frac{h_1^2 K_1}{V_1} + \frac{h_2^2 K_2}{V_2} + \frac{h_3^2 K_3}{V_3} \\
C_2 &= \frac{h_1 \bar{h}_1 K_1}{V_1} + \frac{h_2 \bar{h}_2 K_2}{V_2} + \frac{h_3 \bar{h}_3 K_3}{V_1}.
\end{aligned} \qquad (10.1)$$

Note that if the stop is at the second lens, there is further simplification, since the second term in the formula for C_2 vanishes. Note also that in solving these four equations, it is not desirable to make P exactly zero, since exact correction of field curvature results in higher lens powers and increased higher-order aberrations. A typical value for P is about 0.35 K.

The four equations for the power and the three shape-independent aberrations can be rewritten in terms of the five parameters, K_1, K_2, K_3, d_1, and d_2. The extra variable permits a range of solutions that can be used in the way outlined below.

This leaves the four shape-dependent aberrations (S_1, S_2, S_3, and S_5) and only three shape factors to control them. Of course, if we bend the outer lenses, which are not in contact with the stop, all four aberrations will change. The simplest procedure is to correct S_3 and S_5 by choice of lens shapes of the outer components, and then to bend the second lens, which will only change S_1 and S_2. In general, we can always find a shape to correct S_2; however, the shape that is best for correcting S_2 will not also correct S_1. However, by returning to the first four equations and using the spare variable there to create a range of different solutions, one can be found that gives an acceptable value of S_1 when the combination of shapes gives zero S_2. In this way, we can design the triplet to correct all of the Seidel aberrations.

As an example, contributions of each of the lenses in a thin-lens design are listed in Table 10.1. We can see that for each of the even aberrations, S_1, S_3, S_4, and C_1, the outer, positive lenses have a positive coefficient, and the inner, negative, lens has a negative coefficient.

Table 10.1. Seidel aberrations for a typical triplet.

	S1	S2	S3	S4	S5	C1	C2
Lens 1	0.254	-0.147	0.478	0.241	-0.820	0.055	-0.028
Lens 2	-0.339	0.211	-0.670	-0.371	0.145	-0.082	0.006
Lens 3	0.103	-0.058	0.164	0.199	0.698	0.028	0.022
Total	0.018	0.006	-0.028	0.069	0.023	0.001	0.000

Of the odd aberrations, distortion and lateral color are balanced by contributions of opposite sign from the first and the third lens; the aperture stop is so close to the second lens, the chief ray height there is small making it only a small contribution to the total S_5 and the total C_2. Mainly, a balance between the contributions of the first and second lenses corrects coma. It should be noted that the signs in this table of the Seidel coefficients of the odd aberrations S_2, S_5, and C_2 are arbitrary, since they depend on the sign of the Lagrange invariant, H.

The analytical design of a triplet is possible, but it is difficult. Triplets are probably the most complex systems that can be designed from the first principles described in this procedure. One of the attractions of this analytical approach is that it leads to the discovery of multiple solutions, in some cases. There are, for some pairs of glass types, two solutions that correct all the Seidel aberrations. The analytic approach will find all of the solutions. These days, in practice, if a triplet design were wanted, we would start from an existing design, and optimize!

10.2 Example of an optimized triplet

In this example (Fig. 10.2), a triplet objective with 50-mm focal length and a speed of *f*/4, has been designed using Schott glass types SK16, F6, and SK16, respectively. We can see from Fig. 10.3 that there is good balance between the third-order astigmatism and higher-order aberrations. An optimization program obtained this aberration balance automatically. Naturally, different programs will have different criteria for optimization, but any good optimization routine should be able to give results comparable to this example.

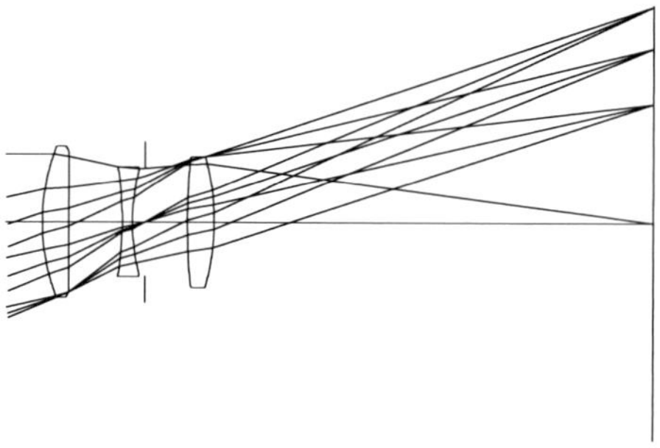

Figure 10.2. Optimized Cooke triplet.

EFL = 49.999

WAVELENGTHS [nm] 587.60 656.30 486.10

SURFACES

# SURF SPACE	RADIUS	SEPN	INDEX1	V	CLR RAD	GLASS
0		0.00000	1.000000			
1 S	18.60900	2.50000	1.620408	60.30	7.001	S-SK16
2 S	460.96700	5.15700	1.000000		6.524	
3 S	-29.93400	1.00000	1.636355	35.33	5.058	S-F6
4 S	18.60500	1.20700	1.000000		4.936	
5#S	Plane	4.00700	1.000000		4.979	
6 S	67.59400	2.50000	1.620408	60.30	5.797	S-SK16
7 S	-22.16200	41.60525	1.000000		6.070	
8 S	Plane				20.178	

LAGRANGE INVARIANT = -2.5251

#	H	U	HBAR	UBAR	D(U/N)	A	ABAR
1	6.24991	0.00000	-5.12787	0.40403	-0.07936	0.33585	0.12847
2	5.92844	-0.12859	-4.24077	0.35484	-0.12103	-0.18753	0.56008
3	4.89504	-0.20039	-1.30500	0.56928	0.16441	-0.36392	0.61287
4	4.83617	-0.05887	-0.97406	0.33094	0.10506	0.32903	0.45586
5	4.91956	0.06909	-0.36065	0.50822	0.00000	0.06909	0.50822
6	5.19640	0.06909	1.67579	0.50822	-0.06094	0.14596	0.53301
7	5.22941	0.01320	2.43615	0.30414	-0.13315	-0.36096	0.31471
		-0.125000		0.424639			

#	S1	S2	S3	S4	S5	C1	C2
1	0.055944	0.021399	0.008185	0.131189	0.053312	0.013328	0.005098
2	0.025233	-0.075362	0.225080	-0.005296	-0.656417	0.007059	-0.021083
3	-0.106585	0.179501	-0.302298	-0.082837	0.648610	-0.019610	0.033025
4	-0.055006	-0.076210	-0.105589	-0.133278	-0.330947	-0.017517	-0.024269
5	0.000000	0.000000	0.000000	0.000000	0.000000	0.000000	0.000000
6	0.006747	0.024637	0.089966	0.036117	0.460411	0.004816	0.017586
7	0.090721	-0.079098	0.068964	0.110157	-0.156171	0.011985	-0.010450
Sum	0.017054	-0.005133	-0.015692	0.056051	0.018798	0.000062	-0.000092

At this aperture, the tangential curves near the axis have slopes of opposite sign to those of the sagittal curves, indicating the presence of astigmatism. Figure 10.4 shows that the tangential-section slopes change sign as the field increases, so there is also substantial field curvature. Spherical aberration is well corrected, and coma is small because of the near front-to-back symmetry. Considering the sagittal curves, the performance clearly drops off above about 10 deg off-axis, while the tangential curve is better. MTF performance shown in Fig. 10.5 is only just acceptable at the *f*/4 aperture shown, although it will obviously improve when the lens is stopped down to *f*/5.6 or *f*/8.

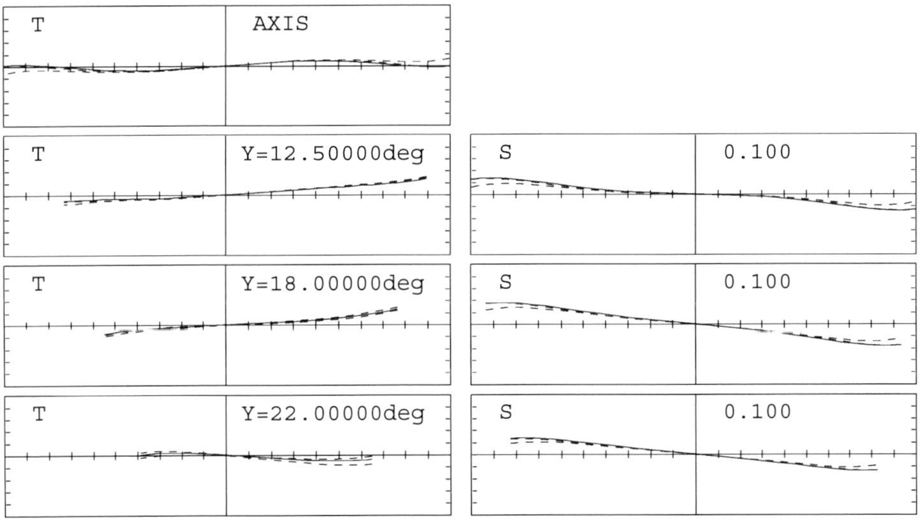

Figure 10.3. Transverse ray aberration curves of a Cooke triplet.

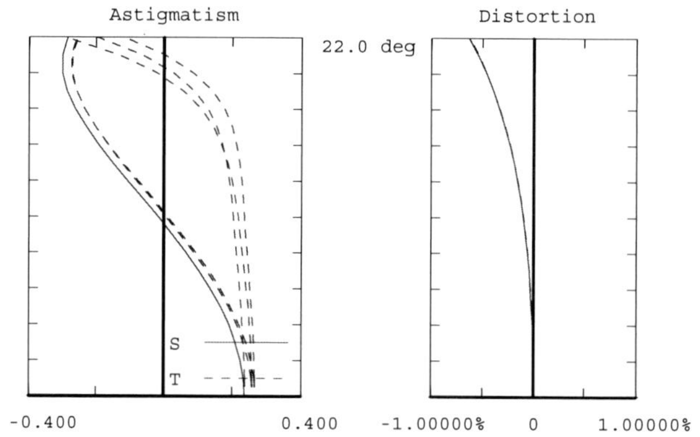

Figure 10.4. Astigmatism and distortion of a Cooke triplet.

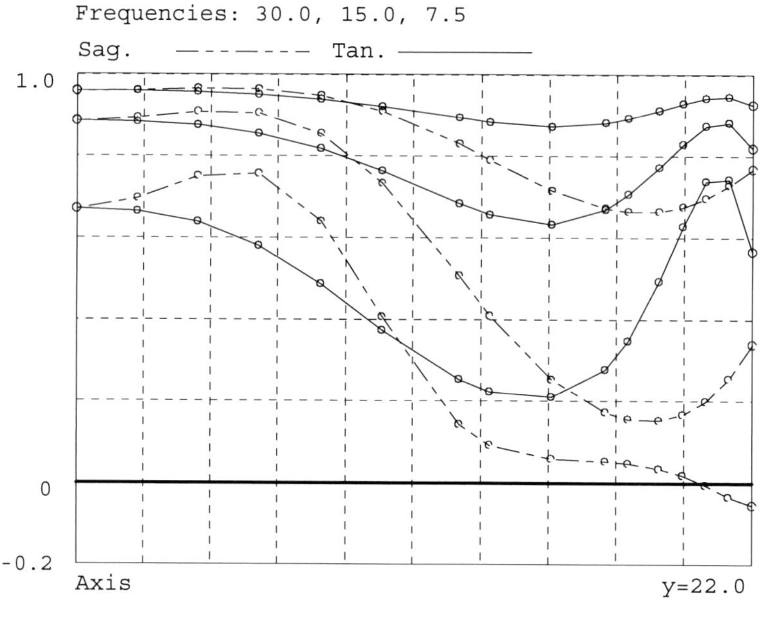

Figure 10.5. MTF as a function of field angle for a Cooke triplet.

10.3 Glass choice

It is highly desirable for the positive lenses to be made of high-index glass, because this reduces the spherical aberration (S_1) and the field curvature (S_4). On the other hand, the positive lenses should have a low dispersion (high V-value), indicating that a high-index crown is needed. As far as the negative lens is concerned, it should have a low index to introduce extra negative S_4, but a high dispersion to correct longitudinal chromatic aberration. Since these are incompatible, a compromise is needed. We will now investigate the effect of choosing different glass types for the second lens, whilst keeping the glass of the first and third lenses unchanged (SK16, with $n = 1.620409$, $V = 60.30$).

If the refractive index of the negative lens is reduced and the lens reoptimized, the resulting design becomes asymmetric, as shown in Fig. 10.6. In this example, the F6 has been replaced by LF7, which has a lower refractive index ($n = 1.575008$) and a slightly lower dispersion ($V = 41.47$). The improved color correction is offset by the presence of higher-order spherical aberration and some coma, as may be seen in Fig. 10.7. Note that the first two lenses in Fig. 10.6 have moved together to form what is almost a Petzval lens with a central stop.

If a high-index (SF18, with $n = 1.721505$, $V = 29.23$) flint is chosen as the negative lens, the difference between the V-values is increased and, to keep the longitudinal color correct, the power of the negative lens in Fig. 10.8 must be reduced accordingly. In this case, it is difficult to obtain sufficient negative S_1 to balance the positive S_1 of the two positive lenses, as shown in the curves in Fig. 10.9.

Although good color correction has been achieved, the Seidel aberrations have deteriorated to the point where the lens is useless.

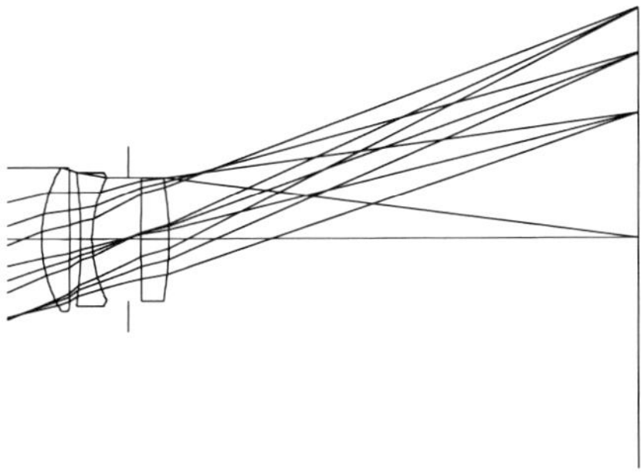

Figure 10.6. Triplet with a lower-index central flint.

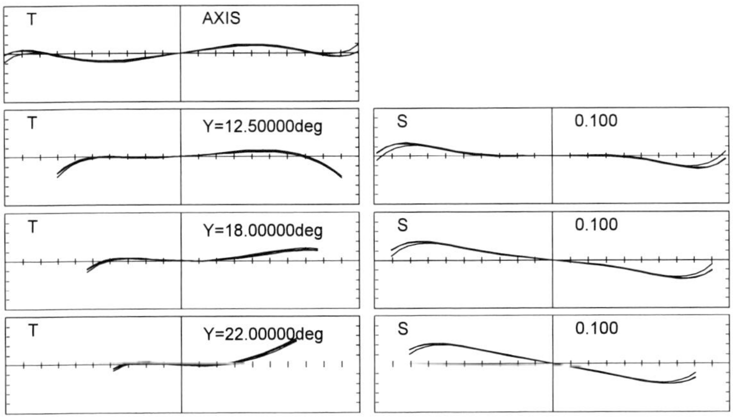

Figure 10.7. Transverse ray aberration curves of triplet with a lower-index central flint.

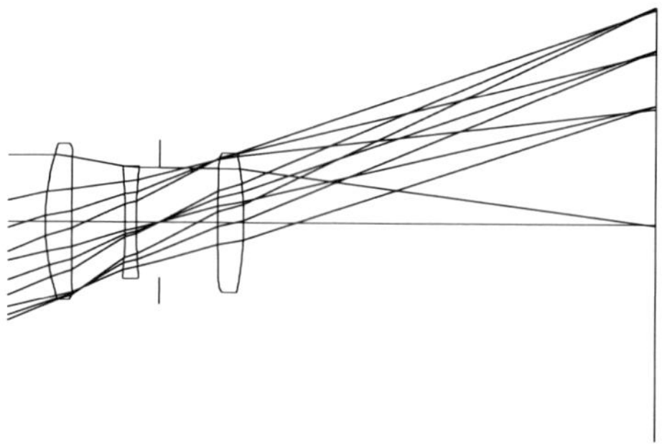

Figure 10.8. Triplet with a higher-index central flint.

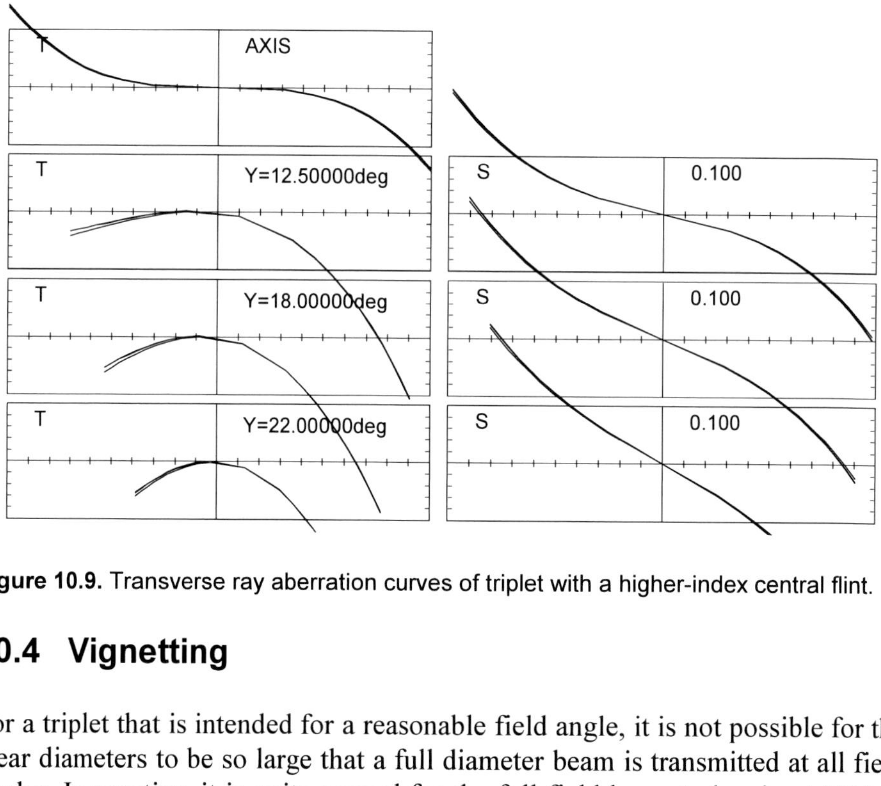

Figure 10.9. Transverse ray aberration curves of triplet with a higher-index central flint.

10.4 Vignetting

For a triplet that is intended for a reasonable field angle, it is not possible for the clear diameters to be so large that a full diameter beam is transmitted at all field angles. In practice, it is quite normal for the full-field beam to be about 60% of the width of the axial beam. Allowing for the fact that the illumination in the corner of the image formed by a perfect lens will also fall off as the fourth power of the cosine of the field angle, the illumination in the corner of the image may only be about 40% of that on-axis. The effect of this vignetting is shown, as before, by a shortening of the aberration curves along the horizontal pupil height axis, particularly in the tangential section.

To eliminate vignetting as shown in Fig. 10.10, the triplet would have to be larger and considerably more expensive. Although the image illumination would be much more uniform, it is clear from Figs. 10.11 and 10.12 that the sagittal image surface is much more curved, and this affects the MTF curves of Fig. 10.13..

It is interesting to see the tangential MTF in Fig. 10.13 at 30 cycles/mm starts at about 30% on-axis; as the tangential image surface approaches the chosen focal plane, the MTF approaches 70% at about 20 deg off-axis, and it then falls very quickly.

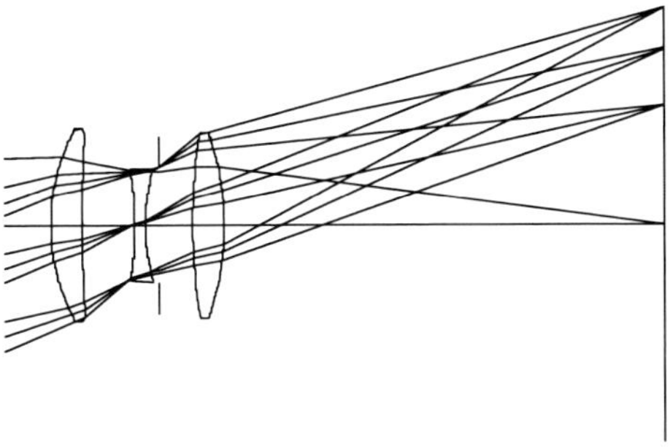

Figure 10.10. An unvignetted triplet.

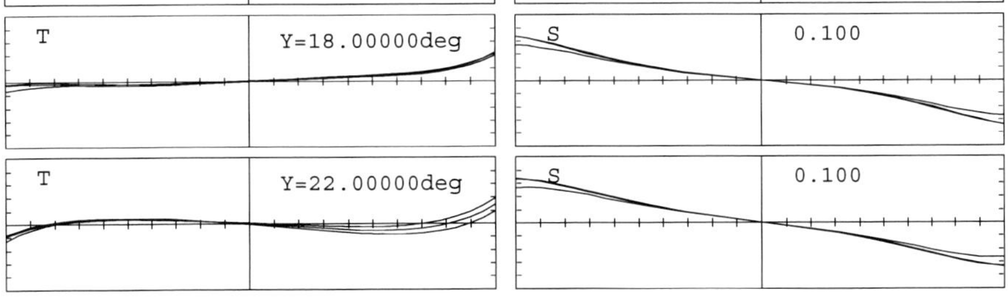

Figure 10.11. Transverse ray aberration curves of an unvignetted triplet.

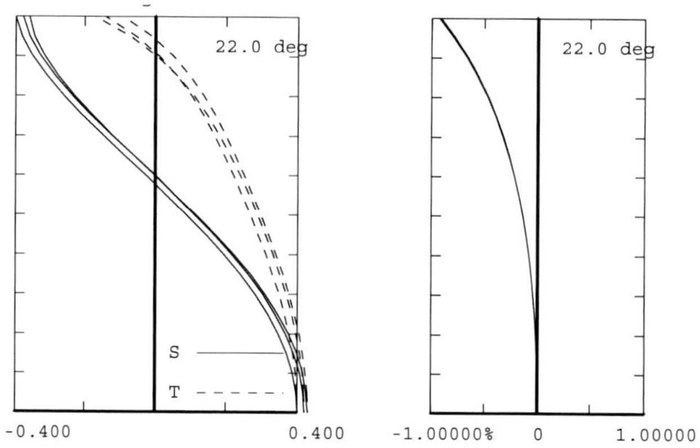

Figure 10.12. Astigmatism and distortion of an unvignetted triplet.

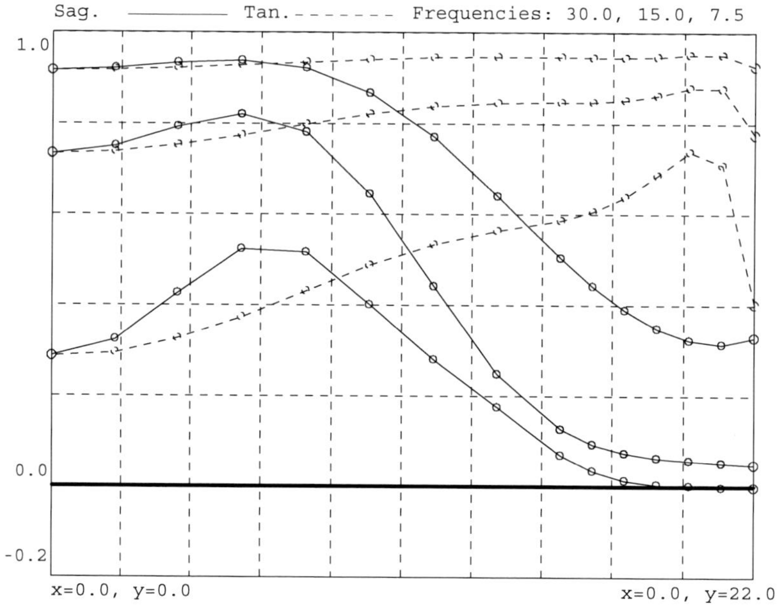

Figure 10.13. MTF as a function of field angle for an unvignetted triplet.

CHAPTER 11
EYEPIECES AND AFOCAL SYSTEMS

In this chapter we will be considering the design of eyepieces and afocal systems for the visual waveband. The most common afocal system is a telescope. Since the major problem in designing visual afocal systems is the eyepiece, these subjects are naturally connected, and it is logical to consider them together.

11.1 Eyepieces—design considerations

In most instruments, eyepieces are just one part of a more complex optical system, but, as we will see, it is usually sensible to consider the design of the eyepiece more or less independently of the rest of the system.

Figure 11.1 shows a complete afocal system. Following the normal convention with light travelling from left to right, this diagram shows the complete system the "right" way round. The aperture stop is at the objective lens, the first element on the left. The real image of the objective is the exit pupil of the system, shown as the last surface on the right. After the objective is a block of material that represents a series of prisms that invert the image and present an upright image to the observer. To the right of the prism assembly is the eyepiece lens, which will be the subject of the first section of this chapter.

Although no physical element is located at the exit pupil, when the eye of the observer is located there, he or she will see a bright, magnified image. If, because the user wears spectacles or for any other reason, the pupil of the eye cannot be located at and centered on the exit pupil, the field of view may be reduced, or the image could vanish entirely. The distance between the last surface of the eyepiece and the exit pupil is called the *eye relief*.

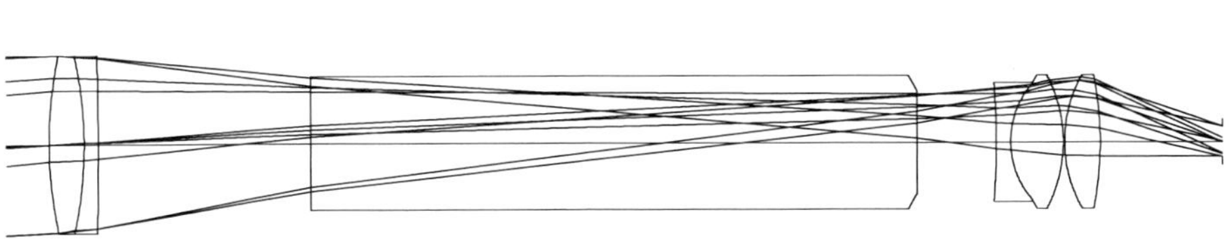

Figure 11.1. An afocal system.

Because the eyepiece comes at the end of the afocal system, one cannot know what will be delivered by the front end of the system. In the cases of both telescopes and microscopes, the objective lens delivers a real intermediate image to a plane just in front of the eyepiece. So the eyepiece is a lens that operates with a finite object distance and an infinite image distance (parallel rays out). The easiest way to analyze an eyepiece, considered on its own, is to trace the rays in the reverse direction, i.e., trace from the eye-pupil to the intermediate image surface, like the eyepiece shown in Fig. 11.2.

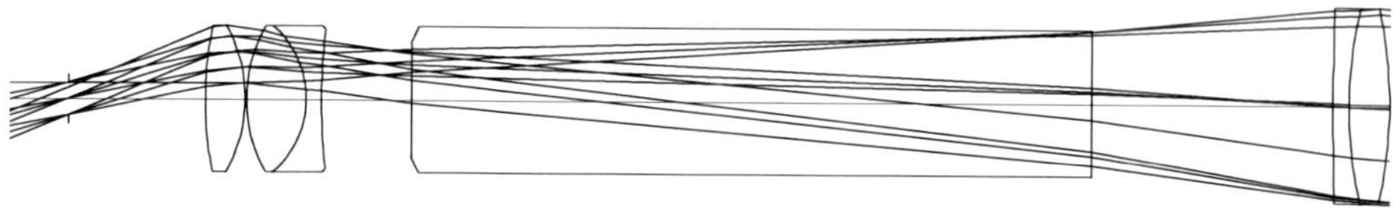

Figure 11.2. A reversed afocal system.

11.1.1 Specification of an eyepiece

The specification of the eyepiece will usually consist of the items in the following sections.

11.1.1.1 Focal length

In a simple afocal system such as a telescope as shown above, the angular magnification of the system will be determined by the ratio of the focal lengths of the eyepiece and the objective, so if we are to design the eyepiece independently, its focal length must be known. Typical eyepiece focal lengths are of the order of 25 mm (one inch!). Note that, because the human eye is always required to view the image, there is not a very large variation in focal lengths of typical eyepieces for simple instruments such as binoculars and similar instruments.

11.1.1.2 Field angle

Typically, eyepieces have semi-field angles in the eye-space between 20 and 30 deg. Anything much larger than +/− 30 degrees is very difficult, although there are some complex eyepieces with much larger fields.

11.1.1.3 Pupil diameter

This is related, of course, to the objective diameter and the angular magnification of the complete instrument. For example, in the case of 8 × 30 binoculars, the objective diameter is 30 mm and the angular magnification is 8. Therefore, the pupil diameter will be 30/8 = 3.75 mm. This is really the **exit** pupil of the instrument, but for us it will appear to be the **entrance** pupil diameter, because we are ray tracing in reverse when we design the eyepiece.

This diameter is obviously related to the eye pupil diameter, which is typically taken to be between 3 mm and 7 mm (when the eye pupil is dilated). However, some instruments, for example an 8 × 20 telescope, and many microscopes, will have pupil diameters smaller than the eye pupil diameter, but other instruments may need pupil diameters larger than the eye pupil diameter. If the pupil diameter is too small, and the eye moves too far away from the optical axis, then parts of the image will disappear. This latter situation would occur if the instrument were to be used in a vehicle, for example, where vibration might cause the eye to be moving with respect to the instrument.

11.1.1.4 Exit pupil position ("eye relief")

The distance from the (real) exit pupil to the eyepiece is obviously an important parameter, and it is probably obvious that a large eye relief will make the design more difficult. Typically, eye relief varies between 0.5 and 0.8 of the eyepiece focal length. It is also important to ensure that the pupil positions of the objective and eyepiece match; i.e., the chief rays in the diagrams on the previous page must cross the axis close to the objective. Again, because the eye is ultimately the detector, eye relief does not vary much between instruments. An eye relief of about 20 mm is generally comfortable, but if the user is expecting to wear spectacles or a visor, a larger eye relief is needed. This will inevitably make the eyepiece larger and more expensive.

11.1.2 Aberration considerations

11.1.2.1 Prism aberrations

In many visual instruments there are erecting prisms between the objective and the eyepiece, and these will in almost all cases introduce aberrations (except if the prism assembly is in a collimated beam, which is unlikely).

Usually we can represent the prism assembly as a parallel-sided glass block, and we can include the block in our tracing. However, if we are designing the objective and the eyepiece separately, we have to decide whether to treat the prism as part of the eyepiece or as part of the objective. One of the major aberrations will be lateral color, and it is common practice to design the eyepiece to correct the lateral color of the prism, since it is not possible to correct this aberration in a simple telescope objective.

11.1.2.2 Pupil spherical aberration

The exit pupil is of course an image of the entrance pupil, and as such it will have aberrations. In this context, the most important is pupil spherical aberration. If the exit pupil has a large amount of pupil spherical aberration, the beams from different field points will not overlap correctly at the eye position, and the eyepiece will be uncomfortable or even impossible to use.

11.1.2.3 Distortion

In many visual instruments, distortion cannot be easily corrected, if we define distortion in the normal way: the image size is proportional to the tangent of the semi-field angle. However, in an afocal system, it is usually preferable to design the system so that the chief ray angles in the object and image space are proportional. This is a condition that is similar to one imposed on laser scanning lenses, which are described as "f-theta" lenses, and one that tends to be met when the exit pupil is external to the eyepiece.

Since the objective operates at a very small field angle, and it is effectively close to the stop, the objective will in any case have very little distortion.

11.1.2.4 Field curvature

All simple eyepieces consist primarily of positive elements, and therefore have a positive Petzval sum. An optimization will often try to balance this by introducing astigmatism, but in the case of a visual system this is not desirable, because the eye can accommodate for an image that is defocused due to field curvature but cannot accommodate for an astigmatic image. The answer to this is sometimes to optimize for a curved image surface, either by letting the program vary the image curvature by selecting a non-zero curvature for the image surface, or by setting non-zero longitudinal aberration targets for the optimization.

11.1.2.5 Special factors in optimization

As pointed out earlier, since the eyepiece is intended to image the objective at the exit pupil, the chief ray through the eyepiece should cross the optical axis at the objective. If this is not done, there is a distinct possibility that the chief ray will not only fail to cross the axis near to the correct position, it may even diverge, which is obviously useless.

11.1.2.6 General comments on eyepieces

The most important difference between eyepieces and any of the lens types discussed earlier is that the aperture stop is external to the lens. Remember that in the case of doublets and Petzval objectives, although there is no particular restriction on the pupil position, it is almost always at the front element. It is generally desirable for a lens to have at least approximate front-back symmetry. In the case of an eyepiece, the exit pupil must be coincident with the eye pupil, and therefore external to the lens. This means that eyepieces **cannot** use the symmetry principle, by which the system is symmetrical about the stop.

Although the designs may in fact be approximately symmetric, the ray paths through the eyepiece are not symmetric. This in turn means that correction of odd aberrations (coma, distortion and lateral color) is not automatic, and can be quite difficult.

11.2 Simple eyepiece types

11.2.1 The Ramsden eyepiece

The simplest eyepiece, beyond a single lens, is the Ramsden, with two singlets, as shown in Fig. 11.3. The color correction is limited to the choice of the two glasses. Naturally the performance is poor, and the example is included only as an introduction to eyepieces. Realistically, this type of eyepiece would rarely be used at these apertures. Remember that in these drawings the eye position is to the **left** of the drawing, and the intermediate image is on the right. This can often cause confusion!

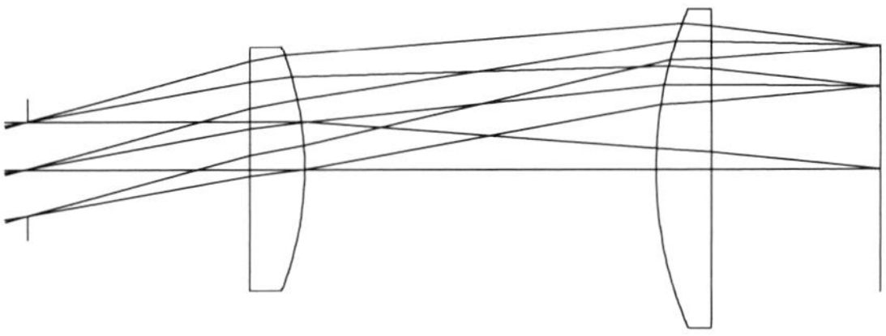

Figure 11.3. The Ramsden eyepiece.

EFL = 24.997

WAVELENGTHS [nm]
587.60 656.30 486.10

SURFACES

#	SURF SPACE	RADIUS	SEPN	INDEX1	V	CLR RAD	GLASS
0			0.00000	1.000000			
1#S		Plane	12.00000	1.000000		2.500	
2 S		Plane	3.00000	1.516798	64.14	6.500	S-BK7
3 S		-17.66000	18.99000	1.000000		6.500	
4 S		21.37500	3.00000	1.516798	64.14	8.500	S-BK7
5 S		Plane	9.12765	1.000000		8.500	
6 S		Plane				6.601	

LAGRANGE INVARIANT = -0.6698

#	H	U	HBAR	UBAR	D(U/N)	A	ABAR
1	2.49966	0.00000	0.00000	0.26795	0.00000	0.00000	0.26795
2	2.49966	0.00000	3.21539	0.26795	0.00000	0.00000	0.26795
3	2.49966	0.00000	3.74535	0.17665	-0.07315	-0.21469	-0.05374
4	1.11055	-0.07315	6.75234	0.15835	0.02968	-0.02119	0.47425
5	0.91276	-0.06593	6.74263	-0.00324	-0.05653	-0.10000	-0.00491
		-0.100000		-0.004910			

#	S1	S2	S3	S4	S5	C1	C2
1	0.000000	0.000000	0.000000	0.000000	0.000000	0.000000	0.000000
2	0.000000	0.000000	0.000000	0.000000	0.027186	0.000000	0.003558
3	0.008428	0.002109	0.000528	0.008655	0.002298	0.002851	0.000713
4	-0.000015	0.000331	-0.007414	0.007151	0.005895	-0.000125	0.002798
5	0.000516	0.000025	0.000001	0.000000	0.000000	0.000485	0.000024
Sum	0.008929	0.002466	-0.006885	0.015806	0.035380	0.003211	0.007093

Note that the eye relief (12 mm) is about half the focal length, which is reasonable but not as good as more complex eyepieces to be described later in these notes. Transverse ray aberrations for this design are shown in Fig. 11.4, in which dashed lines show the y component of the rays in the x (sagittal) section.

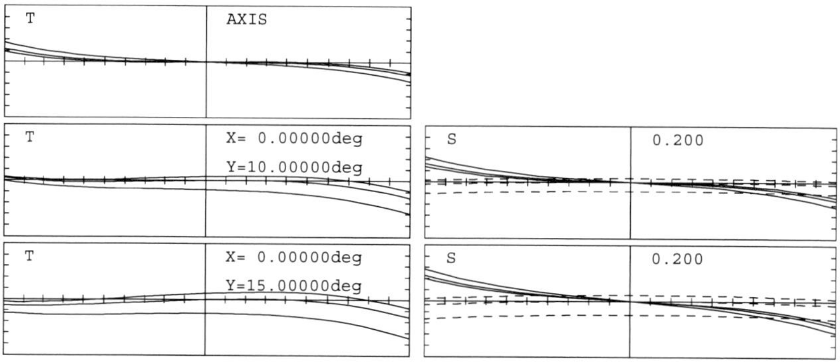

Figure 11.4. Transverse ray aberrations of the Ramsden eyepiece.

With two single lenses, spherical aberration and axial color are not correctable. And with this configuration, lateral color (evident in the separation of the tangential curves at the origin for 10 and 15 deg) is not correctable. Also, as we have noted in the previous chapter, the Petzval sum is not correctable in any type of simple eyepiece with two positive lenses.

11.2.2 The achromatized Ramsden, or Kellner eyepiece

In the Kellner a cemented doublet that provides a considerable improvement in performance replaces the first lens of the Ramsden, sometimes called the "eye-lens" since it is next to the eye (Fig. 11.5). This allows the usable semi-field angle to be increased from 15 deg for the simple Ramsden to 20 deg for the Kellner (Fig. 11.6). In this particular example, the eye relief is only 9.4 mm.

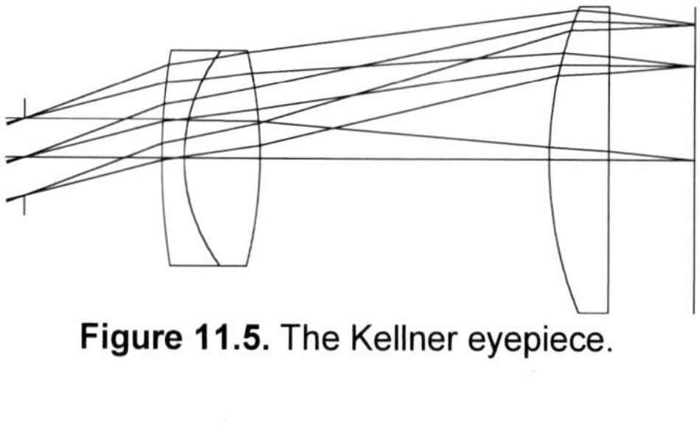

Figure 11.5. The Kellner eyepiece.

EFL = 25.021

WAVELENGTHS [nm]

587.60	656.30	486.10

SURFACES

# SURF SPACE	RADIUS	SEPN	INDEX1	V	CLR RAD	GLASS
0		0.00000	1.000000			
1#S	Plane	9.40000	1.000000		2.500	
2 S	43.28068	1.50000	1.672693	32.20	7.000	S-SF5
3 S	11.60995	5.00000	1.620408	60.30	7.000	S-SK16
4 S	-28.10963	19.40000	1.000000		7.000	
5 S	25.96998	4.00000	1.620408	60.30	10.000	S-SK16
6 S	Plane	5.76060	1.000000		10.000	
7 S	Plane				10.000	

LAGRANGE INVARIANT = -0.9099

#	H	U	HBAR	UBAR	D(U/N)	A	ABAR
1	2.50000	0.00000	0.00000	0.36397	0.00000	0.00000	0.36397
2	2.50000	0.00000	3.42132	0.36397	-0.01389	0.05776	0.44302
3	2.46516	-0.02323	3.70003	0.18580	0.00332	0.31631	0.84387
4	2.37951	-0.01713	4.71044	0.20208	-0.06970	-0.16492	0.05592
5	0.82222	-0.08027	9.04621	0.22349	0.04222	-0.04861	0.57183
6	0.57558	-0.06166	9.06443	0.00456	-0.06186	-0.09992	0.00738
		-0.099915		0.007384			

#	S1	S2	S3	S4	S5	C1	C2
1	0.000000	0.000000	0.000000	0.000000	0.000000	0.000000	0.000000
2	0.000116	0.000888	0.006814	0.007693	0.111269	0.001804	0.013834
3	-0.000818	-0.002183	-0.005824	-0.001376	-0.019207	-0.004789	-0.012776
4	0.004511	-0.001530	0.000519	0.011277	-0.004000	0.002492	-0.000845
5	-0.000082	0.000965	-0.011351	0.012207	-0.010062	-0.000254	0.002985
6	0.000355	-0.000026	0.000002	0.000000	-0.000000	0.000365	-0.000027
Sum	0.004082	-0.001885	-0.009840	0.029802	0.078000	-0.000382	0.003172

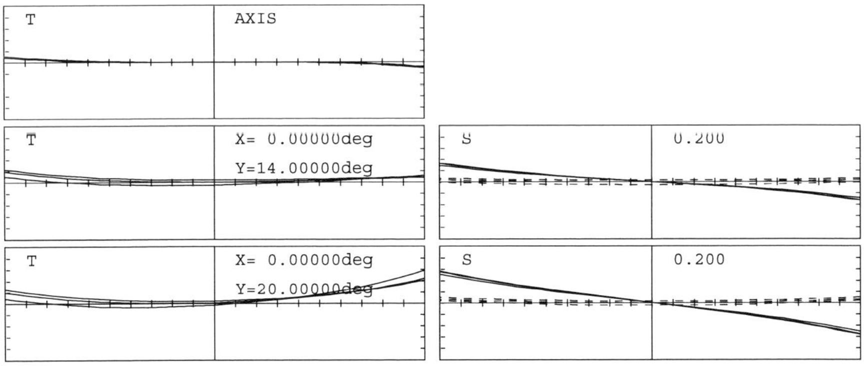

Figure 11.6. Transverse ray aberrations of the Kellner eyepiece.

This eyepiece still has poorly corrected astigmatism, but it is better than the Ramsden eyepiece. The spherical aberration and the axial color and lateral color are all at acceptable levels, although the semi-field of 20 deg is perhaps a little large for this type of eyepiece.

There is significant high-order coma at the edge of the field of view. If this were third-order (Seidel), it would be proportional to the first power of the field angle, so that the coma at 20 deg would be about 50% more than at 14 deg. It is clearly worse than this, suggesting a higher-order dependence on field angle.

11.2.3 The Ploessl eyepiece

In this eyepiece, both lenses are cemented doublets and are achromatic (Fig. 11.7). It is usually possible to make both lenses identical, without losing any performance. By careful choice of glass types, the nearly plane surfaces can be made planar and the other surfaces can all have the same curvature. Typical semi-field angles would be up to 20 deg, but the performance shown in Fig. 11.8 will be significantly better than the simpler eyepieces.

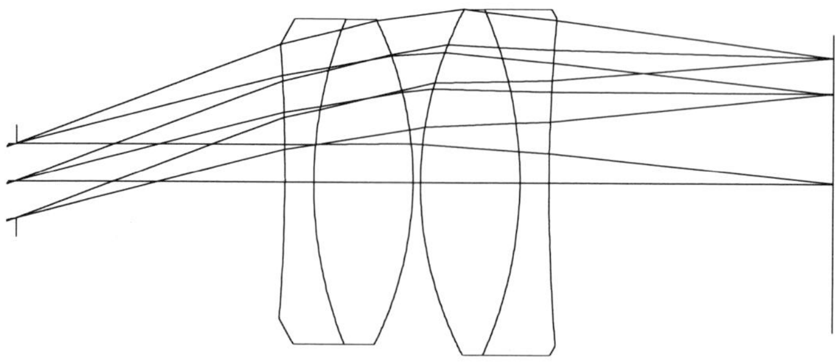

Figure 11.7. The Ploessl eyepiece.

```
EFL =     24.972

WAVELENGTHS  [nm]
      587.60       656.30       486.10

SURFACES
  # SURF SPACE    RADIUS        SEPN        INDEX1      V       CLR RAD  GLASS
  0                           0.00000     1.000000
  1#S              Plane     18.70000     1.000000               2.500
  2 S          -125.00000     2.00000     1.666796    32.99      9.200   S-SF19
  3 S            28.51033     6.80000     1.622988    58.04     10.900   S-SK15
  4 S           -24.50980     0.50000     1.000000              10.900
  5 S            24.50980     6.80000     1.622988    58.04     11.600   S-SK15
  6 S           -28.51033     2.00000     1.666796    32.99     11.600   S-SF19
  7 S           125.00000    19.52340     1.000000              11.000
  8 S              Plane                                        10.000

LAGRANGE INVARIANT =     -0.9099

  #     H         U         HBAR       UBAR       D(U/N)      A         ABAR
  1   2.50000   0.00000   0.00000   0.36397    0.00000   0.00000    0.36397
  2   2.50000   0.00000   6.80624   0.36397    0.00480  -0.02000    0.30952
  3   2.51600   0.00800   7.28654   0.24015    0.00173   0.16043    0.82627
  4   2.58807   0.01060   9.01053   0.25353   -0.05511  -0.15417   -0.18518
  5   2.56378  -0.04858   9.10175   0.18244    0.00540   0.05602    0.55380
  6   2.08720  -0.07009   8.89686  -0.03013    0.00339  -0.23256   -0.55537
  7   1.95457  -0.06632   8.85458  -0.02114   -0.06033  -0.08448    0.08284
             -0.100114             0.012002
```

#	S1	S2	S3	S4	S5	C1	C2
1	0.000000	0.000000	0.000000	0.000000	0.000000	0.000000	0.000000
2	-0.000005	0.000074	-0.001150	-0.002650	0.058801	-0.000606	0.009383
3	-0.000112	-0.000577	-0.002972	-0.000470	-0.017730	-0.002225	-0.011458
4	0.003390	0.004072	0.004891	0.012967	0.021450	0.002639	0.003170
5	-0.000043	-0.000429	-0.004245	0.012967	0.086222	0.000950	0.009391
6	-0.000383	-0.000915	-0.002185	-0.000470	-0.006342	-0.002675	-0.006389
7	0.000841	-0.000825	0.000809	-0.002650	0.001805	0.002002	-0.001963
Sum	0.003688	0.001400	-0.004852	0.019694	0.144206	0.000085	0.002133

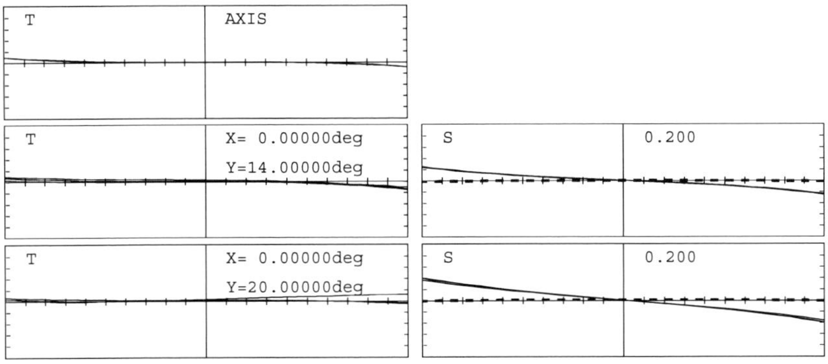

Figure 11.8. Transverse ray aberrations of the Ploessl eyepiece.

The main uncorrected aberration is field curvature. In this example, the Petzval sum has been balanced by the introduction of astigmatism. Alternatively, eyepieces can be balanced for zero astigmatism with a curved field.

11.2.4 The Erfle eyepiece

This eyepiece has an extra positive element, and the form of the lenses is different (Fig. 11.9). It has a concave surface near to the intermediate image. This surface reduces Petzval sum, and it also helps to achieve a longer eye relief by moving the chief rays away from the axis. The semi-field angle for this design is 30 deg (Fig. 11.10). In this example, the lateral color is not corrected, because the eyepiece is intended for use with a prism. This is a typical eyepiece for use in afocal systems of reasonable quality.

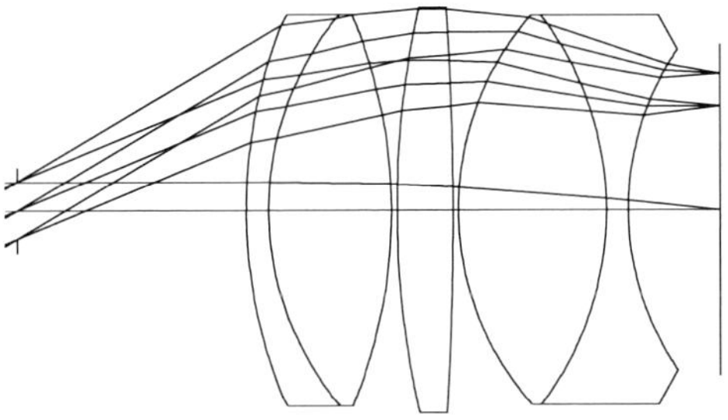

Figure 11.9. The Erfle eyepiece.

EFL = 25.066

WAVELENGTHS [nm]
587.60 656.30 486.10

SURFACES

#	SURF SPACE	RADIUS	SEPN	INDEX1	V	CLR RAD	GLASS
0			0.00000	1.000000			
1#S		Plane	20.50000	1.000000		2.500	
2 S		42.66940	2.00000	1.616589	36.61	16.200	S-F4
3 S		26.21026	11.00000	1.516798	64.14	17.100	S-BK7
4 S		-43.72923	0.50000	1.000000		17.100	
5 S		79.42181	5.00000	1.617648	55.12	17.700	User- SSK4
6 S		-291.63021	0.50000	1.000000		17.700	
7 S		25.23978	13.30000	1.608808	58.90	17.000	S-SK3
8 S		-27.66022	2.00000	1.648306	33.83	17.000	S-SF12
9 S		24.30016	8.16869	1.000000		14.000	
10 S		Plane				14.463	

LAGRANGE INVARIANT = -1.4434

#	H	U	HBAR	UBAR	D(U/N)	A	ABAR
1	2.50000	0.00000	0.00000	0.57735	0.00000	0.00000	0.57735
2	2.50000	0.00000	11.83568	0.57735	-0.01382	0.05859	0.85473
3	2.45531	-0.02235	12.33837	0.25134	0.00218	0.11531	1.16732
4	2.26111	-0.01765	15.62573	0.29885	-0.04186	-0.10521	-0.08870
5	2.23436	-0.05350	15.76004	0.26863	0.02641	-0.02537	0.46706
6	2.01529	-0.04381	16.21152	0.09030	-0.04806	-0.08205	0.05614
7	1.97772	-0.07514	16.26738	0.11173	0.02768	0.00321	0.75625
8	0.96213	-0.07636	13.94722	-0.17445	0.00275	-0.17881	-1.09187
9	0.81473	-0.07370	13.63085	-0.15818	-0.05503	-0.06621	0.66386
		-0.099739		0.102922			

#	S1	S2	S3	S4	S5	C1	C2
1	0.000000	0.000000	0.000000	0.000000	0.000000	0.000000	0.000000
2	0.000119	0.001731	0.025248	0.018623	0.639992	0.001526	0.022260
3	-0.000071	-0.000722	-0.007309	-0.003235	-0.106732	-0.001446	-0.014633
4	0.001048	0.000883	0.000745	0.016232	0.014313	0.001264	0.001065
5	-0.000038	0.000699	-0.012875	0.010016	0.052651	-0.000393	0.007229
6	0.000652	-0.000446	0.000305	0.002728	-0.002075	0.001146	-0.000784
7	-0.000001	-0.000133	-0.031309	0.031236	-0.017204	0.000041	0.009610
8	-0.000085	-0.000517	-0.003158	-0.001122	-0.026134	-0.000895	-0.005465
9	0.000197	-0.001971	0.019758	-0.033720	0.139990	0.000627	-0.006289
Sum	0.001820	-0.000476	-0.008594	0.040757	0.694800	0.001870	0.012994

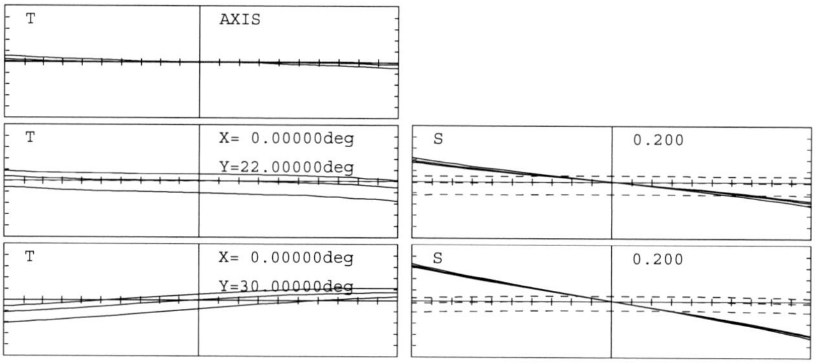

Figure 11.10. Transverse ray aberrations of the Erfle eyepiece.

11.3 Afocal systems for the visible waveband

We will now consider the design of a typical simple binocular telescope, one arm of which is shown in Fig. 11.1. When we are designing the complete system, it can be designed either with the objective first, as in Fig. 11.1, or with the eyepiece first, as in Fig. 11.2. The objective can often be a simple doublet, similar to those described in Chapter 8, and an eyepiece, rather similar to the Ploessl type. Although it is possible to design the objective and eyepiece separately and then reoptimize the eyepiece as part of the complete system, this approach is not usually very helpful.

It is important to appreciate how the various aberrations are corrected in the two parts of the system (Table 11.1).

Table 11.1. Aberration correction in a telescopic system.

	OBJECTIVE	EYEPIECE
S_1	Should be corrected	Should be corrected
S_2	Should be corrected	Should be corrected
S_3	Uncorrectable, but small because K is small	Can be large, because K is large. Must be corrected.
S_4	Uncorrectable, but small because K is small	Usually uncorrectable.
S_5	Automatically small	Potentially large, but should be corrected.
C_1	Should be corrected	Should be corrected
C_2	Automatically small	Must be corrected

The diameter of the objective is primarily determined by the specification, but one does have a free choice of the powers of the lenses, as long as they are in the correct ratio to give the correct angular magnification. It might be useful to consider the effect of increasing the focal length of the objective, keeping the magnification constant by increasing the focal length of the eyepiece. The main effects of increasing the two focal lengths are:

1. Spherical aberration will be improved because the objective will have a lower numerical aperture.
2. Astigmatism will be improved because the powers of the lenses (mainly the eyepiece) are reduced.
3. Field curvature will be improved for the same reason.
4. The length and weight and probably the cost will increase.
5. Because the eyepiece focal length will be larger, the eye relief will probably also be larger. In general, eye relief is a fixed fraction (e.g. 80%) of the eyepiece focal length, although, as we have seen above, it does vary from one eyepiece configuration to another.

There has to be a compromise between weight and performance. It will be essential to determine whether the image has to be inverted with a prism system. If this is to be done, the geometry of the prisms must be worked out so that the effective length of the prism is known, as this must be included in the ray tracing and optimization.

We also have to consider whether we need to correct the intermediate image. This will be important in any instrument with a graticule, or reticle, in the image plane, but does not matter in normal binoculars, other than for the apparent sharpness of the field stop. If the intermediate image does have to be corrected, either the objective or the eyepiece must be designed to work on its own; if not, the complete system can be optimized, varying all parameters (or at least all curves) together. For simple systems, little improvement is obtained by optimizing the complete system. It is very difficult, if not impossible, to decide from first principles how much astigmatism and field curvature is acceptable. However, it seems that an eyepiece with one diopter of these aberrations over most of the field is acceptable for most purposes, although the aberration can be much worse at the edge.

11.3.1 Simple example of a complete telescopic system

As stated above, it is usual to design the objective and eyepiece separately, but we would usually complete the design by ray tracing through the complete system. Very often we could also try reoptimizing the complete system, perhaps changing only the eyepiece, in order to maintain good correction at the intermediate image. The example shown in Fig. 11.1 and tabulated below is quite a simple design, derived from a paper by Williamson.[1] This has a performance that is acceptable for some applications, but not for all.

```
WAVELENGTHS  [nm]
      546.10      656.30      486.10

SURFACES
  # SURF SPACE     RADIUS        SEPN       INDEX1       V       CLR RAD  GLASS
  0                             0.00000   1.000000
  1 S             55.03517      5.00000   1.620669    49.99     12.000  S-SSKN8
  2 S            -51.99181      2.00000   1.746197    28.42     12.000  S-SF3
  3 S           -424.86298     30.00000   1.000000              12.000
  4 S              Plane       85.00000   1.518721    64.38      9.000  S-BK7
  5 S              Plane       11.30000   1.000000               7.000
  6 S            -76.07976      2.00000   1.746197    28.42      8.000  S-SF3
  7 S             11.64547      7.50000   1.566050    61.01      8.000  S-SK11
  8 S            -18.37722      0.20000   1.000000               9.000
  9 S             18.08014      5.00000   1.645140    58.18      9.000  S-LAKN6
 10 S            -44.71992     17.50000   1.000000               9.000
 11#S              Plane        0.00000   1.000000               2.002
 12 S              Plane                                       295.786
```

LAGRANGE INVARIANT = -0.8391

#	H	U	HBAR	UBAR	D(U/N)	A	ABAR
1	12.00000	0.00000	0.00000	0.06993	-0.05152	0.21804	0.06993
2	11.58248	-0.08350	0.21573	0.04315	0.01631	-0.49638	0.06320
3	11.45951	-0.06149	0.29642	0.04034	-0.09228	-0.15447	0.06923
4	7.63468	-0.12749	2.39423	0.06993	0.07222	-0.12749	0.06993
5	0.49905	-0.08395	6.30792	0.04604	-0.07222	-0.12749	0.06993
6	-0.94163	-0.12749	7.09810	0.06993	0.08265	-0.11512	-0.02337
7	-1.09823	-0.07830	7.25793	0.07991	-0.01784	-0.30141	1.22785
8	-1.83441	-0.09816	8.46393	0.16080	-0.03454	0.00260	-0.46945
9	-1.85386	-0.09722	8.46215	-0.00888	0.08574	-0.19975	0.45915
10	-1.94827	-0.01888	7.51746	-0.18894	0.00852	0.04061	-0.58738
11	-2.00007	-0.00296	0.18006	-0.41928	0.00000	-0.00296	-0.41928
		-0.002960		-0.419280			

#	S1	S2	S3	S4	S5	C1	C2
1	0.029395	0.009427	0.003023	0.004900	0.002541	0.020045	0.006429
2	-0.046553	0.005927	-0.000755	-0.000601	0.000173	-0.042398	0.005398
3	0.025232	-0.011309	0.005068	0.000708	-0.002589	0.026615	-0.011928
4	-0.008962	0.004916	-0.002696	0.000000	0.001479	-0.005164	0.002832
5	0.000586	-0.000321	0.000176	0.000000	-0.000097	0.000338	-0.000185
6	0.001031	0.000209	0.000043	-0.003955	-0.000794	0.001630	0.000331
7	-0.001780	0.007250	-0.029533	-0.003983	0.136534	-0.003016	0.012286
8	-0.000000	0.000077	-0.013963	0.013849	0.020476	0.000028	-0.005102
9	0.006342	-0.014578	0.033509	0.015272	-0.112129	0.002496	-0.005738
10	0.000027	-0.000396	0.005726	0.006174	-0.172144	0.000533	-0.007714
11	0.000000	0.000000	0.000000	0.000000	0.000000	0.000000	0.000000
Sum	0.005318	0.001203	0.000599	0.032365	-0.126551	0.001108	-0.003390

Since the system is afocal with an image at infinity, the aberrations are plotted not as transverse ray aberrations but as **angular ray aberrations**. These are the differences (in radians) between the angle of the ray being considered and the relevant chief ray. The interpretation of angular aberration curves is essentially the same as for transverse aberration curves.

In Fig. 11.11 we see that on-axis we have some defocus. Since the angular aberration of the marginal ray is about 0.002 rad, and since the exit pupil radius is 2 mm, this means that the rays are diverging from a point about 1000 mm in front of the eye. For a normal user there is no difficulty accommodating to an image at this distance (1 diopter is quite reasonable), and, in fact, it has been claimed that a normal, relaxed eye does focus at this distance in all instances. This defocus has been introduced deliberately in order to counter the effect of field curvature. The system has a positive Petzval sum, which means that at off-axis field positions up to 3 deg, the image moves away from the eye toward infinity.

At 2 deg (12 deg semi-field in the eye space), the aberrations are very small; and at larger field angles, the aberrations get progressively worse. At 4 deg off-axis there is significant astigmatism. If we consider the tangential section, there is about 5 diopters of aberration, with the image in front of the eye; in the sagittal case there is about 2 diopters of aberration, with the image behind the eye. This amount of aberration would be very noticeable but, as we have mentioned, this is what would be expected from such a simple design.

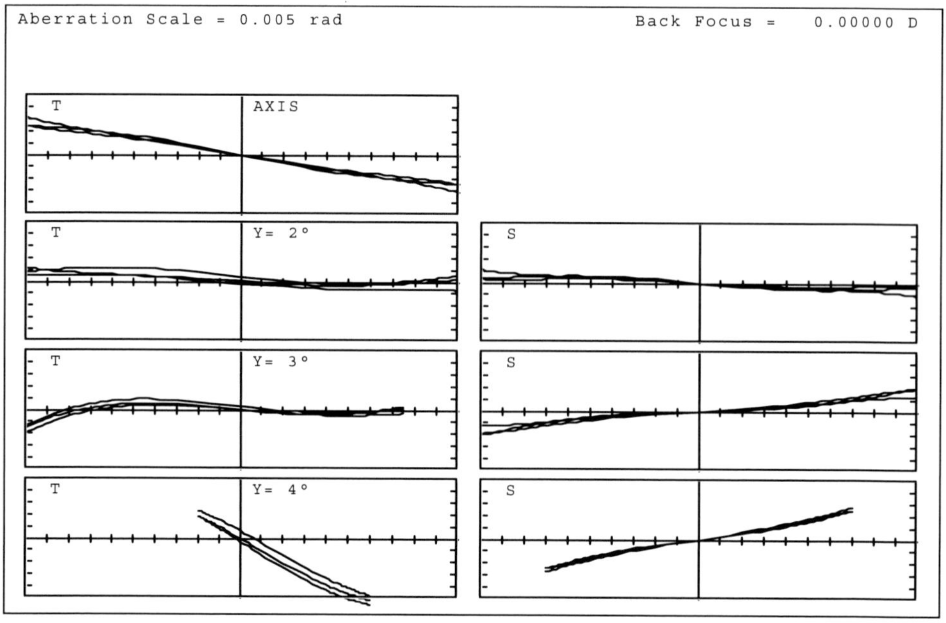

Figure 11.11. Angular ray aberrations of a simple telescopic system.

11.3.2 More complex example of a telescopic system

By using a longer focal length objective and eyepiece, and by inserting field-correcting elements both before and after the intermediate image, enough variables are added to allow a substantial reduction in residual aberrations, most notably S_3, S_4, S_5 and C_2. Rather than regarding this as a redesign exercise of the eyepiece alone, it is better in this case to permit simultaneous rebalancing of the optimization of the telescope as a whole. Williamson[1] has described an example based on an eyepiece design by Scidmore et al.[2] This is shown in Fig. 11.12, with aberrations in Fig. 11.13. The power of the doublet in front of the intermediate image is a compromise—more negative power further reduces S_4 and increases eye relief. However, this sends the chief rays further from the axis and therefore makes the eyepiece lenses bigger.

Figure 11.12. More complex telescopic system.

WAVELENGTHS [nm]
546.10 656.30 486.10

SURFACES

#	SURF SPACE	RADIUS	SEPN	INDEX1	V	CLR RAD	GLASS
0			0.00000	1.000000			
1	S	89.81175	7.00000	1.609935	56.87	12.000	S-SK2
2	S	-31.90963	3.00000	1.616690	44.51	12.000	S-KZFSN4
3	S	692.42487	78.00000	1.000000		12.000	
4	S	Plane	96.00000	1.571243	56.35	10.540	S-BAK4
5	S	Plane	7.00000	1.000000		11.490	
6	S	-52.33766	3.00000	1.609935	56.87	11.610	S-SK2
7	S	23.82229	13.70000	1.711348	30.52	12.880	User-SFN64
8	S	-73.78005	23.00000	1.000000		12.880	
9	S	-45.48018	3.00000	1.711348	30.52	12.800	User-SFN64
10	S	52.56712	2.00000	1.000000		13.750	
11	S	93.39074	12.00000	1.518721	64.38	14.420	S-BK7
12	S	-36.92367	0.20000	1.000000		16.200	
13	S	112.57458	8.30000	1.518721	64.38	16.900	S-BK7
14	S	-56.18798	0.20000	1.000000		17.050	
15	S	37.34785	6.00000	1.518721	64.38	16.290	S-BK7
16	S	-171.91879	34.00000	1.000000		15.930	
17#	S	Plane	0.00000	1.000000		2.001	
18	S	Plane				2.549	

LAGRANGE INVARIANT = -0.8391

#	H	U	HBAR	UBAR	D(U/N)	A	ABAR
1	12.00000	0.00000	-0.02007	0.06993	-0.03144	0.13361	0.06970
2	11.64566	-0.05062	0.28457	0.04352	0.00121	-0.66905	0.05571
3	11.49901	-0.04888	0.41469	0.04337	-0.03855	-0.05218	0.07109
4	6.13351	-0.06879	5.91312	0.07049	0.04093	-0.06879	0.07049
5	1.93067	-0.04378	10.22009	0.04486	-0.04093	-0.06879	0.07049
6	1.44915	-0.06879	10.71354	0.07049	0.04876	-0.09648	-0.13421
7	1.35243	-0.03224	11.07756	0.12134	0.00034	0.03950	0.94398
8	0.89086	-0.03369	12.26386	0.08659	-0.04656	-0.07832	-0.13627
9	-0.63282	-0.06625	12.95265	0.02995	0.04025	-0.05233	-0.25485
10	-0.76630	-0.04449	13.36029	0.13588	-0.06052	-0.10109	0.66749
11	-0.93933	-0.08651	14.18695	0.41333	0.05127	-0.09657	0.56524
12	-1.58169	-0.05353	16.83022	0.22027	-0.02383	-0.01624	-0.35772
13	-1.59350	-0.05908	16.84984	0.09809	0.03665	-0.07323	0.24777
14	-1.87624	-0.03406	16.96162	0.01347	-0.01198	-0.00102	-0.43801
15	-1.88312	-0.03441	16.93439	-0.13613	0.03083	-0.08483	0.31729
16	-1.91575	-0.00544	15.46736	-0.24450	0.00110	0.00867	-0.50797
17	-2.00001	-0.00248	1.25525	-0.41800	0.00000	-0.00248	-0.41800
		-0.002478		-0.418003			

#	S1	S2	S3	S4	S5	C1	C2
1	0.006736	0.003514	0.001833	0.002970	0.002506	0.010682	0.005573
2	-0.006284	0.000523	-0.000044	-0.000057	0.000008	-0.014858	0.001237
3	0.001207	-0.001645	0.002240	-0.000388	-0.002524	0.005142	-0.007005
4	-0.001188	0.001217	-0.001247	0.000000	0.001278	-0.002722	0.002790
5	0.000374	-0.000383	0.000393	0.000000	-0.000402	0.000857	-0.000878
6	-0.000658	-0.000915	-0.001273	-0.005097	-0.008861	-0.000931	-0.001296
7	-0.000001	-0.000017	-0.000406	0.001088	0.016293	0.000372	0.008880
8	0.000254	0.000443	0.000770	0.003967	0.008242	0.000950	0.001653
9	0.000070	0.000340	0.001654	-0.006435	-0.023283	0.000451	0.002196
10	-0.000474	0.003129	-0.020661	-0.005568	0.173182	-0.001055	0.006965
11	0.000449	-0.002629	0.015386	0.002575	-0.105127	0.000481	-0.002817
12	-0.000010	-0.000219	-0.004823	0.006513	0.037230	-0.000136	-0.003002
13	0.000313	-0.001060	0.003585	0.002136	-0.019357	0.000619	-0.002095
14	-0.000000	-0.000010	-0.004314	0.004280	-0.014299	-0.000010	-0.004360
15	0.000418	-0.001563	0.005845	0.006439	-0.045945	0.000848	-0.003170
16	0.000000	-0.000009	0.000545	0.001399	-0.113950	0.000088	-0.005163
17	0.000000	0.000000	0.000000	0.000000	0.000000	0.000000	0.000000
Sum	0.001207	0.000717	-0.000516	0.013823	-0.095009	0.000777	-0.000491

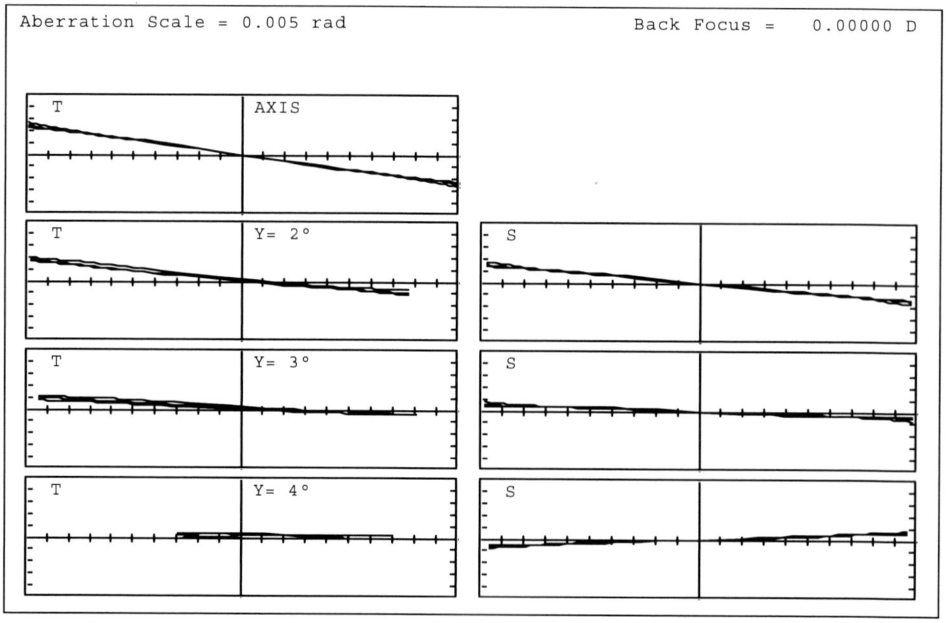

Figure 11.13. Angular ray aberrations of a more complex telescopic system.

11.3.3 Galilean telescopes

A totally different type of afocal system, which we shall now discuss, is the Galilean telescope. In this example, the location of the pupil of the user's eye is shown to the right of Fig. 11.14. We have a positive objective and a negative "eye lens." Since there is no intermediate image, there is no field stop to clearly limit the field of view. Instead, it is the clear aperture of the objective that limits the field. Angular ray aberrations of this design are shown in Fig. 11.15.

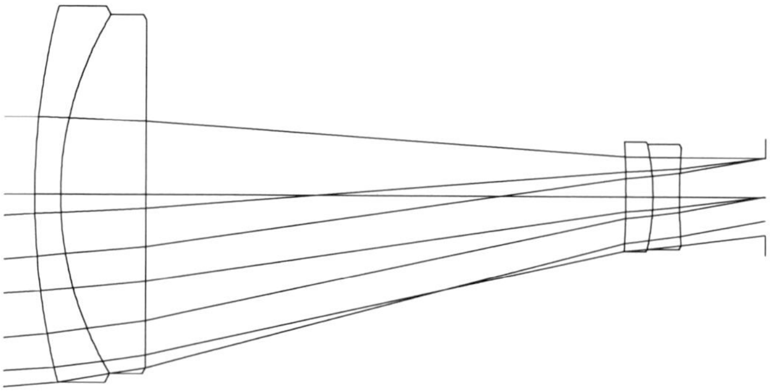

Figure 11.14. The Galilean telescope.

WAVELENGTHS [nm]

587.60	656.30	486.10

SURFACES

#	SURF SPACE	RADIUS	SEPN	INDEX1	V	CLR RAD	GLASS
0			0.17070	1.000000			
1	S	64.82330	2.50000	1.672693	32.20	17.083	S-SF5
2	S	30.93886	8.00000	1.642496	57.94	16.301	S-LAKN6
3	S	Plane	45.00000	1.000000		15.786	
4	S	-71.52104	2.50000	1.672693	32.20	4.986	S-SF5
5	S	-21.97470	2.50000	1.642496	57.94	4.777	S-LAKN6
6	S	60.93983	8.00000	1.000000		4.446	
7#	S	Plane	0.00000	1.000000		3.500	
8	S	Plane				3.504	

LAGRANGE INVARIANT = -0.6124

#	H	U	HBAR	UBAR	D(U/N)	A	ABAR
1	7.00000	0.00000	-12.83976	0.08749	-0.02596	0.10799	-0.11058
2	6.89143	-0.04343	-12.50986	0.13196	0.00153	0.29994	-0.45561
3	6.57038	-0.04013	-11.49423	0.12695	-0.04148	-0.06592	0.20852
4	3.60420	-0.06592	-2.11077	0.20852	0.05447	-0.11631	0.23803
5	3.55635	-0.01914	-1.82879	0.11279	-0.00224	-0.30272	0.32787
6	3.50018	-0.02247	-1.53780	0.11640	0.01368	0.05744	0.14973
7	3.50018	0.00000	-0.13805	0.17497	0.00000	0.00000	0.17497
		0.000000		0.174968			

#	S1	S2	S3	S4	S5	C1	C2
1	0.002119	-0.002170	0.002222	0.002327	-0.004659	0.009442	-0.009669
2	-0.000948	0.001441	-0.002188	-0.000133	0.003527	-0.011864	0.018021
3	0.001184	-0.003746	0.011851	0.000000	-0.037490	0.002924	-0.009250
4	-0.002656	0.005435	-0.011124	-0.002109	0.027082	-0.005236	0.010716
5	0.000729	-0.000789	0.000855	0.000188	-0.001129	0.006179	-0.006692
6	-0.000158	-0.000412	-0.001073	-0.002407	-0.009075	-0.001357	-0.003538
7	0.000000	0.000000	0.000000	0.000000	0.000000	0.000000	0.000000
Sum	0.000270	-0.000241	0.000542	-0.002135	-0.021744	0.000088	-0.000413

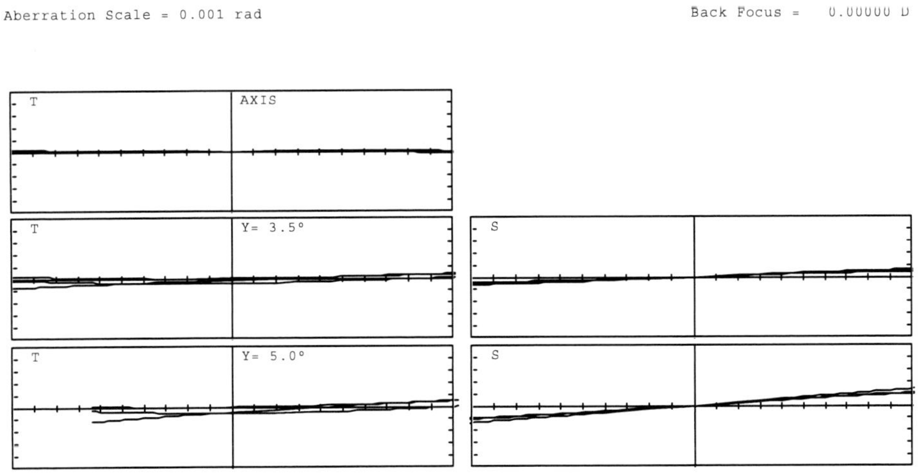

Figure 11.15. Angular ray aberrations of the Galilean telescope.

We would not attempt to design the objective and eye-lens separately, as the conditions for the Galilean telescope are totally different from the normal afocal case of a positive objective and eyepiece. For one thing, the eye pupil is not imaged on the objective, or anywhere else for that matter, so there is no intermediate image, and the objective cannot reasonably be designed on its own. Both components will normally be separately achromatized, in order to correct axial and lateral color. In this particular example we have an angular magnification of 2×; the main uncorrected aberration is field curvature. The Petzval sum is negative because the negative lens has twice the power of the positive lens.

These systems can be quite difficult to design. Invariably the customer wants a telescope that is light and therefore compact, and this increases the powers of the lenses. If the customer also wants a larger magnification than 2, it gets even more difficult to provide a sensible eye relief. In order to achieve a good design, it is essential for the designer to be able to determine what level of aberration is acceptable in any particular case, and this can only be determined experimentally.

Apart from laser beam expanders, Galilean telescopes have three main applications. Traditionally, they are used as "opera glasses" where the light weight and the relatively low magnification are desirable. They are also used by surgeons as operating spectacles, where a magnification of 2 or 3 may be very useful; in this case they are generally fixed to a special pair of spectacles, which restricts the acceptable weight of the optics.

They can also be used to improve the vision of patients with low visual acuity. In this case they are also mounted on a spectacle frame for semi-permanent use. Since the patients, by definition, have impaired visual capabilities, it would be reasonable to suppose that they do not require such good aberration correction as would be the case of operating spectacles. In this case, it may be possible to use single lenses, with obvious savings in cost and weight.

11.3.4 Magnifiers

As we have seen, an eyepiece views the aerial image from the objective lens. A magnifier, on the other hand, directly views an illuminated or self-luminous screen of some sort. There is no pupil imaging function as such, and no exit pupil; like the Galilean telescope, the field of view and optimum viewing position are determined only by the powers and diameters of the lens elements.

There is often a requirement to provide the observer with the largest possible field of view and range of eye movement, and for these so-called "biocular" magnifiers, in which both eyes share a common optic, each needs to see as much of the field of view as possible. The simultaneous need for a short focal length and large lens diameters can push the required f-number close to the *f*/0.5 limit.

Petzval lens derivatives are the natural choice for such extreme f-numbers, and the considerations for S_4 correction that we saw in Chapter 9 apply also to such magnifiers. Sometimes the screen can be curved, or else an attempt can be made to reduce S_4, bearing in mind the defocus range in diopters that the eye can accommodate. Figure 11.16 shows one such design, by Rogers et al.[4] If rays through a magnifier are traced "forwards," as shown in the figure, then the aberrations of Fig. 11.17 are assessed as for an afocal system, since acceptable tolerances for comfortable viewing are expressed in angular terms in eye space.

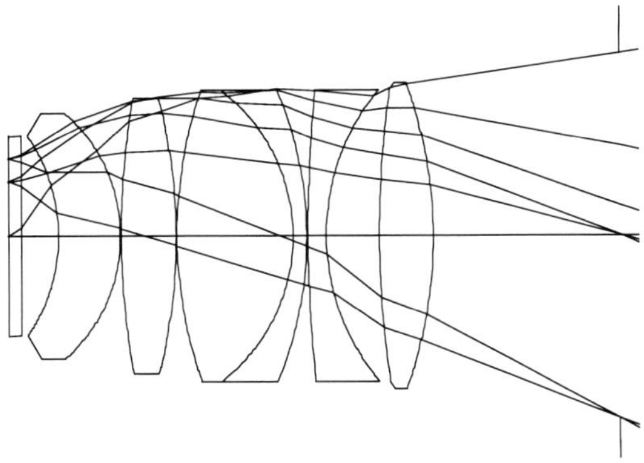

Figure 11.16. A biocular magnifier.

```
EFL =     49.171

OBJECT DISTANCE = 0.0001, CURVATURE = 0, TILT ANGLE = 0

WAVELENGTHS   [nm]
      587.60        656.30        486.10
```

SURFACES

# SURF SPACE	RADIUS	SEPN	INDEX1	V	CLR RAD	GLASS
0		0.00000	1.000000			
1 S	Plane	3.47000	1.620408	60.30	26.000	S-SK16
2 S	Plane	0.05000	1.000000		26.000	
3 S	-45.31000	16.61000	1.850260	32.17	26.000	User-LASFN9
4 S	-41.94000	0.24000	1.000000		32.000	
5 S	195.81000	14.60000	1.744000	44.72	36.000	User-LAF2
6 S	-145.34000	0.24000	1.000000		36.000	
7 S	112.37000	31.14000	1.620408	60.30	38.000	S-SK16
8 S	-47.24000	3.75000	1.748400	27.71	38.000	User-SF63
9 S	-95.12000	0.24000	1.000000		38.000	
10 S	349.66000	4.81000	1.654460	33.65	38.000	User-SF9
11 S	58.28000	14.20000	1.000000		38.000	
12 S	256.57000	14.41000	1.620408	60.30	38.000	S-SK16
13 S	-115.95000	50.00000	1.000000		40.000	
14#S	Plane	5.00000	1.000000		47.710	
15 S	Plane	0.00000	1.000000		100.000	
16 S	Plane				143.623	

LAGRANGE INVARIANT = 16.4000

#	H	U	HBAR	UBAR	D(U/N)	A	ABAR
1	0.00008	0.82000	20.00005	0.44651	-0.50770	0.82000	0.44651
2	1.75606	0.50605	20.95621	0.27555	0.50770	0.82000	0.44651
3	9.83306	0.82000	25.35433	0.44651	-0.52658	0.60298	-0.11307
4	18.85076	0.54291	33.63386	0.49847	0.32893	0.17289	-0.56153
5	19.00013	0.62236	33.69156	0.24043	-0.44147	0.71939	0.41249
6	23.60584	0.31546	34.63261	0.06446	0.24844	0.26690	-0.30316
7	23.70888	0.42932	34.61704	-0.06487	-0.31567	0.64031	0.24319
8	29.44380	0.18417	29.69740	-0.15798	0.01007	-0.71155	-1.27467
9	30.25497	0.21631	29.32091	-0.10040	0.01643	-0.17792	-0.71448
10	30.28861	0.14015	29.22341	-0.40623	-0.10966	0.22678	-0.32266
11	30.53126	0.05045	27.88335	-0.27860	0.39582	0.95019	0.33063
12	36.58494	0.42632	25.78442	-0.14781	-0.29765	0.56891	-0.04732
13	39.58938	0.20850	23.91549	-0.12970	-0.00265	-0.21541	-0.54438
14	45.89045	0.12602	7.00927	-0.33812	0.00000	0.12602	-0.33812
15	46.52055	0.12602	5.31865	-0.33812	0.00000	0.12602	-0.33812
		0.126021		-0.338124			

#	S1	S2	S3	S4	S5	C1	C2
1	0.000028	0.000015	0.000008	0.000000	0.000005	0.000000	0.000000
2	-0.599485	-0.326433	-0.177750	0.000000	-0.096789	-0.009143	-0.004979
3	1.882610	-0.353010	0.066193	-2.727801	0.499081	0.084695	-0.015881
4	-0.185332	0.601955	-1.955132	2.946987	-3.221518	-0.046553	0.151204
5	4.340970	2.489054	1.427190	0.585975	1.154322	0.130399	0.074769
6	-0.417787	0.474539	-0.539001	0.789458	-0.284479	-0.060108	0.068273
7	3.068507	1.165408	0.442618	0.916411	0.516154	0.096393	0.036610
8	-0.150052	-0.268804	-0.481537	-0.257215	-1.323404	-0.190555	-0.341362
9	-0.015739	-0.063206	-0.253824	1.210344	3.841206	0.083139	0.333870
10	0.170818	-0.243039	0.345793	0.304277	-0.924914	0.080737	-0.114872
11	-10.911032	-3.796583	-1.321052	-1.825555	-1.094888	-0.340997	-0.118653
12	3.524415	-0.293122	0.024379	0.401361	-0.035408	0.132156	-0.010991
13	0.004865	0.012294	0.031069	0.888116	2.322915	0.054149	0.136843
14	0.000000	0.000000	0.000000	0.000000	0.000000	0.000000	0.000000
15	0.000000	0.000000	0.000000	0.000000	0.000000	0.000000	0.000000
Sum	0.712787	-0.600932	-2.391047	3.232359	1.352281	0.014312	0.194830

Aberration Scale = 0.2 rad Back Focus = 0.00000 D

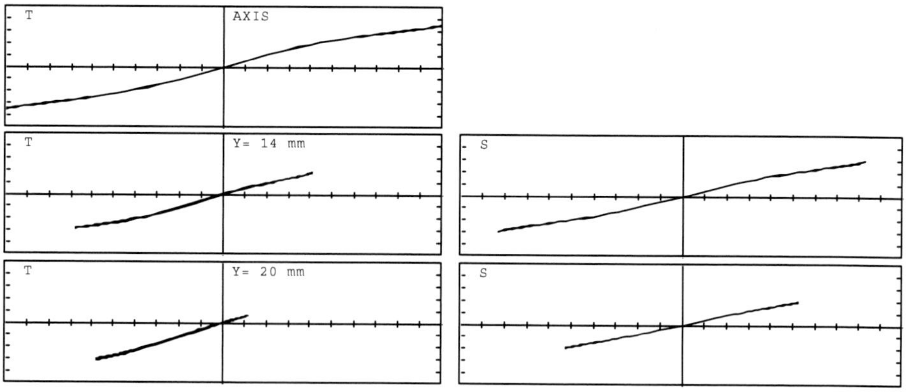

Figure 11.17. Angular ray aberrations of a biocular magnifier.

References

1. D. Williamson, "The eye in optical systems," SPIE Proc. Vol. **531**, p. 136 (1985).
2. Scidmore et al., U.S. Patent 3,384,434 (1968).
3. M. Freeman and D. Freeman, "Innovative binocular design," SPIE Proc. Vol. **1780**, p. 711-720 (1993).
4. P. Rogers and M. Roberts, U.S. Patent 4,183,624 (1980).
5. P. Rogers, "Compact viewing optics using polarization," SPIE Proc. Vol. **656**, p. 90-95 (1986).
6. P. Rogers, "Holographic heads-up display for air and ground applications," SPIE Proc. Vol. **1013**, p. 136 (1988).

CHAPTER 12
THERMAL IMAGING LENSES

In the infrared (IR) region of the spectrum, there are two atmospheric wavelength bands, or "windows," where the absorption due to atmospheric gases is low. One is from about 3 to 5 μm, the other from about 8 to 12 μm. An important difference between design in the visible and in the IR is that in the visible, objects are almost always detected by the light that they reflect; whereas in the IR region, objects are detected by the radiation that they emit.

12.1 Photon detection

The number of photons emitted by a black body at temperature T in degrees kelvin is given by

$$M = \frac{2\pi c}{[\exp(1.4388/\lambda T) - 1]\lambda^4}, \tag{12.1}$$

where

M = photon emittance per second per square centimeter, per centimeter of wavelength
c = the speed of light in vacuo, in centimeter per second
λ = wavelength in centimeter.

Because in most thermal imagers an object is detected by the difference between the number of photons emitted by the object and by its background, the signal detected is proportional to

$$S = (\partial M/\partial T)\Delta T, \tag{12.2}$$

where ΔT is the temperature difference between the object and the background. Table 12.1 indicates the emittance at a range of wavelengths for an object at a typical temperature of 293 K (20°C), and it also lists the signal per degree kelvin ($\partial M/\partial T$).

Table 12.1. Output of a room temperature object at different wavelengths.

Wavelength (μm)	Emittance	Signal (per degree K)
3.0	1.81×10^{18}	1.01×10^{17}
4.0	3.43×10^{19}	1.44×10^{18}
5.0	1.64×10^{20}	5.49×10^{18}
8.0	9.95×10^{20}	2.09×10^{19}
10.0	1.40×10^{21}	2.36×10^{19}
12.0	1.54×10^{21}	2.19×10^{19}

From this table it can be determined that the emittance from an object at this temperature is greater in the 8- to 12-μm atmospheric transmission window than it is in the 3- to 5-μm waveband, by a factor of about 30. The signal is larger by a factor of about 15. For thermal imagers with detectors that detect radiation rather than photons, the ratios are similar.

In most cases, lens designs for the two IR regions are rather different, because of the different materials used to fabricate the lenses in the two cases. Some systems, however, do cover both wavebands, and, as may be expected, the range of suitable materials for these is even more restricted.

12.1.1 8- to 13-μm waveband

Although the published literature describes many materials that are claimed to transmit in this waveband, the number of materials that are widely used is very small. Table 12.2 gives details of the most common materials. Note that index data varies between manufacturers' data sheets, because of different manufacturing techniques.

Table 12.2. Commonly used materials in the 8- to 12-μm region.

Material	*n* (10 μm)	*n* (12 μm)	*n* (8 μm)	V
Germanium	4.0032	4.0023	4.0052	1030
Zinc selenide	2.4065	2.3930	2.4173	58
Zinc sulphide	2.2004	2.1699	2.2232	22

For many purposes germanium is the obvious material to use, because of its high refractive index and low dispersion. Zinc selenide and zinc sulphide can be used as negative elements to correct chromatic aberration. For our examples, we will use germanium exclusively.

The thermal image is detected either by a heat-sensitive image tube or by a detector array, across which the image is scanned. In principle, the detector does not affect the lens design. In this chapter we will not consider the problems associated with this topic, because of its specialized nature.

12.1.2 3- to 5-μm waveband

In this waveband many materials are theoretically available, but only a few are in common use. Table 12.3 lists these materials; all are crystalline and therefore subject to different fabrication techniques.

Table 12.3. Commonly used materials in the 3- to 5-μm region.

Material	n(4 μm)	n(5 μm)	n(3 μm)	V
Silicon	3.4253	3.4220	3.4323	235
Germanium	4.0254	4.0159	4.0442	107
Zinc selenide	2.4331	2.4295	2.4376	177
Zinc sulphide	2.2518	2.2461	2.2572	113

In this waveband, germanium does not have a low dispersion, so that germanium aspheric single lenses that we will describe in the 8- to 12-μm band are not useful for the 3- to 5-μm band, as we must correct chromatic aberration. But chromatic correction is possible in a doublet, since with silicon as the low-dispersion "crown," the obvious material for the "flint" is germanium.

12.2 Single-material lenses

Because in the 8- to 12-μm region germanium is unique in that it has both a high refractive index and a low dispersion, high-performance thermal objectives made of this material alone are feasible. The examples that follow provide an excellent introduction to the effect of simple design changes such as the addition of more lenses and the effect of aspheric surfaces. It also demonstrates some of the approaches that were explored in earlier chapters.

In almost all IR applications, the field of view in the object space is quite small, and it is often possible to use quite simple designs consisting of one or two germanium components. The chromatic aberration can be left uncorrected, since the dispersion of germanium across the 8- to 12-μm band is low. The V value is 1030—remember that a high V-number indicates low dispersion.

In Chapter 4 we discussed the transverse ray aberrations resulting from monochromatic Seidel aberrations. It can easily be shown that the transverse ray aberration resulting from axial color is given by

$$\delta\eta' = C_1/u' = (h^2K)/(u'V) = h/V. \tag{12.3}$$

If, for example, we have an all-germanium lens system with f'= 100 mm and an aperture of f/1.4, then since V = 1030, we will have

$$K = 1/100 = 0.01\ \text{mm}^{-1};\ h = 100/(2 \times 1.4) = 35\ \text{mm}.$$

Therefore,

$$\delta\eta' = 35/1030 = 0.034 \text{ mm}.$$

Since typical detector array elements are about 0.050 mm square, it is clear that the use of germanium optics, uncorrected for axial color, is quite possible.

12.2.1 Single germanium lens

A single germanium lens can be corrected for spherical aberration and coma over the wide apertures needed especially for use with uncooled detectors by the addition of a single aspheric surface (Fig. 12.1). With the stop at the lens, the astigmatism is uncorrected and has a value very close to that given by the formula for the astigmatism of a thin lens:

$$S_3 = H^2K. \tag{7.8}$$

The flat ray-intercept curves in Fig. 12.2, and the spot diagrams in Fig. 12.3, show that spherical and coma have been corrected. They also show that systems of this type are limited by axial color and astigmatism. However, these are quite acceptable under the conditions represented by this example. The use of an aspheric surface on a germanium lens is really only economical if a machine tool using single-point diamond turning is available to generate the surface.

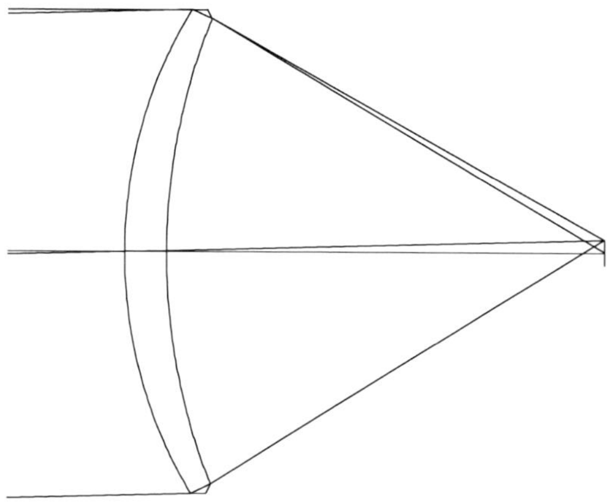

Figure 12.1. Single aspheric IR germanium lens.

```
EFL =    100.000

WAVELENGTHS   [nm]
    10000.00     12000.00      8000.00

SURFACES
  # SURF SPACE     RADIUS       SEPN        INDEX1       V      CLR RAD  GLASS
  0                           0.00000    1.000000
  1#A             94.12232    9.00000    4.003286               50.000  User-GE
  2 S            127.25048   92.82652    1.000000               48.176
  3 S              Plane                                         2.669
ASPHERIC SURFACE  1 CC =  -0.078689  (ELLIPSE)
 A6 = 5.26148e-014

LAGRANGE INVARIANT =      -1.3093

  #      H          U         HBAR       UBAR      D(U/N)       A         ABAR
  1  50.00000   0.00000   0.00000   0.02619  -0.09955   0.53122   0.02619
  2  46.41326  -0.39853   0.05887   0.00654  -0.40045  -0.13526   0.02804
               -0.50000              0.02758

  #     S1         S2         S3         S4         S5         C1         C2
  1  1.404642  0.069240  0.003413  0.013663  0.000842  0.023120  0.001140
  2  0.340043 -0.070487  0.014611 -0.010106 -0.000934  0.005465 -0.001133
     ASPHERIC  SURFACE  1
    -1.771376  0.000000  0.000000  0.000000  0.000000

Sum -0.026692 -0.001247  0.018024  0.003557 -0.000092  0.028585  0.000007
```

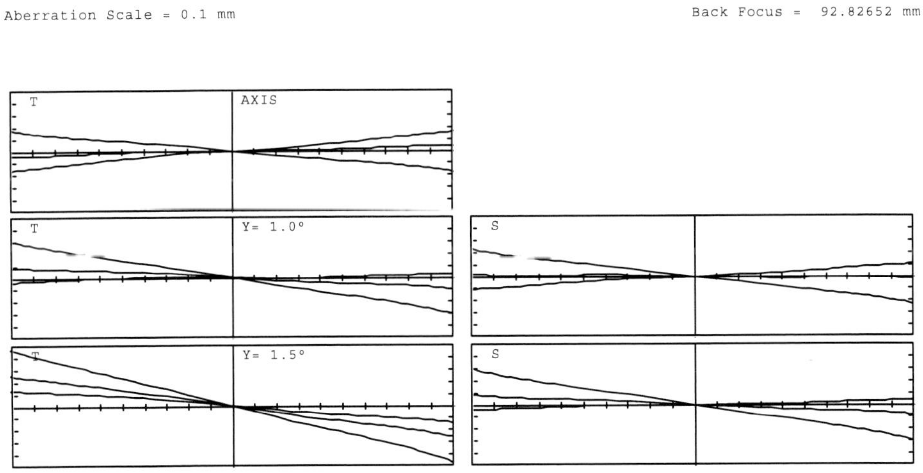

Figure 12.2. Transverse ray aberrations of single aspheric IR germanium lens.

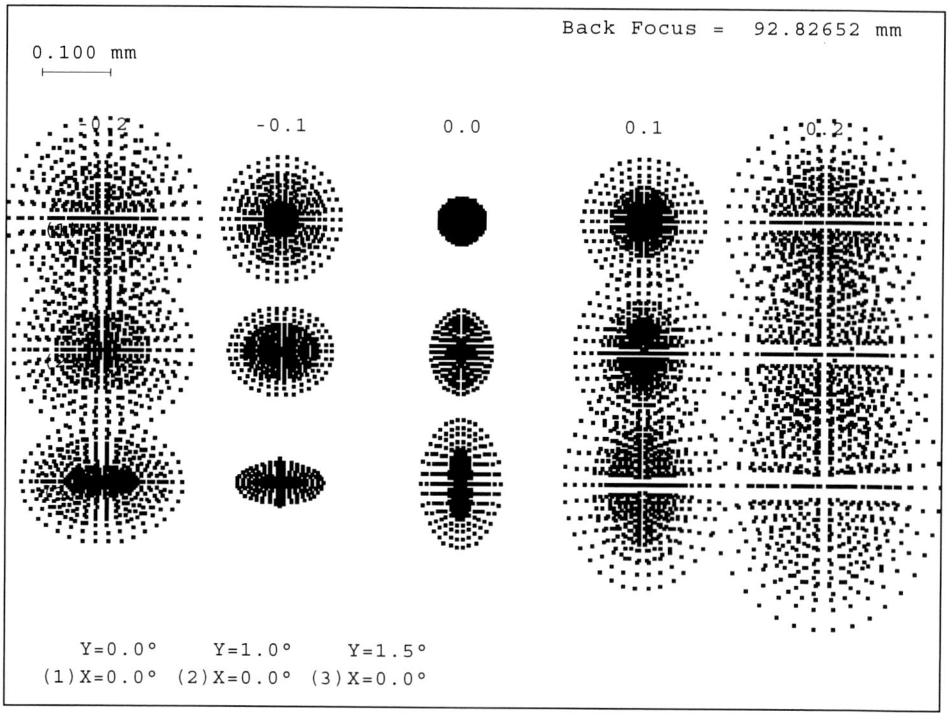

Figure 12.3. Spot diagrams of single aspheric IR germanium lens.

12.2.2 Germanium doublets

In these next examples, two single lenses with spherical surfaces replace the single aspheric lens. The obvious advantage is that the difficulty of manufacturing the aspheric is avoided, but there are two extra air-germanium surfaces to be ground, polished, and coated. Also, the lens will clearly be heavier. Several solutions exist (up to five, we believe), but two solutions are preferable to the others. These are given here.

12.2.2.1 Plus-minus germanium doublet solution

The design of an air-spaced germanium-germanium doublet is easy. The negative element in Fig. 12.4 corrects the positive spherical aberration of the positive element, but being also made of germanium it cannot correct chromatic aberration. Thus, the powers can be chosen freely. The overall aberration shown in Figs. 12.5 and 12.6 is essentially the same as in the aspheric case, except that now the uncorrected spherical aberration is the limiting monochromatic aberration. Note the extreme ray angles in the space between the two elements, giving rise to large opposing S_1 and S_2 surface contributions across the air gap.

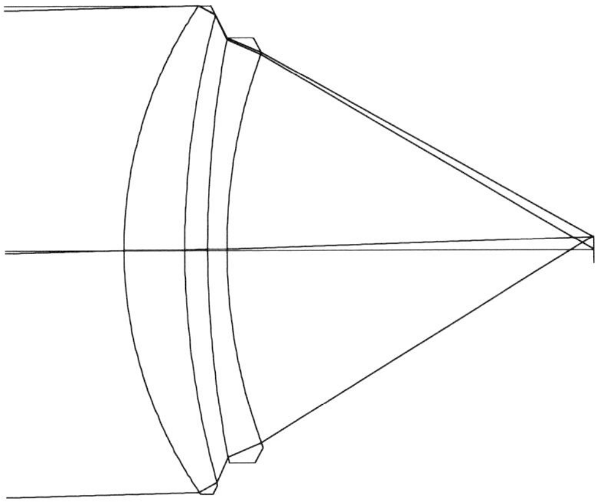

Figure 12.4. Plus-minus germanium doublet solution.

EFL = 100.000

WAVELENGTHS [nm]
10000.00 12000.00 8000.00

SURFACES

# SURF	SPACE	RADIUS	SEPN	INDEX1	V	CLR RAD	GLASS
0			0.00000	1.000000			
1#S		87.41224	12.66040	4.003286		50.000	User-GE
2 S		177.93926	4.81037	1.000000		48.284	
3 S		225.47366	4.00000	4.003286		43.516	User-GE
4 S		115.70070	76.85146	1.000000		40.566	
5 S		Plane				2.820	

LAGRANGE INVARIANT = -1.3093

#	H	U	HBAR	UBAR	D(U/N)	A	ABAR
1	50.00000	0.00000	0.00000	0.02619	-0.10719	0.57200	0.02619
2	44.56718	-0.42912	0.08281	0.00654	-0.85848	-0.71521	0.02805
3	39.92193	-0.96567	0.21550	0.02758	0.87224	-0.78862	0.02854
4	38.42573	-0.37405	0.24019	0.00617	-0.40656	-0.16789	0.03302
		-0.50000		0.03095			

#	S1	S2	S3	S4	S5	C1	C2
1	1.753585	0.080278	0.003675	0.014712	0.000842	0.024895	0.001140
2	19.571162	-0.767537	0.030101	-0.007227	-0.000897	0.027746	-0.001088
3	-21.656048	0.783714	-0.028362	0.005704	0.000820	-0.027405	0.000992
4	0.440336	-0.086616	0.017038	-0.011115	-0.001165	0.005615	-0.001105
Sum	0.109036	0.009839	0.022452	0.002073	-0.000400	0.030852	-0.000061

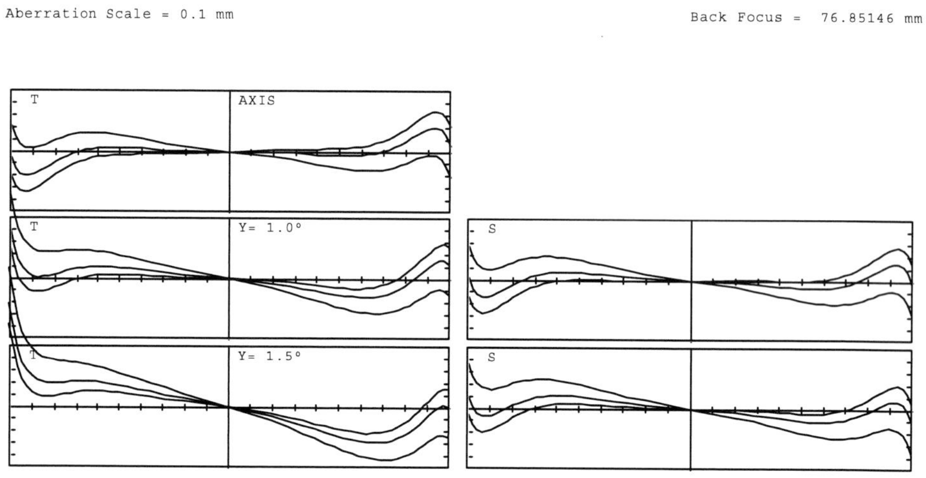

Figure 12.5. Transverse ray aberrations of plus-minus germanium doublet solution.

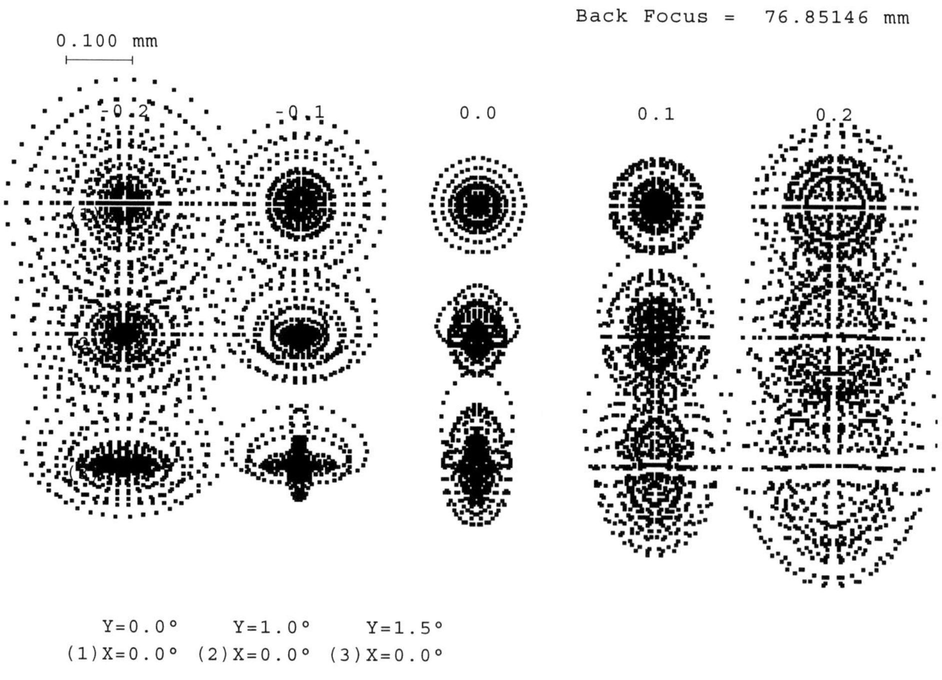

Figure 12.6. Spot diagrams of plus-minus germanium doublet solution.

12.2.2.2 Plus-plus germanium doublet solution

It turns out that it is also possible to correct spherical aberration with two positive lenses, as shown in Fig. 12.7. Because the second lens operates in a converging beam, it can, and does have enough negative spherical aberration to correct the spherical aberration of the first lens. Thus, the performance shown in Figs. 12.8 and 12.9 is close to that of the aspheric single lens. Note that the surface contributions of S_1 and S_2 are very much smaller than the plus-minus solution.

This correction of S_1 with two positive lenses is only possible because of the high index of germanium. With lower-index materials, as we have already seen in Chapter 7, it is possible to correct spherical aberration with all-positive systems, but more elements are needed.

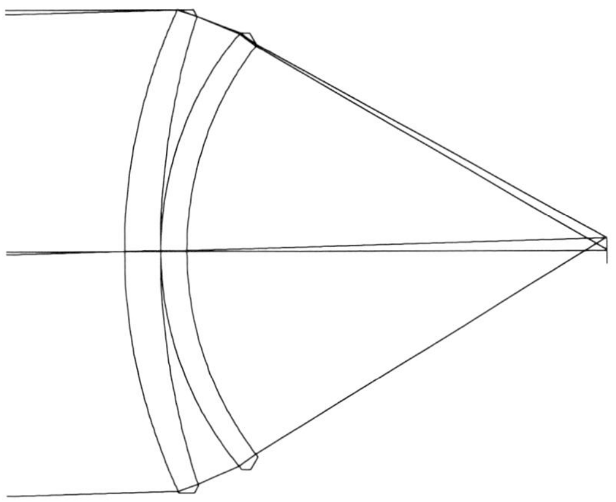

Figure 12.7. Plus-plus germanium doublet solution.

EFL = 100.000

WAVELENGTHS [nm]
10000.00 12000.00 8000.00

SURFACES

# SURF SPACE	RADIUS	SEPN	INDEX1	V	CLR RAD	GLASS
0		0.00000	1.000000			
1#S	115.85208	7.50000	4.003286		50.000	User-GE
2 S	152.38799	0.10000	1.000000		48.558	
3 S	68.44854	5.50000	4.003286		45.130	User-GE
4 S	68.23840	88.35523	1.000000		42.609	
5 S	Plane				2.691	

LAGRANGE INVARIANT = -1.3093

#	H	U	HBAR	UBAR	D(U/N)	A	ABAR
1	50.00000	0.00000	0.00000	0.02619	-0.08088	0.43158	0.02619
2	47.57167	-0.32378	0.04906	0.00654	-0.27775	-0.04645	0.02747
3	47.53581	-0.35862	0.05177	0.02715	0.20610	0.33585	0.02791
4	44.17761	-0.61058	0.08596	0.00622	-0.34748	0.14740	0.02992
		-0.50000		0.02866			

#	S1	S2	S3	S4	S5	C1	C2
1	0.753238	0.045702	0.002773	0.011101	0.000842	0.018784	0.001140
2	0.028506	-0.016861	0.009974	-0.008439	-0.000908	0.001923	-0.001138
3	-1.105100	-0.091833	-0.007631	0.018788	0.000927	0.013897	0.001155
4	0.333529	0.067710	0.013746	-0.018846	-0.001035	-0.005668	-0.001151
Sum	0.010172	0.004717	0.018861	0.002604	-0.000174	0.028936	0.000006

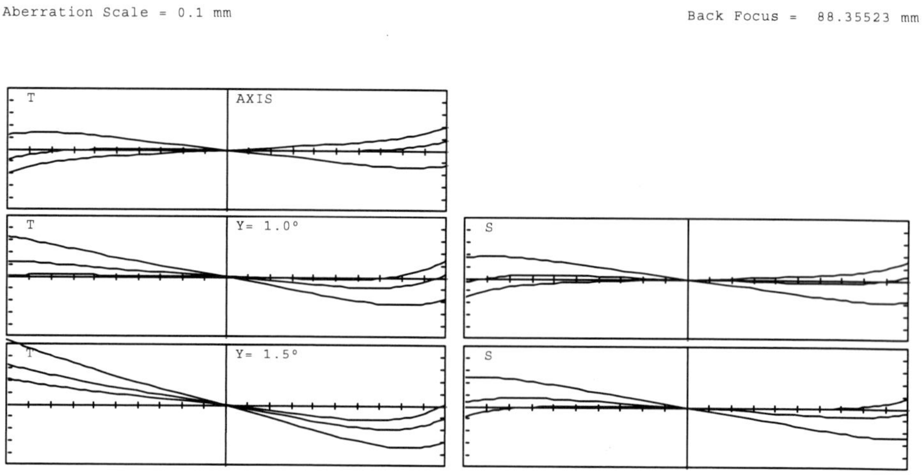

Figure 12.8. Transverse ray aberrations of plus-plus germanium doublet solution.

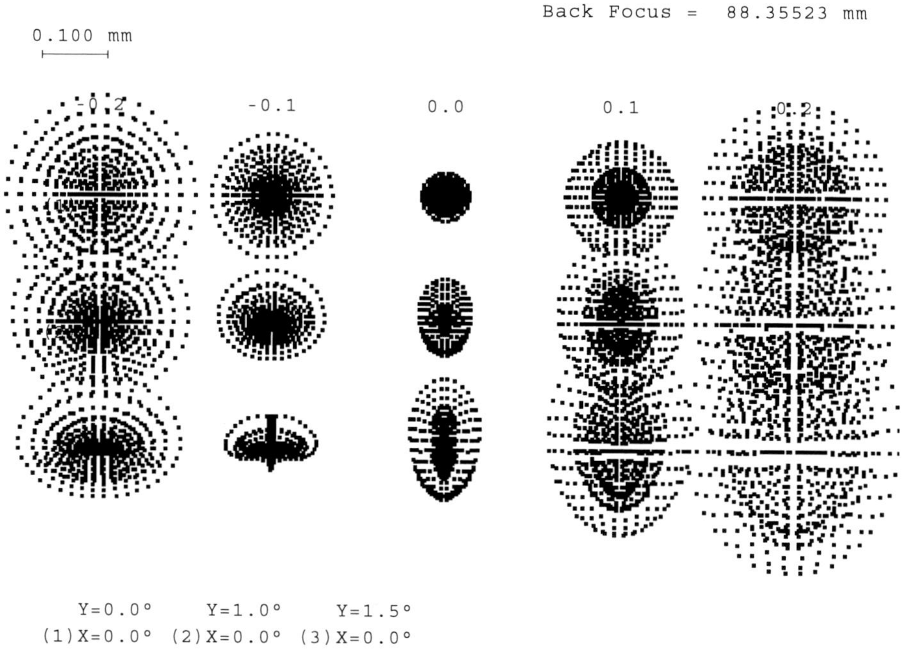

Figure 12.9. Spot diagrams of plus-plus germanium doublet solution.

12.2.3 Germanium Petzval lens

In the three designs shown above, in which the lenses are all essentially at the stop, the astigmatism is fixed and uncorrectable, and this limits the possible field of view. In order to correct astigmatism, separated elements are necessary. The simplest solution, which uses an aspheric on the first surface, is the aspheric Petzval type lens shown in Fig. 12.10. This design, which also operates at *f*/1, gives an improved field of view of 5 deg. Some spherical aberration is evident at all fields in Fig. 12.11, but the main limiting aberrations are axial color C_1 and, at higher field angles, tangential coma, as shown in Fig. 12.12.

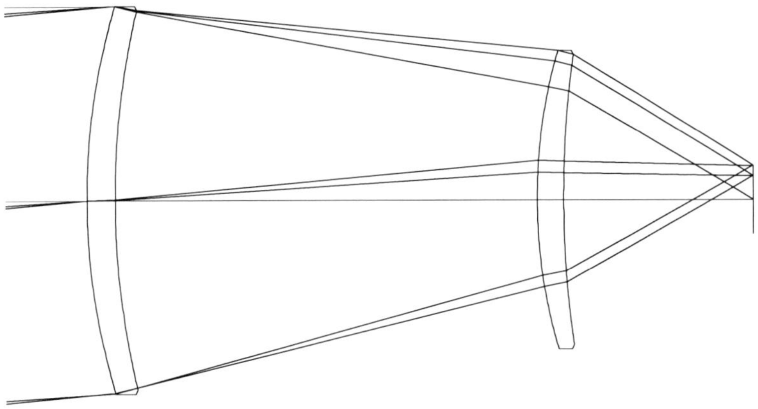

Figure 12.10. Germanium Petzval lens.

```
EFL =    100.000

WAVELENGTHS   [nm]
    10000.00      12000.00       8000.00

SURFACES
  # SURF SPACE     RADIUS         SEPN        INDEX1        V       CLR RAD  GLASS
  0                             0.00000    1.000000
  1#A               165.90967     7.50000    4.003286                50.000  User-GE
  2 S               204.94286   110.92995    1.000000                49.012
  3 S               133.68094     7.00000    4.003286                38.562  User-GE
  4 S               258.53503    50.08480    1.000000                37.555
  5 S                 Plane                                           8.762
ASPHERIC SURFACE   1 CC =   -0.502237   (ELLIPSE)
 A6 = -7.65614e-013

LAGRANGE INVARIANT =      -4.3744

  #      H          U         HBAR       UBAR      D(U/N)       A         ABAR
  1  50.00000   0.00000   0.00000   0.08749  -0.05648   0.30137   0.08749
  2  48.30434  -0.22609   0.16391   0.02185  -0.14076   0.03846   0.09069
  3  26.42535  -0.19723  10.13546   0.08989   0.14788   0.00044   0.16571
  4  25.04240  -0.19756   9.89449  -0.03443  -0.45065  -0.40314   0.01540
               -0.50000            -0.02287

  #      S1         S2         S3         S4         S5         C1         C2
  1  0.256465   0.074453   0.021614   0.086527   0.031394   0.013116   0.003808
  2  0.010059   0.023718   0.055921  -0.070047  -0.033306  -0.001617  -0.003813
  3 -0.000001  -0.000286  -0.107307   0.107388   0.030236   0.000010   0.003812
  4  1.834091  -0.070052   0.002676  -0.055527   0.002019   0.008788  -0.000336
     ASPHERIC  SURFACE   1
    -2.064288   0.000000   0.000000   0.000000   0.000000

Sum 0.036326   0.027832  -0.027096   0.068340   0.030342   0.020297   0.003471
```

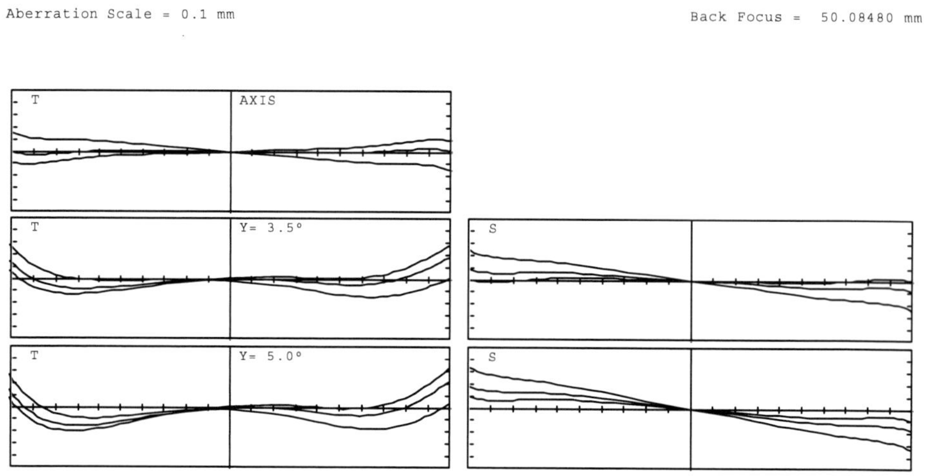

Figure 12.11. Transverse ray aberrations of germanium Petzval lens.

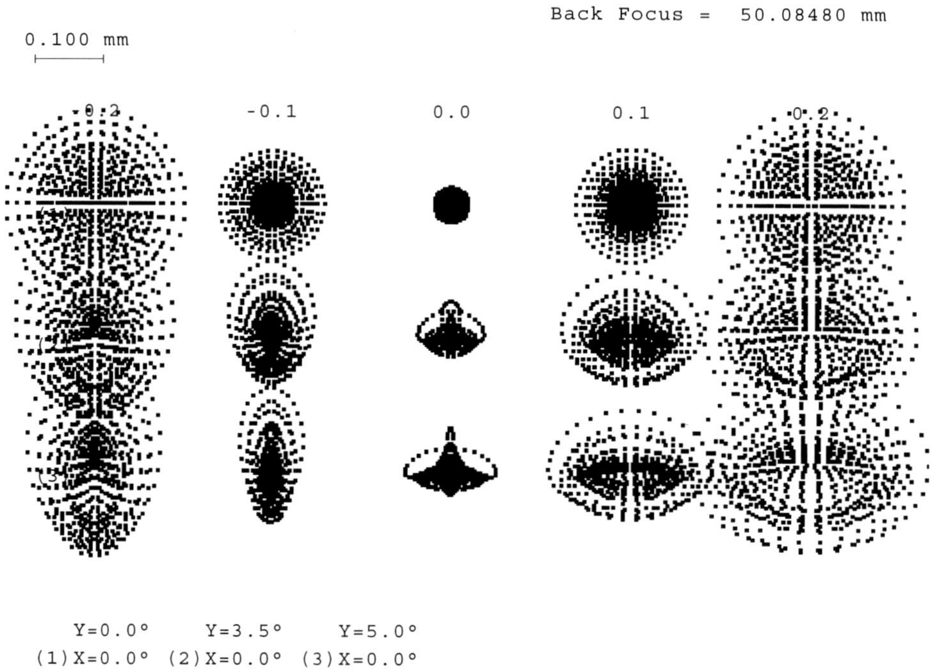

Figure 12.12. Spot diagrams of germanium Petzval lens.

12.2.4 Germanium triplet

Replacing the single aspheric element in the front of the germanium Petzval lens with a germanium-germanium doublet with spherical surfaces gives an all-germanium triplet. Such a design is as shown in Fig. 12.13. Comparison of Figs. 12.14 and 12.15 with the previous lens shows that this lens delivers rather better performance than the aspheric version, especially at the edge of the field.

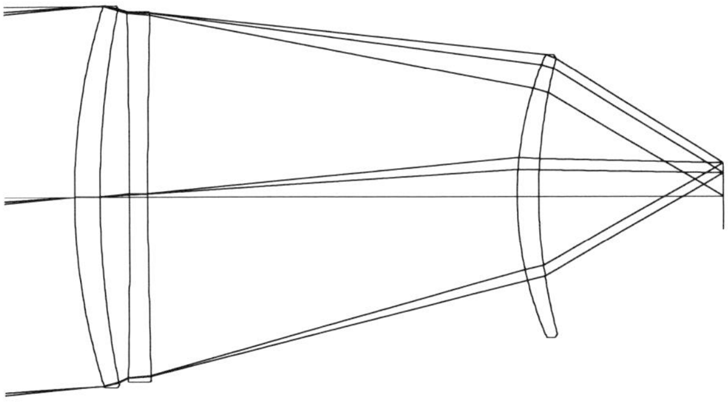

Figure 12.13. All-germanium triplet.

```
EFL =    100.000

WAVELENGTHS   [nm]
    10000.00     12000.00      8000.00

SURFACES
  # SURF SPACE     RADIUS         SEPN        INDEX1      V       CLR RAD  GLASS
  0                             0.00000     1.000000
  1#S             163.19953     6.85558     4.003286              50.233  User-GE
  2 S             247.61351     7.90596     1.000000              49.379
  3 S           -2300.04402     5.00000     4.003286              48.737  User-GE
  4 S            1655.01705    99.49517     1.000000              48.438
  5 S              93.05695     5.74464     4.003286              37.257  User-GE
  6 S             137.57720    49.99240     1.000000              36.269
  7 S               Plane                                          8.720

LAGRANGE INVARIANT =      -4.3744

  #      H          U          HBAR       UBAR      D(U/N)       A          ABAR
  1  50.00000   0.00000    0.00000    0.08749   -0.05741    0.30637    0.08749
  2  48.42429  -0.22984    0.14982    0.02185   -0.27538   -0.13723    0.08991
  3  45.79325  -0.33279    0.85587    0.08931    0.31576   -0.35270    0.08893
  4  45.45228  -0.06819    0.96881    0.02259   -0.17348   -0.16305    0.09277
  5  26.49671  -0.19052   10.14041    0.09218    0.12527    0.09422    0.20115
  6  24.99620  -0.26120    9.80307   -0.05872   -0.43475   -0.31831    0.05017
               -0.50000              -0.02109

  #     S1         S2         S3         S4         S5         C1         C2
  1  0.269455   0.076946   0.021973   0.087964   0.031394   0.013334   0.003808
  2  0.251120  -0.164532   0.107800  -0.057976  -0.032644   0.005784  -0.003790
  3 -1.798756   0.453555  -0.114364  -0.006241   0.030411  -0.014059   0.003545
  4  0.209642  -0.119272   0.067858  -0.008674  -0.033671   0.006451  -0.003670
  5 -0.029466  -0.062908  -0.134304   0.154267   0.042621   0.002173   0.004639
  6  1.101087  -0.173537   0.027350  -0.104346   0.012135   0.006926  -0.001092

Sum 0.003082   0.010253  -0.023687   0.064994   0.050245   0.020610   0.003440
```

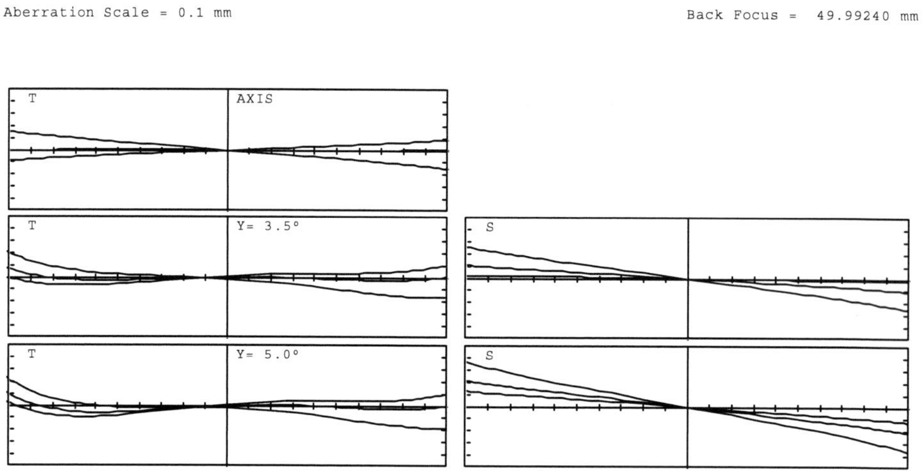

Figure 12.14. Transverse ray aberrations of all-germanium triplet.

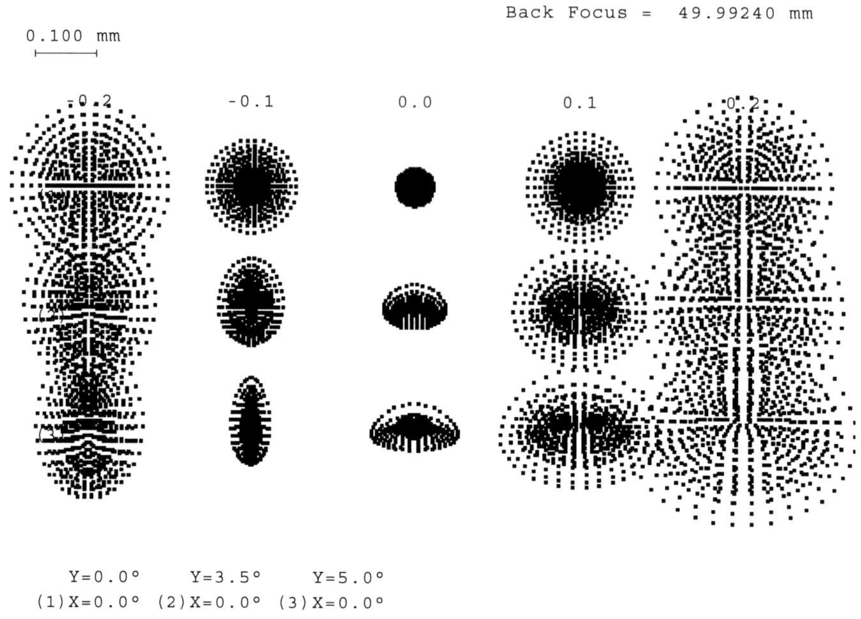

Figure 12.15. Spot diagrams of all-germanium triplet.

12.3 Multiple-material lenses

In the 3- to 5-μm waveband, it is possible to design fairly conventional split color-corrected doublets using silicon in the positive element and either germanium, zinc selenide or zinc sulphide in the negative element. Since both zinc selenide and the "multispectral" type of zinc sulphide are even more expensive than germanium, a silicon-germanium doublet is the obvious starting point. If larger fields are required, then a triplet with silicon-germanium-silicon is possible. Note that the overall shapes and spacing of the example shown in Fig. 12.16 are quite similar to the all-germanium triplet shown in Fig. 12.13. Residual aberrations of this design are shown in Figs. 12.17 and 12.18.

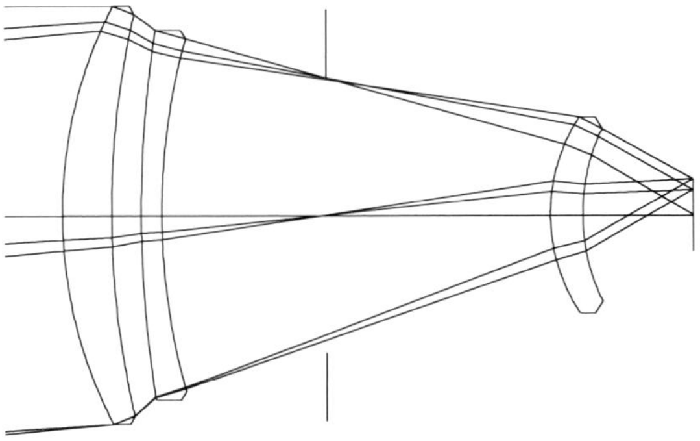

Figure 12.16. 3- to 5-μm waveband silicon-germanium-silicon triplet.

EFL = 100.000

WAVELENGTHS [nm]
4000.00 3000.00 5000.00

SURFACES

#	SURF SPACE	RADIUS	SEPN	INDEX1	V	CLR RAD	GLASS
0			0.00000	1.000000			
1	S	106.99578	12.00000	3.425300		50.000	User-SI
2	S	207.37758	7.30872	1.000000		48.087	
3	S	274.05482	5.00000	4.025400		44.189	User-GE
4	S	152.40613	40.00000	1.000000		42.106	
5#	S	Plane	55.01290	1.000000		32.841	
6	S	41.26623	8.00000	3.425300		23.442	User-SI
7	S	45.93428	26.86392	1.000000		20.586	
8	S	Plane				8.713	

LAGRANGE INVARIANT = -4.3744

#	H	U	HBAR	UBAR	D(U/N)	A	ABAR
1	49.99999	0.00000	-8.54623	0.08749	-0.09660	0.46731	0.00761
2	46.02944	-0.33088	-7.56107	0.08210	-0.49845	-0.37308	0.15632
3	41.68043	-0.59504	-6.15208	0.19278	0.52993	-0.44296	0.17033
4	40.36979	-0.26213	-5.82827	0.06476	-0.18868	0.01109	0.10676
5	30.21804	-0.25379	-0.02828	0.14500	0.00000	-0.25379	0.14500
6	16.25612	-0.25379	7.94858	0.14500	0.15073	0.14014	0.33762
7	13.43196	-0.35302	7.19616	-0.09405	-0.39694	-0.20758	0.21446
		-0.50000		0.05780			

#	S1	S2	S3	S4	S5	C1	C2
1	1.054747	0.017186	0.000280	0.126632	0.002068	0.070261	0.001145
2	3.193499	-1.338061	0.560641	-0.065335	-0.207531	0.051639	-0.021637
3	-4.333789	1.666498	-0.640829	0.052478	0.226242	-0.129799	0.049912
4	0.000937	0.009017	0.086810	-0.094366	-0.072735	-0.003147	-0.030299
5	0.000000	0.000000	0.000000	0.000000	0.000000	0.000000	0.000000
6	-0.048121	-0.115932	-0.279298	0.328333	0.118134	0.006850	0.016504
7	0.229745	-0.237357	0.245221	-0.294967	0.051394	0.008384	-0.008662
Sum	0.097016	0.001352	-0.027174	0.052776	0.117572	0.004189	0.006962

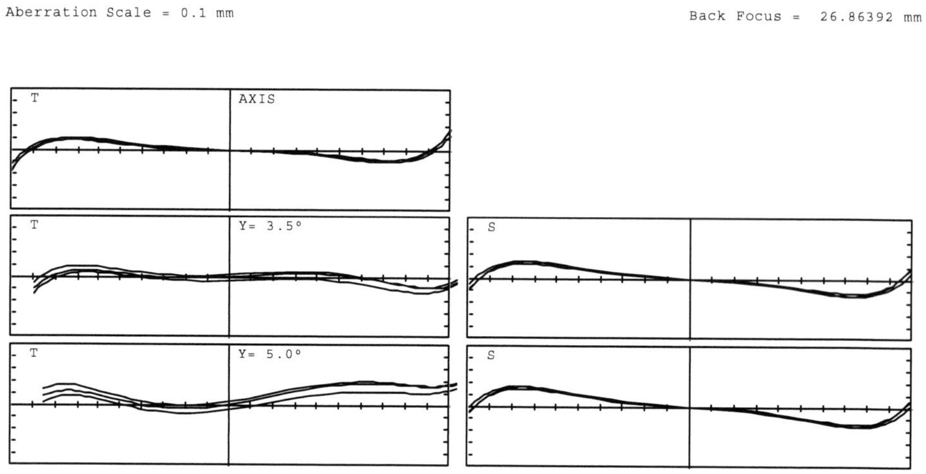

Figure 12.17. Transverse ray aberrations of 3- to 5-μm waveband silicon-germanium-silicon triplet.

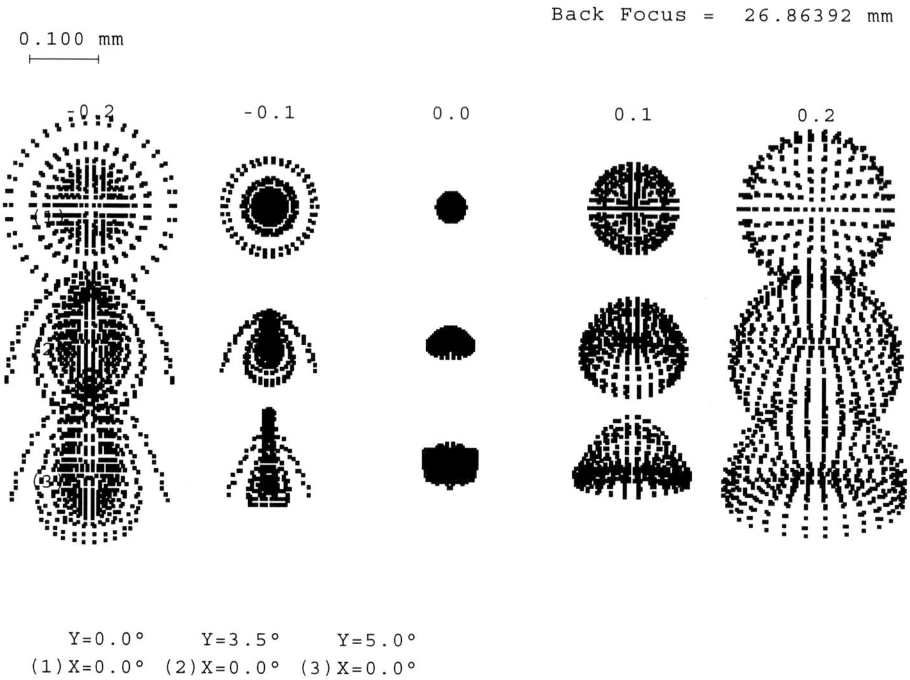

Figure 12.18. Spot diagrams of 3- to 5-μm waveband silicon-germanium-silicon triplet.

12.4 Infrared afocal systems

A large number of these systems have been designed in the last few years, for a wide range of applications.[14] The designs described above are all for use with infrared detectors that are planar arrays and uncooled, in the sense that the detector is essentially at the same temperature as the metal around the optics. Another class of systems employs a much smaller array of detectors cooled to a low temperature (below −190°C). These have a scanning system that scans the image over the array. The scan can be in a raster pattern to build up a television-type picture. Alternatively, in the case of long linear arrays, only one scan motion is needed. In either case, the scanner, which generally consists of small optical apertures scanning over a wide field of view, requires an afocal system, or "telescope," to convert to the narrower field required for applications such as remote surveillance.

Because cooled detectors are almost invariably sold in packages with a "cold shield" extending some distance in front of the substrate, the best performance of the system can only be achieved if this aperture is accurately matched to the acceptance cone of the optics. This means that the scanner and telescope are designed so that the detectors do not "see" the mounts of any of the optics. This can only be done if the entrance pupil of the scanner matches exactly the exit pupil of the telescope.

Because the "telescope" accepts collimated radiation from the object and converts it to a smaller parallel beam, it is convenient to refer to it consisting of an "objective" and an "eyepiece."

12.4.1 The objective

The objective commonly consists of an air-spaced doublet, in which the first element is germanium. The second (negative) element in published designs often consists of a different material, such as zinc selenide, to introduce negative longitudinal chromatic aberration.

12.4.2 The eyepiece

The "eyepiece" commonly consists of two or more positive lenses. The addition of a field lens, made rather thick and meniscus-shaped, is a useful way to introduce some negative S_4, remembering that the eyepiece, because of its high power, will produce most of the field curvature in the system.

Combining the objective and eyepiece gives a complete telescope such as the example in Fig. 12.19, which is based on a design reported by Rogers.[4] Note that since the scanner has such stringent requirements for pupil matching, tracing is usually carried out backwards from scanner space to object space, as shown in Fig. 12.19, and the angular ray aberrations of Fig. 12.20.

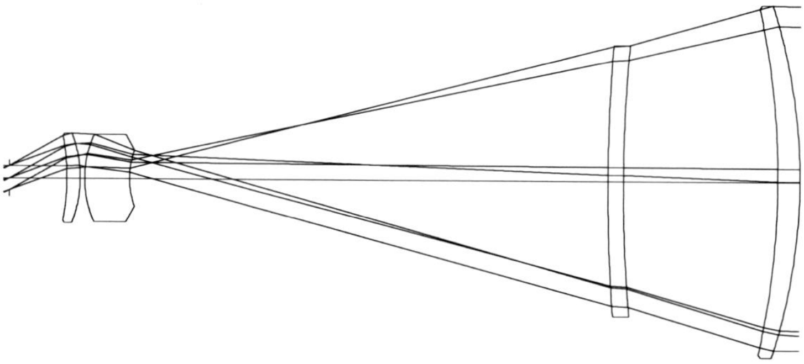

Figure 12.19. Infrared afocal system.

WAVELENGTHS [nm]

10000.00	12000.00	8000.00

SURFACES

#	SURF SPACE	RADIUS	SEPN	INDEX1	V	CLR RAD	GLASS
0			0.00000	1.000000			
1	#S	Plane	26.70000	1.000000		5.357	
2	S	-98.61933	6.00000	4.003286		18.794	G-GE
3	S	-54.79452	2.00000	1.000000		19.903	
4	S	45.78755	21.00000	4.003286		19.276	G-GE
5	S	34.56619	218.00000	1.000000		12.453	
6	S	800.00000	7.00000	2.406527		59.927	G-ZNSE
7	S	649.35065	70.00000	1.000000		60.357	
8	S	-352.11268	10.00000	4.003286		76.905	G-GE
9	S	-250.78993	0.00000	1.000000		78.324	
10	S	Plane				83.184	

LAGRANGE INVARIANT = -2.8916

#	H	U	HBAR	UBAR	D(U/N)	A	ABAR
1	5.35700	0.00000	0.00000	0.53978	0.00000	0.00000	0.53978
2	5.35700	0.00000	14.41213	0.53978	0.01018	-0.05432	0.39364
3	5.60151	0.04075	15.87894	0.24447	-0.15406	-0.24611	-0.18144
4	5.31375	-0.14388	16.09565	0.10835	0.11315	-0.02783	0.45988
5	2.73067	-0.12300	11.12594	-0.23665	-0.22444	-0.17616	0.34116
6	-52.89492	-0.25516	15.33092	0.01929	0.22716	-0.32128	0.03845
7	-53.36662	-0.06739	15.30862	-0.00319	-0.24976	-0.35995	0.04907
8	-72.80985	-0.27776	17.09320	0.02549	0.22168	-0.07098	-0.02305
9	-75.05495	-0.22451	17.52107	0.04279	0.05611	0.29930	-0.10840
		-0.03853					

#	S1	S2	S3	S4	S5	C1	C2
1	0.000000	0.000000	0.000000	0.000000	0.000000	0.000000	0.000000
2	-0.000161	0.001166	-0.008450	-0.063606	0.522165	-0.000253	0.001836
3	0.052269	0.038534	0.028408	0.114478	0.105340	0.001200	0.000885
4	-0.000466	0.007695	-0.127164	0.136997	-0.162490	-0.000129	0.002127
5	0.019020	-0.036834	0.071333	-0.181470	0.213294	0.000419	-0.000811
6	1.240287	-0.148443	0.017766	0.006109	-0.002857	0.171774	-0.020559
7	-1.726886	0.235416	-0.032093	-0.007526	0.005401	-0.194161	0.026469
8	0.081319	0.026408	0.008576	-0.017815	-0.003000	0.004499	0.001461
9	0.377239	-0.136623	0.049480	0.025012	-0.026978	0.019554	-0.007082
Sum	0.042622	-0.012681	0.007856	0.012178	0.650873	0.002902	0.004326

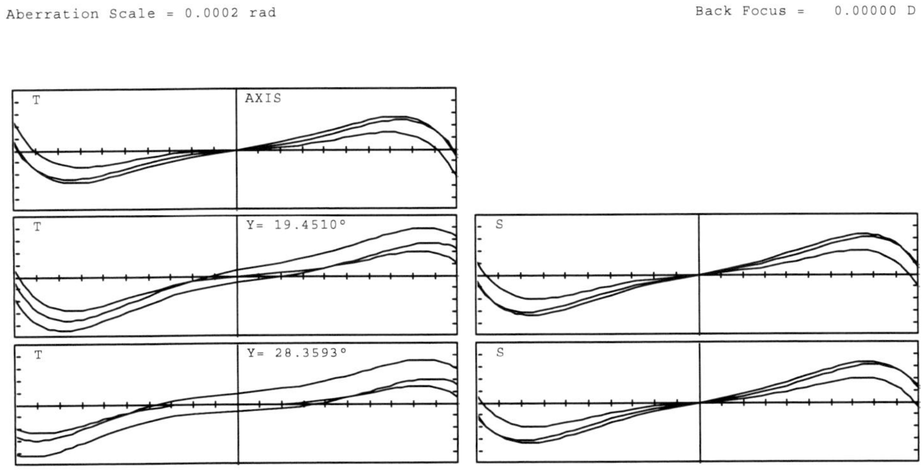

Figure 12.20. Angular ray aberrations of infrared afocal system.

It is possible to correct spherical aberration, coma, and astigmatism, but the Petzval sum, S_4, is positive and can be the limiting factor. Remember that these systems are not used visually and the scanner will not usually accommodate a curved image in the same way that the eye can in a visual system.

12.4.3 Optimization and analysis

Whilst it is important that the detector does not "see" the telescope lens mount, it is not necessary for the entrance pupil to correspond exactly to the front component. Having said that, one must remember that all infrared materials commonly used outside the laboratory are extremely expensive, and any increase in the diameter of the front component may dramatically increase the material costs.

As with visual systems, the angular ray error is the difference between the exiting chief ray angle for a particular field angle and that for other rays at the same entering field angle. In terms of resolution response, spatial frequency is specified, as in other afocal systems, in cycles per radian. Because systems engineers find it more helpful, the resolution is usually defined in object space.

12.5 Other aspects of thermal imaging

12.5.1 Narcissus effect

This problem is unique to infrared systems with cooled detectors. Narcissus is the effect by which a reflection of the detector and its surround occurs over only the central part of the field of view.[15] Its appearance is that of cold circular shadow with a sharp edge overlaid on the scene. Since it is important for small temperature differences in the scene to be detected, this effect can be serious, from both a functional and an aesthetic point of view. Perfect antireflection coatings would solve the problem, but in practice it is necessary to make significant changes to the design to ensure that there are no reflected images in or near focus.

12.5.2 Thermal effects

It is well known that nearly all infrared materials have large thermal coefficients of refractive index; the effect of this is that a large temperature change can alter the focus of the image by a significant amount.[8] If the system can be refocused this is not a problem, but this is not always possible. To reduce the problem, it is necessary to be aware that the design must be such that either the various thermal effects cancel, or are compensated. This compensation is called "athermalization," and it is done either actively, using temperature sensors and motor drives, or passively, using different materials in the mount. This is a large topic and it is not covered here in detail.

12.5.3 Special optical surfaces

Because of the high cost of materials in this application, the use of nonspherical surfaces, which can be costly to generate, may be economical if they permit fewer lenses to be used. Conventional aspheres have already been mentioned, but diffractive optical elements made by single-point diamond turning are also commonly employed.[9,12,13] They have the benefit that in the thermal waveband, the number of diffracting steps, or "zones," can be quite small, so the scatter problems inherent in diffractive surfaces intended for visible wavelengths are avoided.

References

1. P. Rogers, "A comparison between optimized spheric and aspheric optical systems for the thermal infrared," SPIE Proc. Vol. **144**, p.141 (1978).
2. I. Neil, "Use of chalcogenide glass in thermal infrared telescopes," SPIE Proc. Vol. **237**, p.422 (1980).
3. P. Rogers, "Theoretical and practical advantages of using aspheric surfaces in the thermal band," SPIE Proc. Vol. **235**, p.2 (1980).
4. P. Rogers, "Infra-red optical systems," U.S. Patent 4,199,217 (1980).
5. P. Rogers, "Thermal infrared lenses of high aperture and performance," *Opt. Eng.*, Vol. **14**, p. 141 (1975).
6. R. Fischer, "Optical design for the infrared," SPIE Proc. Vol. **531**, p. 82 (1985).

7. M. Roberts and P. Rogers, "Wide waveband infrared optics," SPIE Proc. Vol. **1013**, p. 84 (1988).
8. P. Rogers, "Athermalized FLIR optics," SPIE Proc. Vol. **1354**, p. 742 (1991).
9. P. Rogers, "Use of hybrid optics in the 3 to 5-u infrared band," SPIE Proc. Vol. **2540**, p. 13 (1995).
10. P. Rogers, *Selection of IR optical materials for low-mass applications*, SPIE Proc. Vol. **2774**, p. 301 (1996).
11. P. Conway et al., "Arsenic sulphide athermal singlets for 3- to 5-um imaging," SPIE Proc. Vol. **3061**, p. 396 (1997).
12. C. Puech et al., "Diffractive optics in the Thomson group," SPIE Proc. Vol. **3482**, p. 454 (1998).
13. A. Wood et al., "Hybrid optics in dual-waveband infrared systems," SPIE Proc. Vol. **3482**, p. 602 (1998).
14. I. Neil, "Compact refractor telescopes for the thermal infrared," SPIE Proc. Vol. **330**, p.18 (1982).
15. D. Freeman, "Guidelines for narcissus reduction and modeling," SPIE Proc. Vol. **892**, p. 27 (1988).

CHAPTER 13
CATADIOPTRIC SYSTEMS

In this chapter we will be considering systems with one or two mirrors as well as some lenses. Two-mirror systems are more common and more useful than single-mirror systems, because the image position can usually be outside the optical system, whereas in the case of a single-mirror system the image lies within the incoming beam unless some decentrations are introduced. In this chapter we will only consider centered systems with spherical or aspheric surfaces.

The main advantages of catadioptric systems compared with all-refracting lenses are that they are

1. Very compact,
2. Quite simple for relatively high apertures,
3. Very good color correction because most of the power is in the mirrors, instead of the lenses.

13.1 General considerations

Because the image in a two-mirror system is formed close to the primary mirror, there almost always needs to be a hole in the primary mirror. This restricts the maximum image diameter to essentially the size of the hole, which is considerably less than the diameter of the primary mirror. For example, if the size of the hole is 30% of the diameter of the primary mirror, the linear obstruction ratio of 0.30, gives a reduction in effective aperture of 0.09 (= 0.3^2), which is quite common. These systems are therefore always narrow-field systems. A typical application is a mirror-telephoto system with an aperture of *f*/8, and focal lengths on the order of 500 mm to 1000 mm. A 1000-mm system like this for a 35-mm camera has a diameter of 1000/8 mm = 125 mm. The image diameter is about 40 mm, so the linear obstruction ratio will be about 40/125 = 0.3.

13.1.1 Reminder of Seidel theory—spherical aberration, S_1

A spherical surface has a spherical aberration given by

$$S_1 = -A^2 h\delta(u/n). \qquad (13.1)$$

In the case of a mirror the Seidel theory is simple if we take $n = 1$ and $n' = -1$ at the reflecting surface, in which case we find that

$$A = ni = i \text{ (angle of incidence)}$$

and, since

$$n'u' = nu - hc\,(n' - n),$$
$$-u' = u + 2\,hc,$$

or

$$u' = -u - 2\,hc$$

Therefore,

$$\delta\,(u/n) = u'/n' - u/n = -u' - 0 = 2hc. \tag{13.2}$$

If the object is at infinity, $u = 0$ and the angle of incidence $i = n(hc - u) = hc$, then $A = ni = i = hc$, so that Eq. (13.1) becomes

$$\begin{aligned} S_1 &= -\,(hc)^2\,h\,2hc \\ &= -\,2\,h^4c^3. \end{aligned} \tag{13.3}$$

Considering the **concave** primary mirror in a two-mirror system, $c < 0$; so we see that *spherical aberration* will be positive and it will be quite large. This must be corrected, either by an aspheric surface somewhere or by some refracting elements.

13.1.2 Correction of field curvature, S_4

The other aberration that can be easily computed is field curvature. The concave primary mirror has positive power:

$$\begin{aligned} K &= (n' - n)\,c \\ &= (-1-1)\,c = -\,2\,c \\ &> 0 \text{ when } c < 0. \end{aligned}$$

In contrast to a lens with positive power, the field curvature for this mirror is negative because

$$\begin{aligned} S_4 &= -\,H^2c\;\delta(1/n) \\ &= -\,H^2c\,(1/n' - 1/n) \\ &= -\,H^2c\,(-1 - 1) \\ &= 2H^2c. \end{aligned} \tag{13.4}$$

That is, a concave mirror produces a backward-curving surface relative to the direction of the light. Thus, if we combine a positive lens with a concave mirror, the field curvature, S_4, can be small.

13.1.3 General topics relating to computations with catadioptric systems

To avoid having ray trace formulae that treat reflecting surfaces as a special case, we can use the convention that $n' = -n$ at a mirror surface. That is, the sign of all refractive indices reverses at every reflective surface.

We are considering here centered systems with rotational symmetry. In most of such "cat" systems we have surfaces that are used more than once. A simple example that we shall see below is the Mangin mirror, in which light passes through the refracting surface twice. In almost all lens design programs the surface needs to be entered twice in the lens data, but great care must be taken to ensure that the two curves remain identical throughout the design. This is done by linking of parameters, often referred to as "pickups."

There are always internal obstructions, caused by the secondary mirror if there is one, or by the hole in the primary mirror or the image surface. In almost all cases we find that the chief rays are not transmitted by the system, since they are cut off by both the secondary mirror and also by the hole in the primary mirror. We still need to include these chief rays in the calculation, however, because the chief ray is used as a reference for all other rays from any object point.

13.1.4 Baffles

It is always necessary to ensure that there is no possibility of stray light passing the secondary mirror and reaching the image surface directly without reflecting off the primary mirror. This requires the use of baffles.

13.2 Simple examples

13.2.1 Cassegrain telescope

The classical Newtonian telescope objective uses a concave paraboloidal primary mirror to form an image that can be viewed by an eyepiece. This has been developed into the Cassegrain form by adding a convex hyperboloidal secondary mirror. By virtue of their conic forms, each mirror is free from spherical aberration but each introduces coma and astigmatism that limit the useful field of view. Unless both mirrors have the same radius of curvature, there is also field curvature. Figure 13.1 shows a hypothetical example by Wynne.[1] In the table of Seidel aberrations the contribution from the base sphere is printed separately from the aspheric contribution, so as to identify the role of the aspheric departure from the sphere.

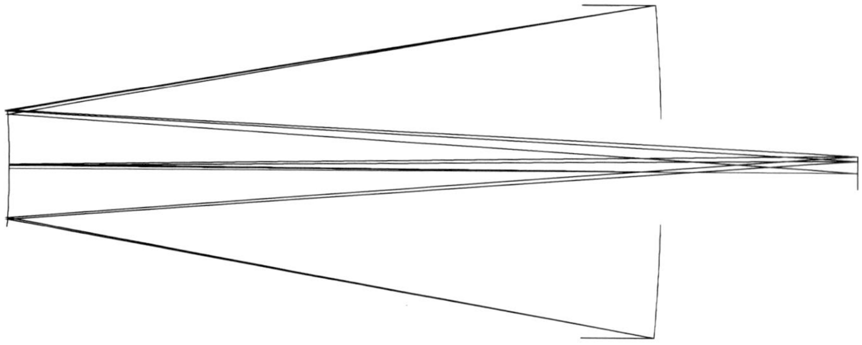

Figure 13.1. Cassegrain telescope.

```
EFL = 31215.733

WAVELENGTHS   [nm]
      587.60       850.00       365.00

SURFACES
  # SURF SPACE     RADIUS          SEPN        INDEX1      V      CLR RAD  GLASS
OBS RAD
  0                               0.00000    1.000000
  1#EMI      -2.28620e+004-7747.00000    1.000000           1905.720           600.000
  2 AM       -1.16250e+00410060.25370    1.000000            670.000
  3 S              Plane                                     192.601
CONIC    SURFACE    1 CC =   -1.000000  (PARABOLA)
ASPHERIC SURFACE  2 CC =    0.000000  (SPHERE) A4 = 3.6991e-013

LAGRANGE INVARIANT =    -11.6415

  #      H          U          HBAR       UBAR      D(U/N)       A          ABAR
  11905.72050   0.00000    0.00000    0.00611   -0.16672   -0.08336    0.00611
  2 614.17849   0.16672   47.32421   -0.00611    0.10567   -0.11388    0.01018
               -0.06105               0.01425

  #     S1         S2         S3         S4         S5         C1         C2
  1  2.207619 -0.161782  0.011856 -0.011856  0.000000  0.000000  0.000000
  2 -0.841668  0.075234 -0.006725  0.023316 -0.001483  0.000000  0.000000
    ASPHERIC SURFACE  1
   -2.207619  0.000000  0.000000  0.000000  0.000000
    ASPHERIC SURFACE  2
    0.842160  0.064891  0.005000  0.000000  0.000385

 Sum 0.000492 -0.021657  0.010131  0.011460 -0.001098  0.000000  0.000000
```

13.2.2 Field corrector for a Cassegrain telescope

Figure 13.2 shows some fifth-order on-axis spherical aberration but the off-axis aberrations are dominated by coma and astigmatism, making the system unusable over this large a field. Wynne showed that a refracting doublet corrector relatively close to focus can correct these aberrations, provided that it is allowed to introduce some spherical aberration, which in turn can be corrected either by changing the asphericity of the secondary mirror, or by shifting it along the optical axis.[1] Figure 13.3 shows one of his designs, and Fig 13.4 shows its residual aberrations. These are clearly smaller than those of Fig. 13.2.

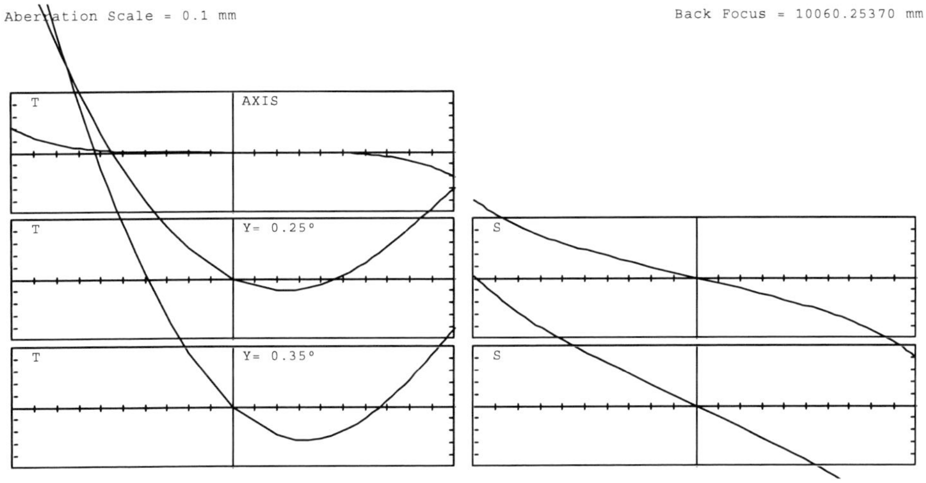

Figure 13.2. Transverse ray aberrations of Cassegrain system.

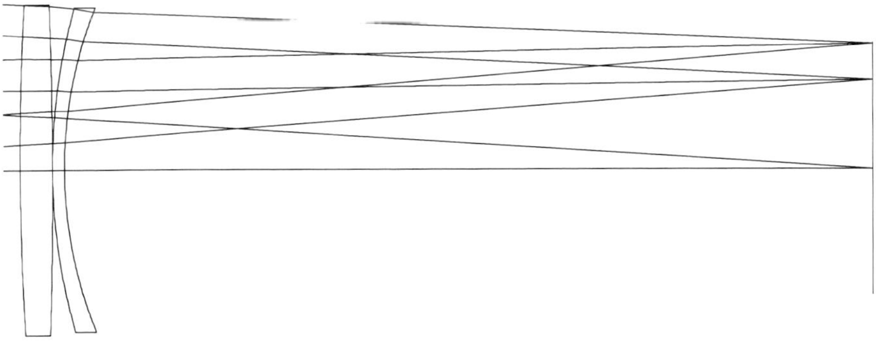

Figure 13.3. Field corrector for a Cassegrain telescope.

```
EFL = 30594.634

WAVELENGTHS  [nm]
      587.60        850.00        365.00

SURFACES
  # SURF SPACE    RADIUS          SEPN       INDEX1     V     CLR RAD GLASS   OBS RAD
  0                             0.00000    1.000000
  1#EMI       -2.28620e+004-7797.80000    1.000000          1905.036             600.000
  2 AM        -1.16250e+004 8378.20000    1.000000           670.000
  3 S            4239.00000   50.80000    1.516800           250.000 S-UBK7
  4 S           -7666.90000    0.00000    1.000000           250.000
  5 S             867.70000   18.80000    1.516800           245.000 S-UBK7
  6 S             640.30000 1253.03001    1.000000           245.000
  7 S            Plane                                       190.231
CONIC    SURFACE   1 CC =   -1.000000   (PARABOLA)
ASPHERIC SURFACE   2 CC =    0.000000   (SPHERE) A4 = 3.6991e-013

LAGRANGE INVARIANT =    -11.6373

  #      H          U         HBAR       UBAR      D(U/N)       A         ABAR
  11905.03610   0.00000    0.00000    0.00611  -0.16666  -0.08333   0.00611
  2 605.49183   0.16666   47.63453   -0.00611   0.10417  -0.11457   0.01021
  3  81.98331  -0.06248  167.47548    0.01430   0.03098  -0.04314   0.05381
  4  79.55585  -0.04778  167.27071   -0.00403  -0.04634  -0.08822  -0.03921
  5  79.55585  -0.07784  167.27071   -0.01739   0.02341   0.01384   0.17539
  6  78.00374  -0.08256  165.82037   -0.07715  -0.00784   0.05956   0.27579
               -0.06226               0.01682

  #     S1         S2         S3         S4         S5         C1         C2
  1  2.204450 -0.161607  0.011847 -0.011847  0.000000  0.000000  0.000000
  2 -0.827932  0.073755 -0.006570  0.023299 -0.001490  0.000000  0.000000
  3 -0.004728  0.005897 -0.007355  0.010885 -0.004403 -0.061482  0.076684
  4  0.028690  0.012751  0.005667  0.006018  0.005193  0.121992  0.054216
  5 -0.000357 -0.004522 -0.057294  0.053178 -0.052147  0.019144  0.242530
  6  0.002168  0.010041  0.046500 -0.072064 -0.118380 -0.080750 -0.373938
     ASPHERIC SURFACE  1
    -2.204450  0.000000  0.000000  0.000000  0.000000
     ASPHERIC SURFACE  2
     0.795517  0.062584  0.004924  0.000000  0.000387

Sum -0.006642 -0.001101 -0.002282  0.009470 -0.170840 -0.001097 -0.000509
```

Aberration Scale = 0.1 mm

Back Focus = 1253.03001 mm

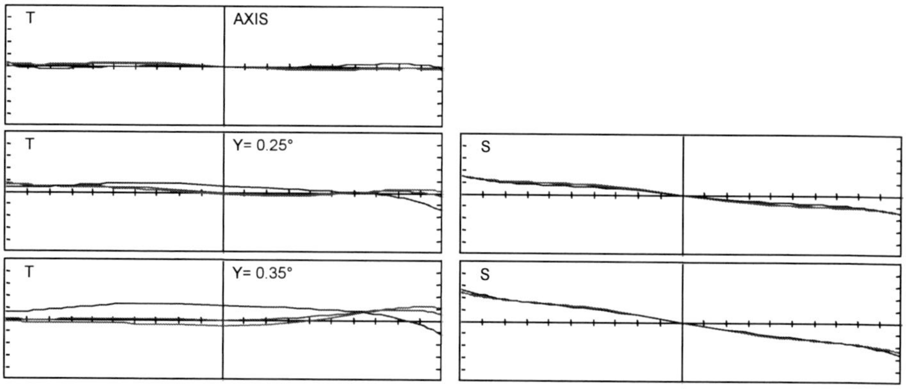

Figure 13.4. Transverse ray aberrations of field corrector for a Cassegrain telescope.

13.2.3 Coma corrector for a paraboloidal mirror

Most large astronomical telescopes are made in a form that can be converted to work at different focal lengths for different types of observations, either by removing the secondary mirror and using only the primary mirror (prime, or Newtonian, focus), or by changing to a more convex secondary (Coudé focus). At the shorter focal length prime focus of the parabolic primary, spherical aberration is zero; but coma is even larger than at the longer focal length Cassegrain focus, since the secondary mirror partially corrects the coma of the primary. This coma of the paraboloid can be corrected without introducing significant spherical aberration or axial color, with a single lens element just before the focus. Figure 13.5 shows such a corrector, by Harmer and Wynne,[2] with the primary mirror off the page to the right and the light coming to prime focus at the left. Residual aberrations are shown in Fig. 13.6.

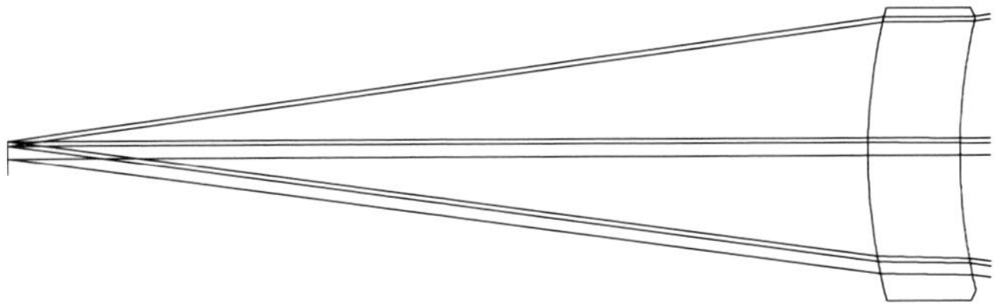

Figure 13.5. Coma corrector for a parabolic mirror.

```
EFL = -13506.684
WAVELENGTHS  [nm]
      587.60       852.00       365.00
SURFACES
  # SURF SPACE    RADIUS          SEPN       INDEX1      V     CLR RAD  GLASS
  0                              0.00000   1.000000
  1#EM      -2.70000e+004-12798.81000   1.000000            1800.846
  2 S            456.70000  -75.33000   1.516800             108.000  S-UBK7
  3 S            450.30000 -691.37472   1.000000             115.000
  4 S             Plane                                       13.489
CONIC   SURFACE   1 CC =  -1.000000  (PARABOLA)

LAGRANGE INVARIANT =    -1.7809

  #      H          U        HBAR        UBAR      D(U/N)       A         ABAR
  11800.84620  0.00000   0.00000    0.00099  -0.13340  -0.06670   0.00099
  2   93.53595  0.13340  12.65675  -0.00099   0.12142  -0.33820  -0.02672
  3   92.16766  0.01816  13.41716  -0.01009  -0.12135  -0.33801  -0.02988
                0.13333              0.00009

  #     S1        S2        S3        S4        S5        C1        C2
  1  1.068674 -0.015845  0.000235 -0.000235  0.000000  0.000000  0.000000
  2 -1.299062 -0.102651 -0.008111 -0.002366 -0.000828 -0.550598 -0.043508
  3  1.277898  0.112979  0.009988  0.002400  0.001095  0.542233  0.047939
     ASPHERIC SURFACE  1
    -1.068674  0.000000  0.000000  0.000000  0.000000

Sum -0.021165 -0.005517  0.002112 -0.000201  0.000267 -0.008365  0.004431
```

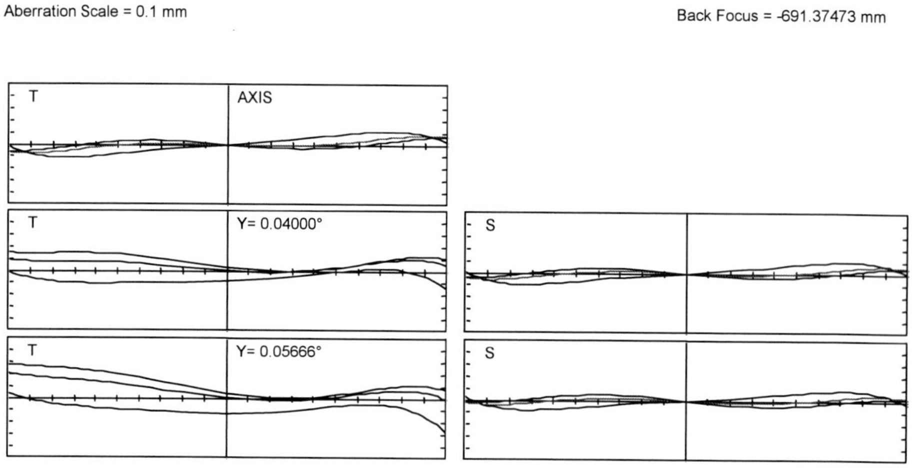

Figure 13.6. Transverse ray aberrations of coma corrector for a parabolic mirror.

13.2.4 Field corrector for a paraboloidal mirror

Wynne (1972)[5] reviewed the correction of the parabolic primary mirror off-axis aberrations over a larger field, including astigmatism and lateral color. He found that if one is restricted to spherical surfaces, such a corrector requires at least three separated lens elements. Furthermore, in order to avoid removal of a secondary spectrum, it is desirable that the corrector be afocal, or nearly so. Figure 13.7 shows one such corrector, and Fig. 13.8 its residual aberrations.

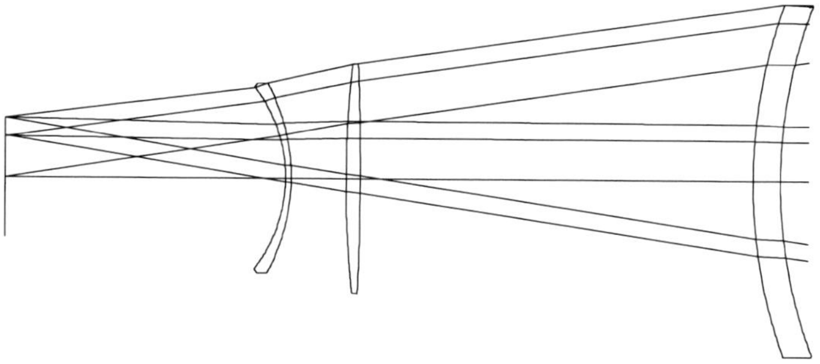

Figure 13.7. Field corrector for a parabolic mirror.

EFL = -17360.570

WAVELENGTHS [nm]

486.10	656.30	365.00

SURFACES

# SURF	SPACE	RADIUS	SEPN	INDEX1	V	CLR RAD	GLASS
0			0.00000	1.000000			
1#EM		-3.38836e+004	-16283.30000	1.000000		2538.098	
2 S		418.36000	-24.68000	1.522370		151.600	S-UBK7
3 S		434.56000	-348.25000	1.000000		152.600	
4 S		-2410.22000	-12.57000	1.522370		99.800	S-UBK7
5 S		894.77000	-48.36000	1.000000		99.250	
6 S		-171.61000	-5.00000	1.528586		82.350	S-UK50
7 S		-127.60000	-251.42202	1.000000		78.700	
8 S		Plane				51.729	

CONIC SURFACE 1 CC = -1.000000 (PARABOLA)

LAGRANGE INVARIANT = -7.3831

#	H	U	HBAR	UBAR	D(U/N)	A	ABAR
1	2538.09805	0.00000	0.00000	0.00291	-0.14981	-0.07491	0.00291
2	98.65171	0.14981	47.36669	-0.00291	0.13832	-0.38562	-0.11031
3	98.21991	0.01750	48.37265	-0.04076	-0.13321	-0.37072	-0.10741
4	47.82752	0.14470	49.73238	-0.00390	0.07779	-0.12486	0.02454
5	46.54715	0.10186	49.67562	0.00452	-0.11533	-0.23426	-0.09139
6	37.73393	0.18224	47.94071	0.03587	0.05450	0.03764	0.24348
7	36.75764	0.19526	47.34035	0.12007	-0.01846	0.14187	0.38357
		0.14620		-0.01257			

#	S1	S2	S3	S4	S5	C1	C2
1	2.133511	-0.082853	0.003217	-0.003217	0.000000	0.000000	0.000000
2	-2.029115	-0.580454	-0.166046	-0.044708	-0.060289	-0.547219	-0.156539
3	1.798184	0.520987	0.150946	0.043041	0.056204	0.523779	0.151754
4	-0.058004	0.011399	-0.002240	0.007760	-0.001085	-0.085900	0.016882
5	0.294616	0.114938	0.044841	0.020904	0.025649	0.156854	0.061193
6	-0.002914	-0.018849	-0.121928	0.109840	-0.078191	0.022156	0.143322
7	0.013658	0.036928	0.099842	-0.147725	-0.129461	-0.081348	-0.219941
	ASPHERIC SURFACE 1						
	-2.133511	0.000000	0.000000	0.000000	0.000000		
Sum	0.016426	0.002098	0.008631	-0.014105	-0.187173	-0.011678	-0.003329

Aberration Scale = 0.1 mm

Back Focus = -251.42199 mm

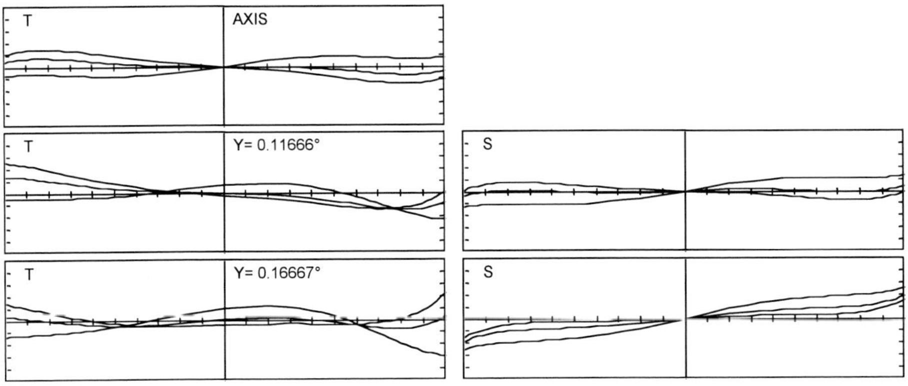

Figure 13.8. Transverse ray aberrations of field corrector for a parabolic mirror.

13.2.5 The Ritchey-Chrétien telescope

From the mid-1960s onward, the Ritchey-Chrétien form became the norm for new telescope designs. In this, the primary mirror becomes a hyperboloid, and the secondary hyperboloid is chosen such that the spherical aberration and coma contributions of the primary and secondary are equal and opposite, thereby making the mirror pair aplanatic.

The main reason for this rather sudden change was the decision to adopt smaller *f*/numbers (higher NAs), following Bowen's recommendation (1961) that the best aperture for detection of faint objects at secondary focus was *f*/8 to *f*/10, rather than the typical *f*/16 of classical Cassegrain systems. The wavefront error due to coma is proportional to the cube of the aperture, while that due to astigmatism is proportional to the square. Therefore, at *f*/8 it is coma rather than astigmatism that limits the useful field size.

Figure 13.9 shows a typical Ritchey-Chrétien design, the Kitt Peak 150-in. *f*/2.8:*f*/8 telescope.[9]

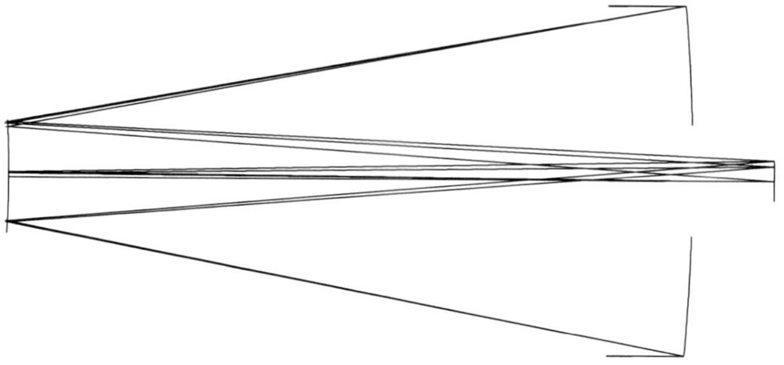

Figure 13.9. The Ritchey-Chrétien telescope.

```
EFL = 30473.313

WAVELENGTHS  [nm]
      546.10      769.90      365.00

SURFACES
  # SURF SPACE     RADIUS        SEPN       INDEX1 V  CLR RAD  GLASS    OBS RAD
  0                          0.00000    1.000000
  1#AMI            Plane  -7665.21000   1.000000   1904.582           600
  2 AM             Plane   8577.11662   1.000000    615.000
  3 S              Plane                            223.844
ASPHERIC SURFACE  1 CC =   0.000000  (SPHERE)
 A2 = -2.3435e-005   A4 = 1.228e-015
ASPHERIC SURFACE  2 CC =   0.000000  (SPHERE)
 A2 = -5.4114e-005   A4 = 7.21e-013   A6 = -1.808e-020
```

LAGRANGE INVARIANT = -13.8507

#	H	U	HBAR	UBAR	D(U/N)	A	ABAR
1	1904.58206	0.00000	0.00000	0.00727	-0.17854	-0.08927	0.00727
2	536.06979	0.17854	55.74365	-0.00727	0.11604	-0.12052	0.01331
		-0.06250		0.01934			

#	S1	S2	S3	S4	S5	C1	C2
1	2.709653	-0.220745	0.017983	-0.017983	0.000000	0.000000	0.000000
2	-0.903471	0.099744	-0.011012	0.041525	-0.003369	0.000000	0.000000
	ASPHERIC SURFACE 1						
	-2.968186	0.000000	0.000000	0.000000	0.000000		
	ASPHERIC SURFACE 2						
	1.162045	0.120836	0.012565	0.000000	0.001307		
Sum	0.000040	-0.000164	0.019537	0.023542	-0.002062	0.000000	0.000000

13.2.6 Field corrector for a Ritchey-Chrétien telescope

As may be seen in Fig. 13.10, the residual astigmatism remains large enough at wider field angles to require refracting correctors close to the secondary focal plane. Figure 13.11 shows an example by Wynne for the Kitt Peak 150-in. telescope,[9] with reduced astigmatism apparent in Fig. 13.12.

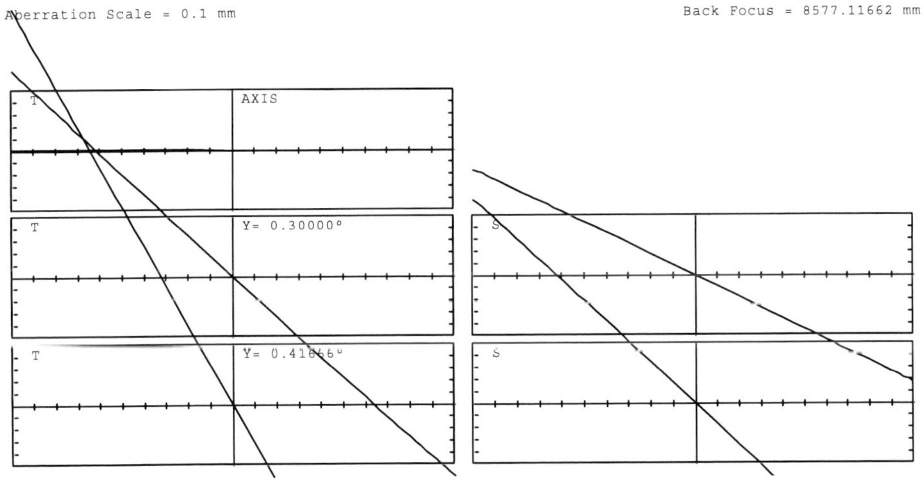

Figure 13.10. Transverse ray aberrations of the Ritchey-Chrétien telescope.

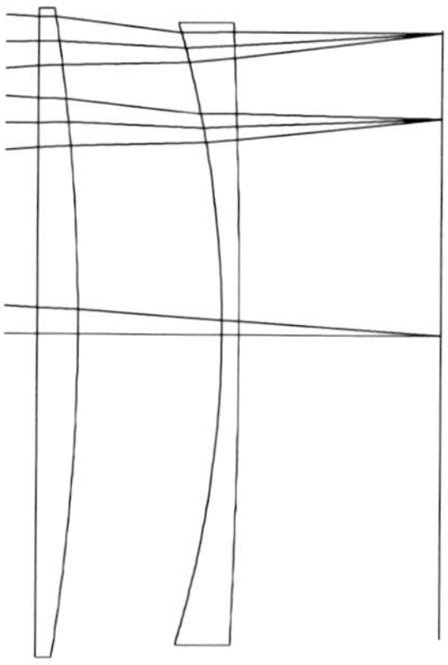

Figure 13.11. Field corrector for a Ritchey-Chrétien telescope.

EFL = 30314.073

WAVELENGTHS [nm]

546.10 769.90 365.00

SURFACES

# SURF SPACE	RADIUS	SEPN	INDEX1	V	CLR RAD	GLASS	OBS RAD
0		0.00000	1.000000				
1#AMI	Plane	-7665.21000	1.000000		1894.630		600
2 AM	Plane	8274.76000	1.000000		615.000		
3 S	Plane	30.48000	1.460031		240.000	G-SILICA	
4 S	-1527.35000	109.22000	1.000000		240.000		
5 S	-795.25000	12.70000	1.460031		230.000	G-SILICA	
6 S	-4810.51000	153.57815	1.000000		230.000		
7 S	Plane				225.221		

ASPHERIC SURFACE 1 CC = 0.000000 (SPHERE)
A2 = -2.3435e-005 A4 = 1.228e-015
ASPHERIC SURFACE 2 CC = 0.000000 (SPHERE)
A2 = -5.4114e-005 A4 = 7.21e-013 A6 = -1.808e-020

LAGRANGE INVARIANT = -13.7783

#	H	U	HBAR	UBAR	D(U/N)	A	ABAR
1	1894.62959	0.00000	0.00000	0.00727	-0.17760	-0.08880	0.00727
2	533.26853	0.17760	55.74365	-0.00727	0.11543	-0.11989	0.01331
3	18.79854	-0.06217	215.76376	0.01934	0.03301	-0.06217	0.01934
4	17.50059	-0.04258	216.16748	0.01325	-0.03828	-0.07890	-0.18730
5	10.13430	-0.06744	211.16844	-0.04577	0.03856	-0.08019	-0.31131
6	9.59863	-0.04218	211.83287	0.05232	-0.03361	-0.06450	0.01209
		-0.06250		0.05613			

#	S1	S2	S3	S4	S5	C1	C2
1	2.653457	-0.217302	0.017796	-0.017796	0.000000	0.000000	0.000000
2	-0.884734	0.098189	-0.010897	0.041092	-0.003351	0.000000	0.000000
3	-0.002399	0.000746	-0.000232	0.000000	0.000072	-0.016598	0.005163
4	0.004170	0.009900	0.023501	0.039163	0.148754	0.019609	0.046549
5	-0.002512	-0.009754	-0.037867	-0.075216	-0.439015	-0.011540	-0.044802
6	0.001342	-0.000252	0.000047	0.012434	-0.002340	0.008791	-0.001648
ASPHERIC SURFACE 1							
	-2.906629	0.000000	0.000000	0.000000	0.000000		
ASPHERIC SURFACE 2							
	1.137946	0.118952	0.012434	0.000000	0.001300		
Sum	0.000641	0.000479	0.004782	-0.000322	-0.294580	0.000263	0.005261

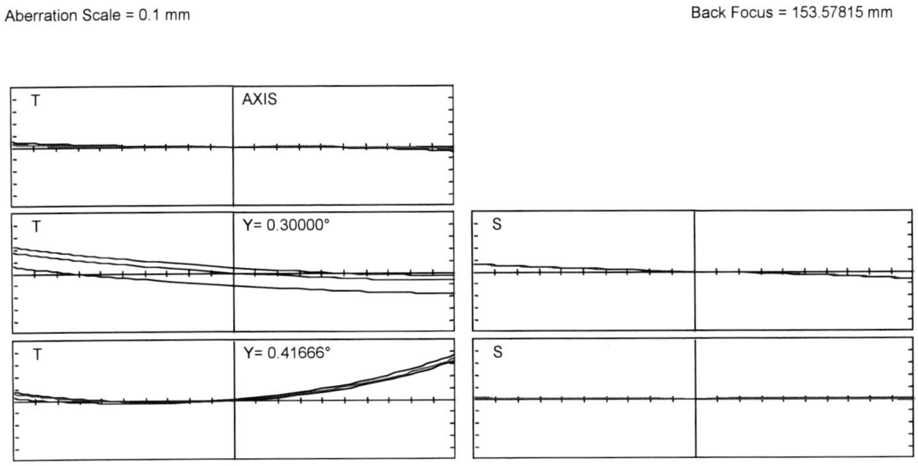

Figure 13.12. Transverse ray aberrations of field corrector for a Ritchey-Chrétien telescope.

13.2.7 Field corrector for a hyperbolic mirror

A drawback of the Ritchey-Chrétien design is that the focus of the hyperbolic primary mirror cannot be used without correctors, even on-axis, because of the large amount of spherical aberration present. Figure 13.13 shows an example of a prime focus hyperboloid corrector designed by Wynne,[8] employing three spherical-surfaced lenses (the plane parallel plate is a filter). Figure 13.14 shows that the aberrations of the primary mirror have been well corrected.

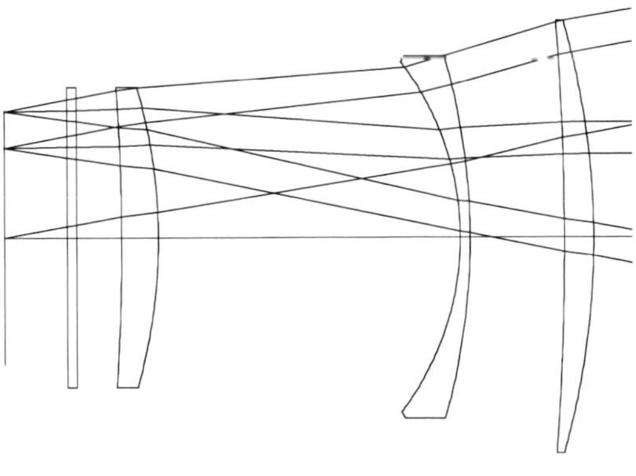

Figure 13.13. Field corrector for prime focus of a hyperbolic mirror.

```
EFL = -10928.056

WAVELENGTHS  [nm]
      440.00       486.10       404.70

SURFACES
  # SURF SPACE    RADIUS        SEPN       INDEX1      V      CLR RAD  GLASS
  0                          0.00000    1.000000
  1#AM           Plane -10295.33000    1.000000             1903.842
  2 S         -506.49000  -19.54000    1.526253              140.850  S-UBK7
  3 S        -1811.44000  -63.53000    1.000000              140.850
  4 S         -417.37000   -6.60000    1.526253              118.150  S-UBK7
  5 S         -184.25000 -202.31000    1.000000              114.000
  6 S         -349.45000  -24.38000    1.526253               97.450  S-UBK7
  7 S        -1382.24000  -30.00000    1.000000               97.450
  8 S            Plane      -6.35000   1.526253               97.450  S-UBK7
  9 S            Plane     -42.04586   1.000000               97.450
 10 S            Plane                                         82.478
ASPHERIC SURFACE  1 CC =   0.000000  (SPHERE)
 A2 = -2.3435e-005   A4 = 1.228e-015

LAGRANGE INVARIANT =    -13.8451

  #      H          U         HBAR       UBAR      D(U/N)       A        ABAR
  11903.84217  0.00000    0.00000   0.00727  -0.17847  -0.08923   0.00727
  2   66.47411  0.17847   74.86962  -0.00727   0.07220  -0.04722   0.15509
  3   63.30503  0.16218   73.96680   0.04620  -0.12288  -0.19420  -0.00820
  4   48.74758  0.22914   70.85191   0.04903   0.10439  -0.11235   0.12073
  5   47.49090  0.19041   70.25357   0.09066  -0.03021   0.10279   0.44359
  6   16.14000  0.15496   82.85578  -0.06229   0.07801  -0.10878   0.29939
  7   13.27637  0.11746   81.85767   0.04094  -0.09726  -0.16461   0.02790
  8    8.04989  0.17422   80.91808   0.03132   0.09943  -0.17422  -0.03132
  9    7.32506  0.11415   80.78778   0.02052  -0.09943  -0.17422  -0.03132
               0.17422               0.03132

  #     S1        S2         S3         S4         S5         C1         C2
  1  2.705444 -0.220485  0.017969 -0.017969  0.000000  0.000000  0.000000
  2 -0.010703  0.035151 -0.115449  0.130494 -0.049411 -0.016122  0.052952
  3  0.293358  0.012383  0.000523 -0.036487 -0.001518  0.063142  0.002665
  4 -0.064228  0.069020 -0.074169  0.158358 -0.090470 -0.028129  0.030227
  5  0.015158  0.065417  0.282312 -0.358718 -0.329735 -0.025072 -0.108200
  6 -0.014898  0.041003 -0.112855  0.189137 -0.209953 -0.009017  0.024819
  7  0.034988 -0.005930  0.001005 -0.047816  0.007934  0.011225 -0.001903
  8 -0.024293 -0.004367 -0.000785  0.000000 -0.000141 -0.007203 -0.001295
  9  0.022105  0.003974  0.000714  0.000000  0.000128  0.006554  0.001178
    ASPHERIC SURFACE  1
   -2.963577  0.000000  0.000000  0.000000  0.000000

Sum -0.006643 -0.003834 -0.000736  0.016998 -0.673165 -0.004623  0.000444
```

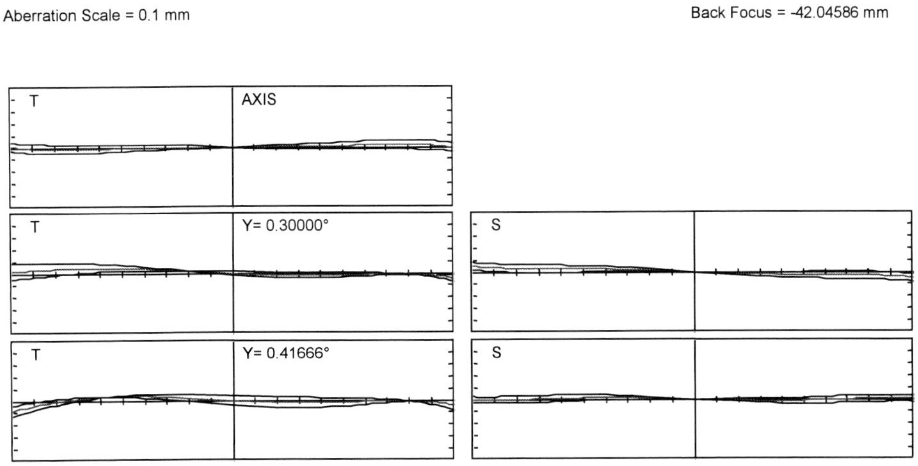

Figure 13.14. Transverse ray aberrations of field corrector for a hyperbolic mirror.

A hyperbolic primary mirror is somewhat more difficult to test during manufacture than a paraboloid, since it is not aberration-free with collimated input light. In practice, in order to reduce the size of the test set, it is more convenient to test at the center of curvature. In this situation, a so-called null element is required to remove the spherical aberration. This is also true if the mirror is a paraboloid. Dall[10] and Ross[11] have described how a simple lens can correct the majority of the spherical aberration of a paraboloid at center of curvature. Offner[12] described an improvement in which a field lens images this null lens onto the test mirror, thereby increasing the accuracy of correction, as well as allowing for the more general case of a hyperboloidal test mirror.

13.2.8 Schmidt camera

The classical Schmidt design uses a primary spherical mirror whose coma, astigmatism, and distortion are zero by virtue of placing the aperture stop at the center of curvature of the mirror. Spherical aberration is corrected, without introducing coma and astigmatism, by means of an aspheric, nearly plane parallel plate, at the pupil. The principal off-axis aberration is field curvature—the image lies on a spherical surface concentric with the mirror and of half its radius. There is also some oblique spherical aberration caused by the off-axis beams seeing the aspheric plate foreshortened. The effect of this is minimized by an appropriate small change of focus for oblique image points, achieved by the choice of a focal surface with a curvature that departs slightly from exact concentricity with the mirror.

A disadvantage of correcting the mirror spherical aberration with a refracting corrector is that since the aberration of the mirror is constant with wavelength, whereas that of the plate varies with the refractive index of the material, there is chromatic variation of spherical aberration. This is known as spherochromatism, and can be balanced against a chromatic difference of focus, which is introduced by adding some vertex power to the plate.

Figure 13.15 shows the original design of the Schmidt camera that forms the U.K. wide-field telescope at Siding Spring Mountain in Australia.[15]

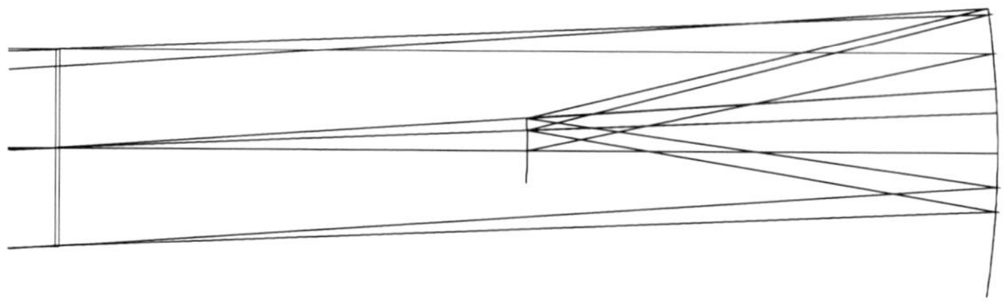

Figure 13.15. The Schmidt camera.

EFL = -3073.148

WAVELENGTHS [nm]

420.00 1200.00 330.00

SURFACES

# SURF	SPACE	RADIUS	SEPN	INDEX1	V	CLR RAD	GLASS
0			0.00000	1.000000			
1#S		Plane	25.40000	1.528385		624.724	S-BK7
2 A		-3.94500e+005	6121.10000	1.000000		625.000	
3 SM		-6121.10000	-3047.95243	1.000000		915.000	
4 S		-3060.60000				205.164	

ASPHERIC SURFACE 2 CC = 0.000000 (SPHERE)

A4 = 2.0296e-012 A6 = 8.163e-020

LAGRANGE INVARIANT = -41.3846

#	H	U	HBAR	UBAR	D(U/N)	A	ABAR
1	624.72414	0.00000	0.00000	0.06624	0.00000	0.00000	0.06624
2	624.72414	0.00000	1.10091	0.04334	-0.00084	-0.00242	0.06624
3	619.60235	-0.00084	406.58217	0.06624	-0.20245	-0.10206	-0.00018
		0.20329		0.06660			

#	S1	S2	S3	S4	S5	C1	C2
1	0.000000	0.000000	0.000000	0.000000	0.103865	0.000000	1.051568
2	0.000003	-0.000084	0.002294	0.001501	-0.103850	0.038420	-1.051500
3	1.306605	0.002303	0.000004	-0.559602	-0.000986	0.000000	0.000000
	ASPHERIC SURFACE 2						
	-1.306786	-0.002303	-0.000004	0.000000	-0.000000		
Sum	-0.000178	-0.000084	0.002294	-0.558101	-0.000971	0.038420	0.000068

13.2.9 The achromatized Schmidt camera

The spherochromatism present in the classical Schmidt design is clearly seen in Fig. 13.16. A modification has been made to the U.K. wide-field telescope by fitting a cemented doublet achromatic corrector in place of the original Schmidt plate.[15] In this design the asphericities of the two components, made from materials with differing dispersions, are approximately in the same ratio as their vertex powers. This means that the spherical aberration of the doublet plate is equal for those pairs of wavelengths at which the paraxial foci are coincident, as shown in Fig. 13.17.

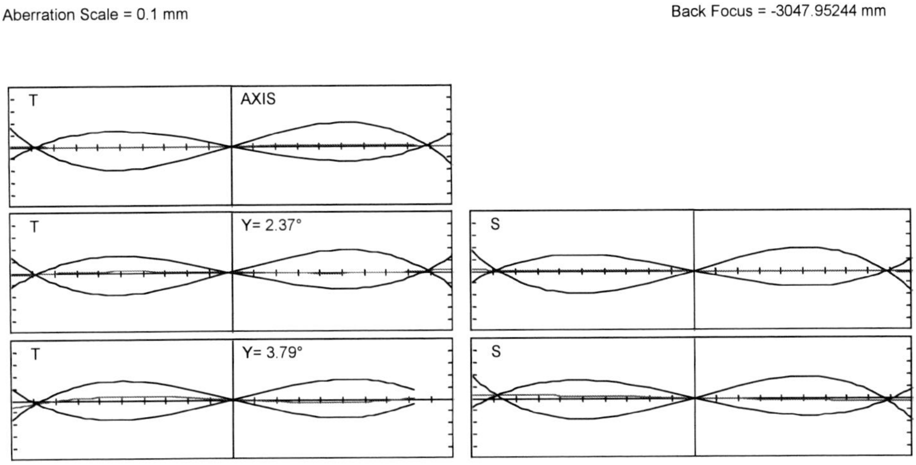

Figure 13.16. Transverse ray aberrations of the Schmidt camera.

```
EFL = -3073.174

WAVELENGTHS   [nm]
      420.00       1200.00        330.00

SURFACES
  # SURF SPACE     RADIUS          SEPN        INDEX1       V       CLR RAD   GLASS
  0                              0.00000     1.000000
  1#S                Plane      12.70000     1.547926              624.724   S-LLF6
  2 A          1.43800e+005     19.00000     1.528364              625.000   S-UBK7
  3 A         -3.57300e+005  6121.10000     1.000000              625.000
  4 SM         -6121.10000  -3047.92076     1.000000              915.000
  5 S          -3060.60000                                         205.166
ASPHERIC SURFACE  2 CC =    0.000000  (SPHERE)
 A4  = -5.6978e-012
ASPHERIC SURFACE  3 CC =    0.000000  (SPHERE)
 A4  = 2.2426e-012    A6  = 8.04e-020

LAGRANGE INVARIANT =     -41.3846

  #      H          U         HBAR       UBAR      D(U/N)        A        ABAR
  1 624.72414   0.00000    0.00000   0.06624    0.00000    0.00000    0.06624
  2 624.72414   0.00000    0.54351   0.04280    0.00004    0.00672    0.06625
  3 624.72519   0.00006    1.36703   0.04334   -0.00088   -0.00259    0.06624
  4 619.59057  -0.00084  406.84540   0.06624   -0.20244   -0.10206   -0.00022
                0.20328              0.06669

  #     S1         S2         S3         S4         S5         C1         C2
  1  0.000000   0.000000   0.000000   0.000000   0.105816   0.000000   1.450636
  2 -0.000001  -0.000010  -0.000100  -0.000098  -0.001953  -0.040867  -0.402606
  3  0.000004  -0.000094   0.002399   0.001657  -0.103843   0.040934  -1.047975
  4  1.306560   0.002859   0.000006  -0.559602  -0.001225   0.000000   0.000000
     ASPHERIC SURFACE  2
     0.135820   0.000118   0.000000   0.000000   0.000000
     ASPHERIC SURFACE  3
    -1.443881  -0.003160  -0.000007   0.000000  -0.000000

Sum -0.001499  -0.000286   0.002299  -0.558043  -0.001205   0.000067   0.000054
```

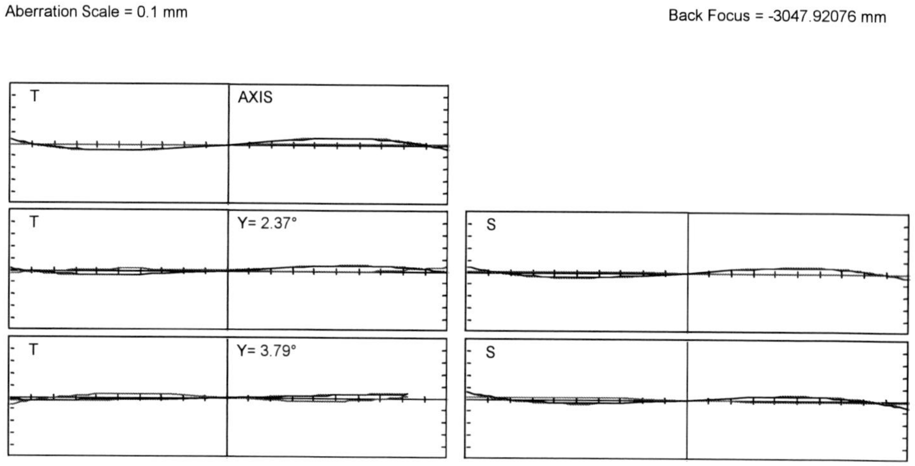

Figure 13.17. Transverse ray aberrations of the achromatized Schmidt camera.

13.2.10 The field-flattened Schmidt camera

The image curvature of the Schmidt camera can sometimes be accommodated in photographic systems by physically bending the photographic plate into a sphere that matches the curvature of the focal surface. However, this is not possible when the Schmidt camera is used with an image tube or solid-state camera. At moderate relative apertures and over small field sizes, field curvature can be corrected by adding a thin plano-convex lens close to the image, as discussed by Linfoot (1955).[14] Wynne (1977)[16] has noted that since this lens is in a converging beam, by making it thicker and bending it, it can be made "self-achromatic." This means that the axial chromatic aberration of its two surfaces can be made equal and opposite. Furthermore, the coma and astigmatism introduced by such a lens can be corrected by moving the aperture stop back toward the mirror, and this also gives a shorter system. Figure 13.18 shows one such design for a spectrograph camera, by Wynne.[16] Figure 13.19 indicates that the system is not corrected for lateral color, but that is unimportant in this particular application.

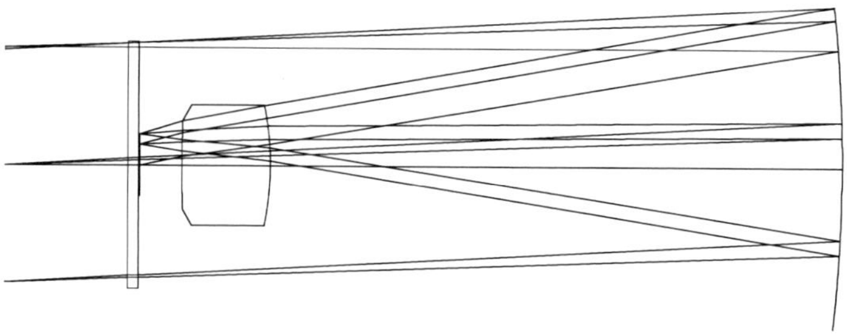

Figure 13.18. The field-flattened Schmidt camera.

EFL = -203.161

WAVELENGTHS [nm]
546.10 850.00 300.00

SURFACES

#	SURF SPACE	RADIUS	SEPN	INDEX1	V	CLR RAD	GLASS
0			0.00000	1.000000			
1	#S	Plane	45.00000	1.000000		40.632	
2	S	Plane	4.00000	1.460031		42.750	G-SILICA
3	A	-7.51930e+005	252.50000	1.000000		42.750	
4	SM	-500.00000	-204.62000	1.000000		56.000	
5	S	-100.10000	-32.30000	1.460031		20.900	G-SILICA
6	S	386.10000	-15.12254	1.000000		15.200	
7	S	Plane				10.754	

ASPHERIC SURFACE 3 CC = 0.000000 (SPHERE)
A4 = 5.253e-009 A6 = 3.45e-014

LAGRANGE INVARIANT = -2.1294

#	H	U	HBAR	UBAR	D(U/N)	A	ABAR
1	40.63230	0.00000	0.00000	0.05241	0.00000	0.00000	0.05241
2	40.63230	0.00000	2.35835	0.05241	0.00000	0.00000	0.05241
3	40.63230	0.00000	2.50193	0.03589	-0.00002	-0.00008	0.05240
4	40.62602	-0.00002	15.73451	0.05241	-0.16250	-0.08128	0.02094
5	7.36935	0.16253	13.57949	0.01053	0.07040	-0.08891	0.12513
6	3.02451	0.13452	11.96587	0.04996	-0.10787	-0.20783	-0.11819
		0.20000		0.08720			

#	S1	S2	S3	S4	S5	C1	C2
1	0.000000	0.000000	0.000000	0.000000	0.000000	0.000000	0.000000
2	0.000000	0.000000	0.000000	0.000000	0.003105	0.000000	0.051539
3	0.000000	-0.000000	0.000003	0.000002	-0.003104	0.000078	-0.051534
4	0.043612	-0.011235	0.002894	-0.018138	0.003927	0.000000	0.000000
5	-0.004101	0.005771	-0.008123	0.014273	-0.008656	-0.015858	0.022318
6	0.014092	0.008014	0.004557	0.003700	0.004696	0.015214	0.008652
ASPHERIC SURFACE 3							
	-0.052695	-0.003245	-0.000200	0.000000	-0.000012		
Sum	0.000908	-0.000694	-0.000868	-0.000162	-0.000045	-0.000566	0.030974

Aberration Scale = 0.1 mm

Back Focus = -15.12254 mm

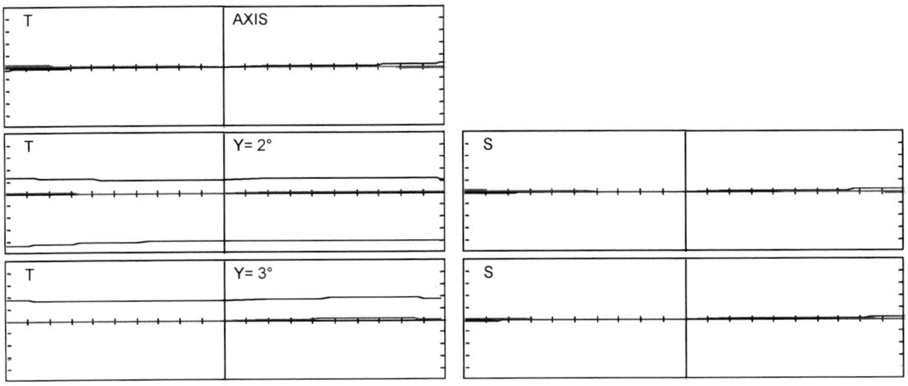

Figure 13.19. Transverse ray aberrations of the field-flattened Schmidt camera.

13.2.11 The Maksutov-Bouwers Cassegrain system

The Maksutov-Bouwers meniscus corrector is another form of "self-achromatic" lens, this time used in collimated space. Generally, it is located away from the aperture stop and has surfaces concentric to it. In the variant of the Cassegrain system shown in Fig. 13.20, the aperture stop is in fact coincident with the meniscus element. This has the effect of reducing the overall length. The design data for this example is taken from *Lens Design Fundamentals* by Rudolf Kingslake, and is intended for an aperture of *f*/10 and a semi-field angle of 1.5 deg. Aberrations of this design are shown in Figs. 13.21 and 13.22.

The meniscus lens, of course, corrects the primary spherical aberration of the mirror; the performance is limited by high-order spherical aberration, chromatic aberration—both longitudinal and transverse—and by field curvature. Because of the curvature of the secondary mirror, the field is curved in the opposite direction from that of the conventional Schmidt system. However, the final negative lens reduces field curvature considerably, as well as giving some telephoto effect.

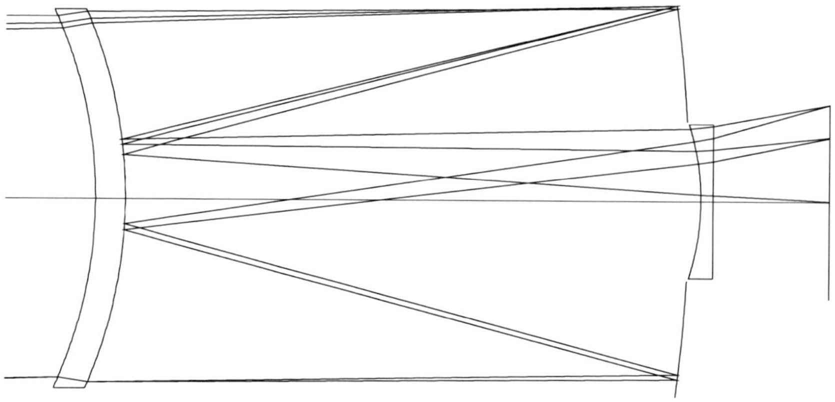

Figure 13.20. The Maksutov-Bouwers Cassegrain system.

EFL = 298.354

WAVELENGTHS [nm]
587.60 656.30 486.10

SURFACES

# SURF SPACE	RADIUS	SEPN	INDEX1	V	CLR RAD	GLASS	OBS RAD
0		0.00000	1.000000				
1#S	-35.47106	2.50000	1.516798	64.14	15.500	S-BK7	
2 S	-37.67245	47.40000	1.000000		15.500		
3 SMI	-119.41068	-47.40000	1.000000		16.000		6.500
4 SM	-37.67245	48.40000	1.000000		5.000		
5 S	-20.00000	1.00000	1.516798	64.14	6.300	S-BK7	
6 S	Plane	9.76894	1.000000		6.300		
7 S	Plane				7.957		

LAGRANGE INVARIANT = -0.3906

#	H	U	HBAR	UBAR	D(U/N)	A	ABAR
1	14.91767	0.00000	0.00000	0.02619	0.09447	-0.42056	0.02619
2	15.27590	0.14329	0.04316	0.01726	-0.08668	-0.39771	0.02445
3	15.64497	0.00779	1.25631	0.02559	-0.26204	-0.12323	0.01507
4	3.59351	0.25425	1.47207	-0.00455	0.19078	-0.15886	0.04363
5	0.52141	-0.06347	5.47491	0.08270	0.04174	-0.08954	-0.19104
6	0.48845	-0.03296	5.62270	0.14779	-0.02827	-0.05000	0.22417
		-0.05000		0.22417			

#	S1	S2	S3	S4	S5	C1	C2
1	-0.249257	0.015520	-0.000966	-0.001466	0.000151	-0.033326	0.002075
2	0.209444	-0.012875	0.000791	0.001380	-0.000133	0.032272	-0.001984
3	0.062256	-0.007615	0.000931	-0.002556	0.000199	0.000000	0.000000
4	-0.017301	0.004751	-0.001305	0.008101	-0.001866	0.000000	0.000000
5	-0.000175	-0.000372	-0.000794	-0.002600	-0.007241	-0.000248	-0.000529
6	0.000035	-0.000155	0.000694	0.000000	-0.003111	0.000130	-0.000582
Sum	0.005002	-0.000746	-0.000649	0.002860	-0.012002	-0.001172	-0.001020

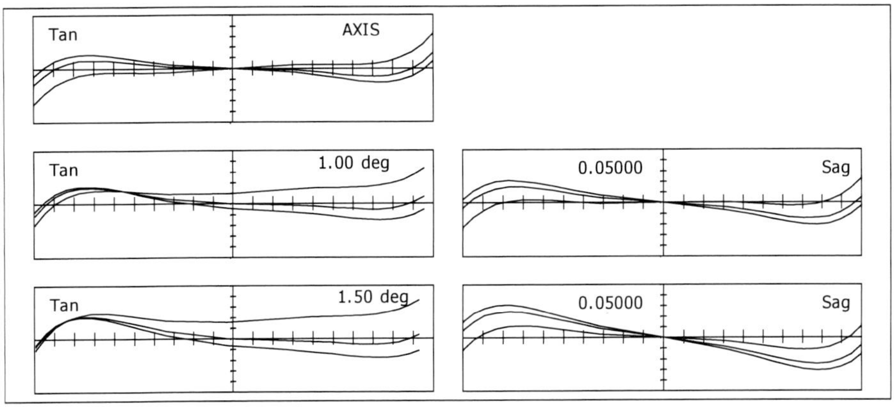

Figure 13.21. Transverse ray aberrations of the Maksutov-Bouwers Cassegrain system.

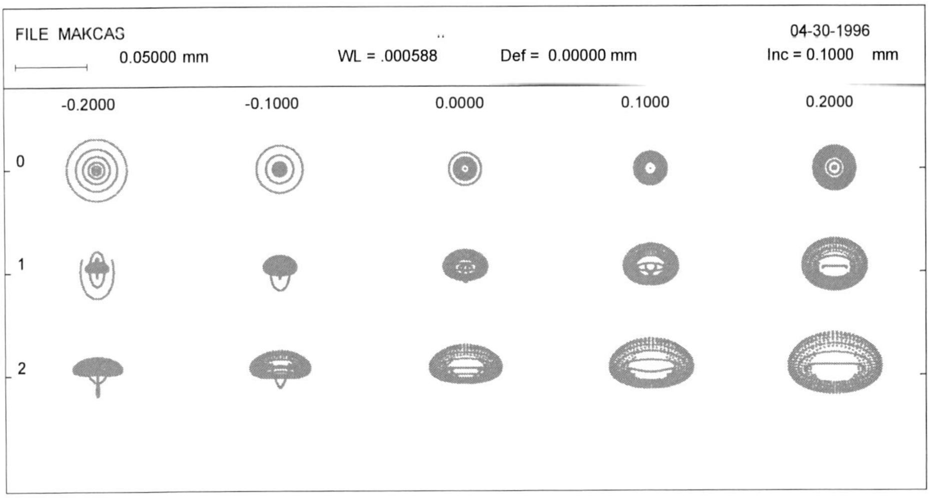

Figure 13.22. Spot diagrams of the Maksutov-Bouwers Cassegrain system.

13.2.12 A simple Mangin mirror system by Wiedemann

The Mangin mirror is a meniscus-shaped negative lens in contact with a mirror. The example shown in Fig. 13.23, designed by Wiedemann,[22] uses a Mangin as the primary mirror. It is remarkably simple yet still obtains good resolution at the extreme aperture of *f*/1.25, at a semi-field of 3 deg, as indicated by the aberrations shown in Figs. 13.24 and 13.25. It is longer than the Maksutov-Bouwers Cassegrain example given above, but even at this aperture it resolves over 100 cycles/mm resolution over an image diameter of 10.5 mm, at least over the visual wavelength range. In the system prescription below, to enable the contributions of each surface to be assessed correctly, the Mangin mirror is represented by a lens, a mirror, and another lens, for a total of five surfaces.

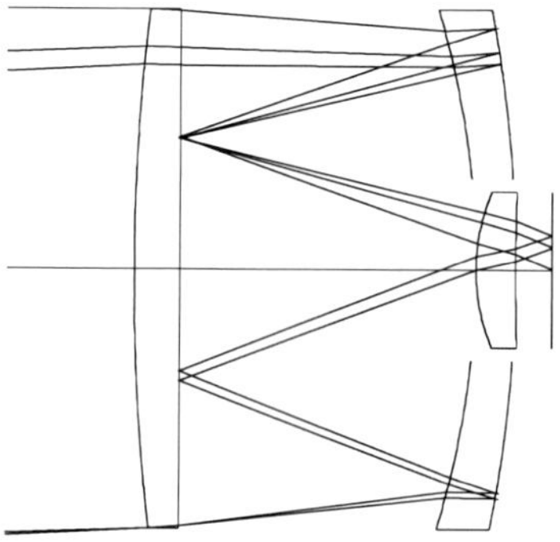

Figure 13.23. Wiedemann Mangin mirror system.

EFL = 100.000

WAVELENGTHS [nm]
587.60 656.30 486.10

SURFACES

# SURF SPACE	RADIUS	SEPN	INDEX1	V	CLR RAD	GLASS	OBS RAD
0		0.00000	1.000000				
1#S	337.98491	7.22700	1.617717	49.75	40.004	S-SSKN8	
2 S	Plane	46.46050	1.000000		40.000		
3 SI	-137.31671	6.19470	1.563838	60.77	40.000	S-SK11	14.000
4 SI	-208.03991	0.00000	1.563838	60.77	40.000	S-SK11	14.000
5 SMI	-208.03991	0.00000	1.563838	60.77	40.000	S-SK11	14.000
6 SI	-208.03991	-6.19470	1.563838	60.77	40.000	S-SK11	14.000
7 SI	-137.31671	-46.46050	1.000000		40.000		14.000
8 SM	Plane	46.46050	1.000000		20.000		
9 S	30.45750	6.19470	1.563838	60.77	12.000	S-SK11	
10S	361.35967	5.72685	1.000000		12.000		
11S	Plane				12.000		

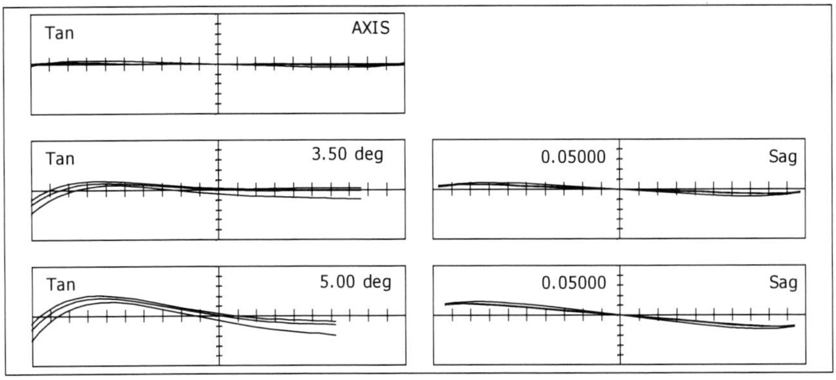

Figure 13.24. Transverse ray aberrations of the Wiedemann Mangin mirror system.

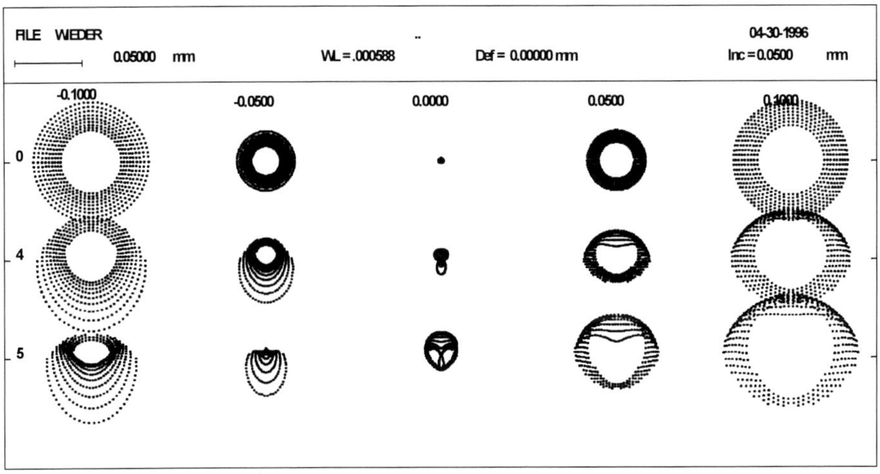

Figure 13.25. Spot diagrams for the Wiedemann Mangin mirror system.

The Seidel aberrations for this system are shown below. The surface contributions have been grouped according to the separate components.

LAGRANGE INVARIANT = -2.0965

#	H	U	HBAR	UBAR	D(U/N)	A	ABAR
1	40.00437	0.00000	0.00000	0.05241	-0.02551	0.10626	0.05241
2	39.68533	-0.04254	0.23571	0.03143	-0.04543	-0.07093	0.05241
3	36.42233	-0.07093	2.64647	0.05241	0.10830	-0.33633	0.03312
4	36.79251	0.06170	2.88254	0.03935	-0.05404	-0.19874	0.04141
5	36.79251	-0.01668	2.88254	0.05568	-0.36413	-0.19874	0.04141
6	36.79251	0.38080	2.88254	-0.02715	0.19764	-0.19874	0.04141
7	34.97794	0.30243	2.94745	-0.01082	-0.15023	-0.07853	0.05332
8	19.64164	0.33340	4.41231	-0.03184	0.00000	-0.33340	0.03184
9	4.30534	-0.33340	5.87718	0.03184	0.15809	-0.19808	0.21656
10	2.97580	-0.26591	5.66747	-0.04194	-0.22473	-0.39369	-0.04525
		-0.40004		-0.05736			

#	S1	S2	S3	S4	S5	C1	C2
1	0.011523	0.005683	0.002803	0.004674	0.003688	0.040600	0.020024
2	0.009071	-0.006702	0.004951	0.000000	-0.003658	0.026886	-0.019864
3	-0.446199	0.043945	-0.004328	-0.012630	0.001670	-0.086466	0.008516
4	0.078536	-0.016365	0.003410	0.008577	-0.002498	0.051613	-0.010755
5	0.529154	-0.110262	0.022976	-0.043501	0.004277	0.000000	0.000000
6	-0.287210	0.059847	-0.012471	0.008577	0.000811	-0.051613	0.010755
7	0.032406	-0.022004	0.014940	-0.012630	-0.001569	0.019388	-0.013165
8	0.000000	0.000000	0.000000	0.000000	0.000000	0.000000	0.000000
9	-0.026705	0.029197	-0.031922	0.047070	-0.016562	-0.004530	0.004953
10	0.103651	0.011914	0.001369	-0.003200	-0.000210	0.006223	0.000715
	0.004227	-0.004746	0.001730	-0.003063	-0.014051	0.002102	0.001179

The following table shows the aberration contributions of each component, once again treating the Mangin mirror as three components. The aberrations of the plane secondary mirror are naturally zero.

	S1	S2	S3	S4	S5	C1	C2
Front lens	.0206	-.0010	.0078	.0047	.0000	.0675	.0002
Meniscus	-.3677	.0276	-.0009	-.0040	-.0008	-.0349	-.0022
Mirror	.5291	-.1103	.0230	-.0435	.0043	.0000	.0000
Meniscus	-.2548	.0378	.0025	-.0040	-.0008	-.0322	-.0024
Field lens	.0769	.0411	-.0306	.0439	-.0168	.0017	.0056
Total	.0042	-.0047	.0017	-.0003	-.0140	.0021	.0012

One can see that the negative meniscus lens overcorrects the positive spherical aberration of the mirror, but the final positive lens gives a good level of correction for the remaining spherical aberration. One can also see that the meniscus lens introduces coma, which is also corrected by the final lens. The last positive lens mainly corrects the astigmatism and field curvature of the mirror. The first positive lens corrects the axial color of the Mangin mirror.

13.3 More complex examples

13.3.1 Canzek Mangin system

This example, shown in Fig. 13.26,[23] has an image plane that lies inside the lens. Whether this is acceptable depends on the dimensions of the detector. In the case of the design shown here, it is an image tube. The basic principles are the same as for the Wiedemann Mangin: the front positive lens corrects the axial color of the Mangin primary on all three passes, and this time there is a more complex field corrector in front of the final image. Residual aberrations of this design are shown in Figs. 13.27 and 13.28.

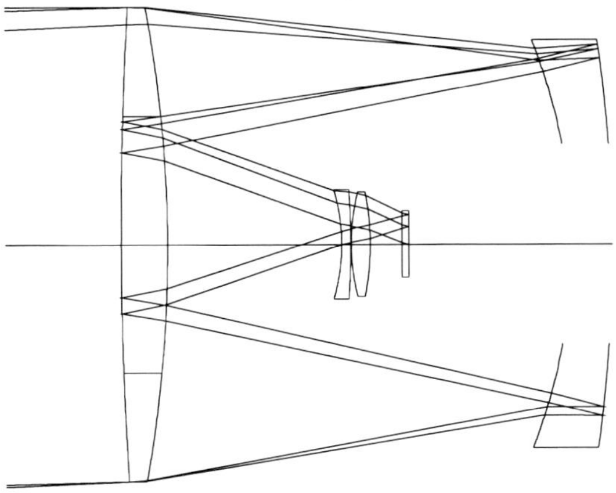

Figure 13.26. The Canzek Mangin system.

EFL = 279.990

WAVELENGTHS [nm]

587.60 656.30 486.10

SURFACES

#	SURF SPACE	RADIUS	SEPN	INDEX1	V	CLR RAD	GLASS	OBS RAD
0			0.00000	1.000000				
1	#SI	1680.23520	20.04240	1.516798	64.14	99.996	S-BK7	54.000
2	SI	-534.45840	174.03120	1.000000		100.000		54.000
3	SI	-230.12080	17.92560	1.516798	64.14	86.000	S-BK7	42.000
4	SMI	-582.25720	-17.92560	1.516798	64.14	86.000	S-BK7	42.000
5	SI	-230.12080	-174.03120	1.000000		86.000		42.000
6	S	-534.45840	-20.04240	1.516798	64.14	54.000	S-BK7	
7	SM	1680.23520	20.04240	1.516798	64.14	54.000	S-BK7	
8	S	-534.45840	75.03160	1.000000		54.000		
9	S	-78.78640	3.92840	1.548138	45.73	23.000	S-LLF1	
10	S	-312.86080	0.20160	1.000000		23.000		
11	S	88.51640	8.25720	1.516798	64.14	22.000	S-BK7	
12	S	-109.80200	13.67520	1.000000		22.000		
13	S	Plane	3.02120	1.516798	64.14	14.000	S-BK7	
14	S	Plane	-0.00202	1.000000		14.000		
15	S	Plane				14.000		

LAGRANGE INVARIANT = -4.3659

#	H	U	HBAR	UBAR	D(U/N)	A	ABAR
1	99.99563	0.00000	0.00000	0.04366	-0.01337	0.05951	0.04366
2	99.58923	-0.02028	0.57692	0.02878	-0.11369	-0.31339	0.04202
3	77.47775	-0.12705	8.07820	0.04310	0.14746	-0.46374	0.00800
4	78.03252	0.03095	8.80200	0.04038	-0.17671	-0.15633	0.03832
5	73.78261	0.23709	8.98383	-0.01014	-0.03761	0.12671	0.07460
6	40.03569	0.19391	15.17265	-0.03556	0.09280	-0.11900	0.06395
7	36.96186	0.15337	15.44868	-0.01377	-0.02901	-0.26599	0.00694
8	33.00624	-0.19736	15.35616	-0.00462	-0.20116	-0.39303	-0.05058
9	8.15018	-0.33127	13.71669	-0.02185	0.21671	-0.43472	-0.19595
10	7.45345	-0.17736	13.90340	0.04753	-0.17307	-0.31145	0.00478
11	7.39546	-0.28763	13.91332	0.04922	0.14384	-0.20408	0.20640
12	5.59460	-0.21810	13.73906	-0.02110	-0.21335	-0.40809	-0.22180
13	0.71064	-0.35714	12.41699	-0.09668	0.20191	-0.35714	-0.09668
14	-0.00072	-0.23546	12.22443	-0.06374	-0.20191	-0.35714	-0.09668
		-0.35714		-0.09668			

#	S1	S2	S3	S4	S5	C1	C2
1	0.004735	0.003473	0.002548	0.003865	0.004705	0.031612	0.023191
2	1.111975	-0.149108	0.019994	0.012151	-0.004311	0.165788	-0.022231
3	-2.456922	0.042379	-0.000731	-0.028222	0.000499	-0.190855	0.003292
4	0.337015	-0.082598	0.020243	-0.043165	0.005618	0.000000	0.000000
5	0.044550	0.026229	0.015442	-0.028222	-0.007524	-0.049662	-0.029238
6	-0.052617	0.028275	-0.015195	0.012151	0.001635	-0.025308	0.013600
7	0.075854	-0.001980	0.000052	-0.014958	0.000389	0.000000	0.000000
8	1.025616	0.131996	0.016988	0.012151	0.003750	0.068909	0.008869
9	-0.333791	-0.150456	-0.067818	-0.085660	-0.069180	-0.027431	-0.012365
10	0.125131	-0.001921	0.000029	0.021571	-0.000332	0.017973	-0.000276
11	-0.044306	0.044810	-0.045320	0.073370	-0.028369	-0.008017	0.008108
12	0.198785	0.108041	0.058721	0.059147	0.064062	0.012128	0.006592
13	-0.018301	-0.004954	-0.001341	0.000000	-0.000363	-0.001348	-0.000365
14	-0.000019	-0.000005	-0.000001	0.000000	-0.000000	-0.000001	-0.000000
Sum	0.017704	-0.005819	0.003612	-0.005820	-0.029419	-0.006214	-0.000823

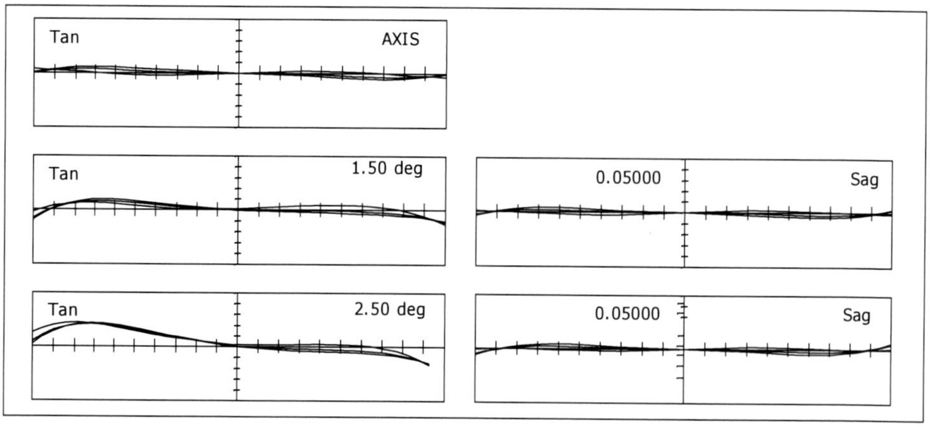

Figure 13.27. Transverse ray aberrations of the Canzek Mangin system.

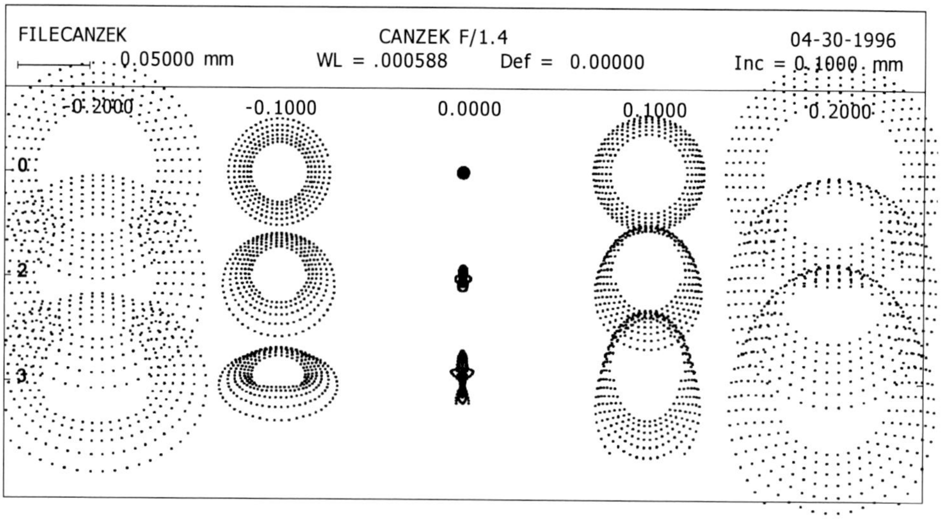

Figure 13.28. Spot diagrams for the Canzek Mangin system.

13.3.2 Mirror telephoto lens

A final, more complex, example of a Mangin system is shown in Fig. 13.29.[24] This is a 500 mm, *f*/6 "mirror-telephoto" lens for 35-mm single-lens reflex cameras, based on a design by Iizuka. As we can see from the diagram, the lens is very compact, and it too has a positive lens in front of the Mangin primary mirror. This time, the secondary mirror is separated from the front positive component, although for ease of mounting it is probably cemented to it. The rear lens group is more complex than before, and cannot now be described as a field group, since for 35-mm photography a large back-focus is needed to clear the reflex mirror. This design gives acceptable resolution for this application, as may be seen by the aberration plots of Figs. 13.30 and 13.31, and the MTF curves of Fig. 13.32.

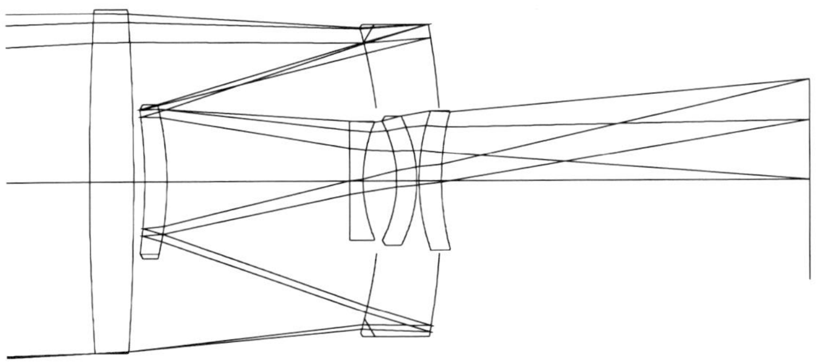

Figure 13.29. Mirror telephoto lens.

EFL = 499.996

WAVELENGTHS [nm]
587.60 656.30 486.10

SURFACES

#	SURF SPACE	RADIUS	SEPN	INDEX1	V	CLR RAD	GLASS	OBS RAD
0			0.00000	1.000000				
1	S	621.74787	10.00000	1.516798	64.14	38.003	S-BK7	
2	S	-621.74787	56.00000	1.000000		37.709		
3	SI	-126.51246	13.70000	1.516798	64.14	33.783	S-BK7	16.000
4	#SMI	-204.41335	-13.70000	1.516798	64.14	34.602	S-BK7	16.000
5	SI	-126.51246	-48.50000	1.000000		30.604		16.000
6	S	-71.44318	-5.00000	1.516798	64.14	17.088	S-BK7	
7	SM	-117.36421	5.00000	1.516798	64.14	15.985	S-BK7	
8	S	-71.44318	41.50000	1.000000		16.010		
9	S	-1661.93460	3.00000	1.516798	64.14	13.145	S-BK7	
10	S	33.80459	7.50000	1.000000		12.915		
11	S	-28.57529	4.50000	1.516798	64.14	13.186	S-BK7	
12	S	-31.19198	0.50000	1.000000		14.351		
13	S	46.36416	5.00000	1.749498	34.94	15.406	S-LAFN7	
14	S	60.05254	83.69040	1.000000		15.135		
15	S	Plane				22.132		

LAGRANGE INVARIANT = -1.6239

#	H	U	HBAR	UBAR	D(U/N)	A	ABAR
1	37.19292	0.00000	-3.51680	0.04366	-0.01344	0.05982	0.03800
2	36.98910	-0.02038	-3.20968	0.03071	-0.04822	-0.12115	0.05441
3	33.53613	-0.06166	-0.45156	0.04925	0.09440	-0.32674	0.05282
4	34.21655	0.04967	-0.02337	0.03125	-0.22071	-0.17856	0.04758
5	30.31052	0.28511	0.40795	-0.03148	-0.12067	-0.06905	0.05265
6	15.34148	0.30864	2.80486	-0.04942	0.12625	-0.09390	0.08868
7	13.95825	0.27665	2.90089	-0.01921	0.15682	-0.23922	0.06662
8	13.76433	-0.03878	3.24409	0.06864	-0.13282	-0.35106	0.03524
9	7.19098	-0.15839	6.59089	0.08065	0.09052	-0.16272	0.07668
10	6.88213	-0.10295	6.75444	0.05452	0.01693	0.15264	0.38576
11	6.50004	-0.05095	8.14911	0.18596	0.07990	-0.27842	-0.09922
12	6.69766	0.04392	9.13805	0.21976	-0.07331	-0.25908	-0.11103
13	6.67548	-0.04436	9.22902	0.18193	-0.00539	0.09962	0.38099
14	6.24030	-0.08704	9.32260	0.01872	-0.02464	0.02953	0.30434
		-0.07439		0.14910			

#	S1	S2	S3	S4	S5	C1	C2
1	0.001788	0.001136	0.000722	0.001445	0.001377	0.011818	0.007508
2	0.026181	-0.011759	0.005281	0.001445	-0.003021	0.023805	-0.010692
3	-0.337998	0.054641	-0.008833	-0.007102	0.002576	-0.058207	0.009410
4	0.240793	-0.064163	0.017097	-0.017010	-0.000023	0.000000	0.000000
5	0.017441	-0.013297	0.010137	-0.007102	-0.002314	0.011118	-0.008476
6	-0.017079	0.016129	-0.015232	0.012576	0.002509	-0.007652	0.007227
7	-0.125264	0.034885	-0.009715	0.029626	-0.005545	0.000000	0.000000
8	0.225312	-0.022616	0.002270	0.012576	-0.001490	0.025668	-0.002576
9	-0.017235	0.008122	-0.003827	-0.000541	0.002058	-0.006216	0.002929
10	-0.002715	-0.006861	-0.017339	-0.026578	-0.110990	-0.005580	-0.014103
11	-0.040257	-0.014347	-0.005113	-0.031442	-0.013028	-0.009613	-0.003426
12	0.032958	0.014124	0.006053	0.028804	0.014938	0.009218	0.003950
13	0.000357	0.001366	0.005224	0.024366	0.113164	0.008155	0.031187
14	0.000134	0.001382	0.014240	-0.018812	-0.047124	-0.002260	-0.023289
Sum	0.004416	-0.001258	0.000965	0.002252	-0.046914	0.000254	-0.000350

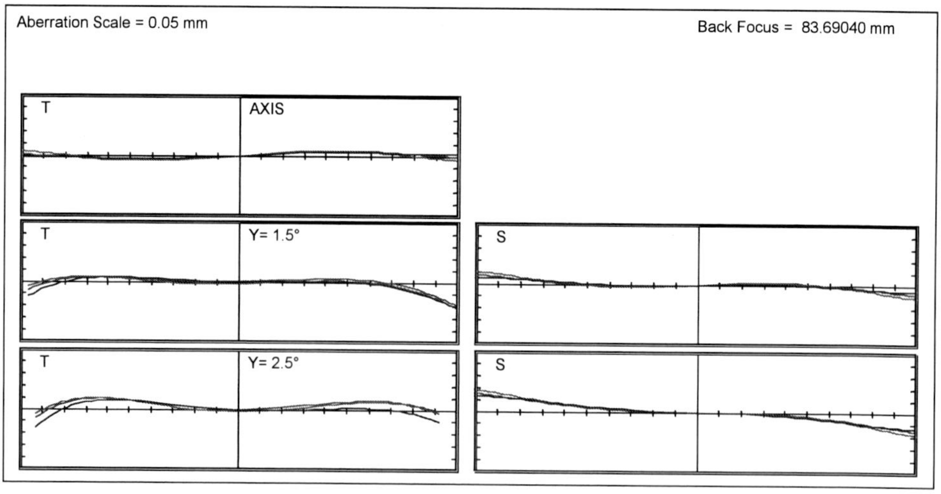

Figure 13.30. Transverse ray aberrations of the mirror telephoto lens.

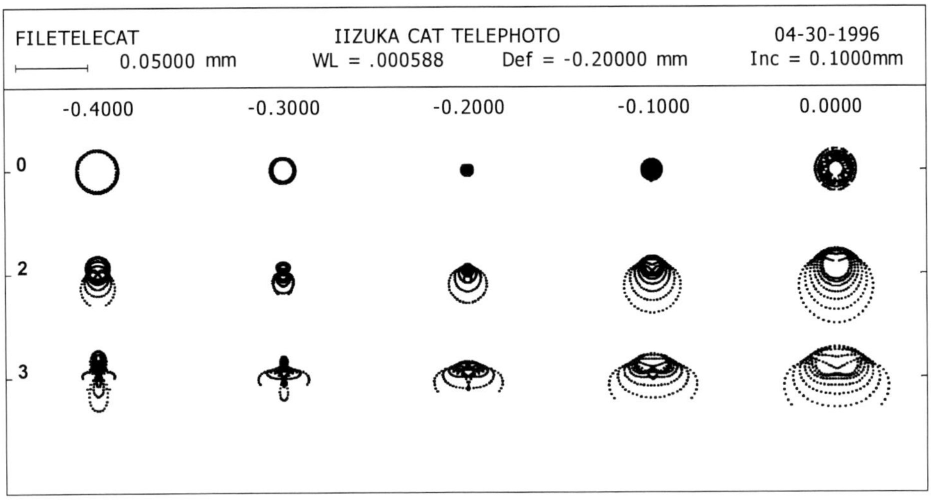

Figure 13.31. Spot diagrams for the mirror telephoto lens.

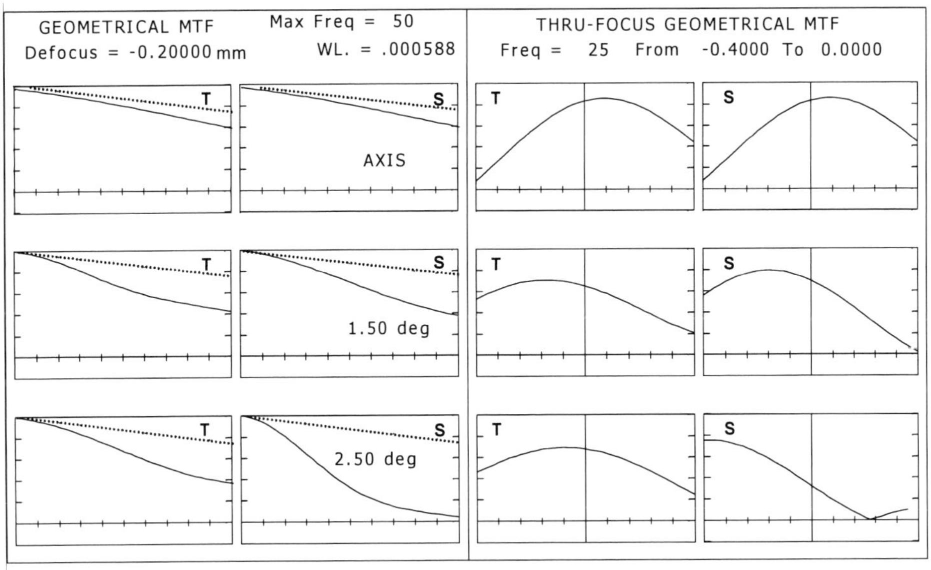

Figure 13.32. MTF curves for the mirror telephoto lens.

References

1. C.G. Wynne, "Wide field Cassegrain telescopes," *Mon. Not. R. Astr. Soc.*, **163**, p. 357-367 (1973).
2. D.L. Harmer and C.G. Wynne, "A single-lens, small field, paraboloid field corrector," *The Observatory*, Vol., **96** No. 1015, p. 239-241 (1976).

3. C.G. Wynne, "Afocal correctors for paraboloidal mirrors," *Applied Optics*, Vol. **6**, pp. 1227-1231 (1967).
4. C.G. Wynne, "Field correctors for parabolic mirrors," *Proc. Physical Soc. B*, Vol. **LXII**, p. 772-787 (1949).
5. C.G. Wynne, "Improved three-lens field correctors for paraboloids," *Mon. Not. R. Astr. Soc.*, **160**, p. 13P-18P (1972).
6. C.G. Wynne, "Data for some four-lens paraboloid field correctors," *Mon. Not. R. Astr. Soc.*, **165**, p. 1P-8P (1973).
7. C.G. Wynne, "A new wide-field triple lens paraboloid field corrector," *Mon. Not. R. Astr. Soc.*, **167**, p. 189-197 (1974).
8. C.G. Wynne, "Field correctors for astronomical telescopes," *Progress in Optics*, Vol. **X**, edited by E. Wolf (1972).
9. C.G. Wynne, "Ritchey-Chrétien telescopes and extended field systems," *Astrophysical Journal*, Vol. **152**, No. 3, part 1, p. 675-694 (1968).
10. H.E. Dall, "A null test for paraboloids," *Amateur Telescope Making* (Book 3), Scientific American, New York (1953).
11. F.E. Ross, *Astrophys. J.* **98**, p. 341-346 (1943).
12. A. Offner, "A null corrector for paraboloidal mirrors," *Applied Optics*, Vol.2, No. 2 (1963).
13. J. Sasian, "Design of null lens correctors for the testing of astronomical optics," *Optical Engineering*, Vol. **27**, No. 12 (1988).
14. E. H. Linfoot, *Recent Advances in Optics*, Oxford University Press, pp. 208-228 (1955).
15. C.G. Wynne, "The optics of the achromatized UK Schmidt telescope," *Q. Jl R. Astr. Soc.*, **22**, pp. 146-153 (1981)
16. C.G. Wynne, "Shorter than a Schmidt," *Mon. Not. R. Astr. Soc.*, **180**, pp. 485-490 (1977).
17. C.G. Wynne, "Maksutov spectrograph cameras," *Mon. Not. R. Astr. Soc.*, **153**, pp. 261-277 (1971).
18. C.G. Wynne, "Five spectrograph camera designs," *Mon. Not. R. Astr. Soc.*, **157**, pp. 403-418 (1972).
19. C.G. Wynne, "New wide-aperture catadioptric systems," *Mon. Not. R. Astr. Soc.*, **107**, pp. 356-368 (1947).
20. J. Maxwell, *Catadioptric imaging systems*, Adam Hilger (1971).
21. S. Rosin and M. Amon, "Color-corrected Mangin mirror," *Applied Optics*, Vol. **6**, pp. 963-968 (1967).
22. E. Wiedemann, "Ueber einfache Spiegelobjektive," *Optica Acta*, Vol. **26**, No. 11, pp. 1389-1396 (1979).
23. L. Canzek, "Lichtstarkes katadioptrisches Objectiv," *Optica Acta*, Vol. **18**, No. 12, pp. 931-937 (1971).
24. Y. Iizuka, "Catadioptric telephoto lens," U.S. Patent 4,666,259 (1987).
25. J. Rayces, "All spherical solid catadioptric optical systems," U.S. Patent 3,926,505 (1975).

INDEX

B

C

D

E

T

U

V

W

Z

Michael Kidger was born in Birmingham, England, on July 6, 1937. He achieved scholarships at the age of 17 to Imperial College, London, where he graduated in 1958, and was awarded an MSc in Applied Optics in 1959. He spent a short time in industry with the optical firm Taylor,Taylor and Hobson of Leicester. In 1963 he joined the optical design team at Imperial College under Professor Charles Wynne. In 1966 he accompanied Wynne to work on bubble chamber optics at the Brookhaven Laboratories. In 1967 Kidger was appointed lecturer in the applied optics section of Imperial College—a post he held for 20 years. He retained a part-time teaching post at IC and left in 1987 after an association lasting 33 years. His PhD dissertation, "The Application of Electronic Computers to the Design of Optical Systems, Including Aspheric Lenses," was published in 1971. In 1982 he formed the company Kidger Optics Ltd. with his wife Tina. He was a regular participant and exhibitor at SPIE meetings and served on several SPIE committees, including the Scholarship Committee. He gave optical design courses worldwide up until his death. Michael Kidger died in Australia on February 2, 1998, at the age of 60.